SOIL PHYSICS

SOIL PHYSICS

SIXTH EDITION

William Jury
Robert Horton

WILEY

JOHN WILEY & SONS, INC.

Published by John Wiley & Sons, Inc., Hoboken, New Jersey
Published simultaneously in Canada

For general information on our other products and services or for technical support, please contact our
Customer Care Department within the United States at (800) 762-2974, outside the United States at
(317) 572-3993 or fax (317) 572-4002.

Wiley also publishes its books in a variety of electronic formats. Some content that appears in print may
not be available in electronic books. For more information about Wiley products, visit our web site at
www.wiley.com.

Library of Congress Cataloging-in-Publication Data:

Jury, William A., 1946–
    Soil physics / William Jury, Robert Horton.—6th ed.
        p.   cm.
    Includes bibliographical references (p.   ).
    ISBN 0-471-05965-X (cloth)
    1. Soil physics.   I. Horton, Robert, 1954–   II. Title.

    S592.3.J86   2004
    631.4'3—dc22   2003057625

Printed in the United States of America

SKY10029167_081821

To Connie and Nannet: For all that you are, and all that you do

CONTENTS

PREFACE

The discipline of soil physics has changed in several significant respects since the fifth edition of this book was published in 1991. Computer speed and storage capacity have grown enormously in the last 13 years. As a result, numerical methods have become standard tools for analysis of transport problems, even in two or three dimensions. A subset of this analysis has been the emergence of new methods for model parameter estimation, model validation, and hypothesis testing. In addition, soil physics has adapted or developed a number of high-technology devices for measuring and monitoring subsurface processes of interest, including a number of noninvasive devices that use electromagnetic radiation to penetrate the soil without disturbing the soil matrix. A significant resurgence of fundamental soil physics has occurred in the last decade, increasing the profile of research areas such as relating soil structure and matrix properties to transport processes. In addition, study of preferential flow of water and chemicals has accelerated, particularly in model development.

The application of soil physics to practical problems in agriculture and environmental protection has continued to expand, producing a shift in focus from the laboratory to the field regime. This transition has given rise to new areas of research, characterizing the effects of spatial and temporal variations in soil properties on the mean behavior of large-scale soil systems. This edition of *Soil Physics* has been altered significantly to reflect the changes in the discipline. As in the fifth edition, the book focuses heavily on transport processes and problem solving, teaching the reader to simplify the general theory for specific applications. This approach is developed systematically using physical principles rather than empirical laws and is illustrated throughout the text with over 70 completely worked examples that provide the reader with a model for using the theory in a practical manner. In addition, this edition provides detailed solutions to the 67 end-of-chapter problems. Other significant changes from the fifth edition include extensive use of numerical solutions to illustrate transport phenomena, major new sections on preferential flow, considerably more coverage of experimental methods for monitoring and measuring in the subsurface, and expanded emphasis on soil structure and its characteristics.

The book should be suitable as an advanced undergraduate- or graduate-level instruction text in soil physics as well as a reference text for professional scientists. In Chapter 1 we describe the important physical and chemical properties of the soil solid phase, emphasizing those characteristics that are most relevant to the transport, retention, or transformation of water, heat, gases, and solutes as well as emphasizing properties of agricultural significance. In Chapter 2 we characterize water in soil, first describing the molecular and fluid properties of water and then developing the thermodynamic description of water potential energy. The relationship of the potential components to soil water measurement devices is established quantitatively by equilibrium analysis.

In Chapters 3 and 4 we introduce the theory of water transport through saturated and unsaturated soil and provide numerous approximate models of water flow. Chapters 5 to 7 deal with the transport of heat, gases, and dissolved chemicals in soil, with emphasis on practical problems encountered in the field. The Appendix deals with methods of assessing the properties of spatially variable soil, including a description of the most modern methods. We provide problem solutions following the Appendix.

The material in the book has formed the basis of soil physics courses taught for many years by the senior author at the University of California–Riverside. Numerous students have contributed to the text by making helpful suggestions for improvement of draft sections used in the course.

WILIAM A. JURY
Riverside, California

ROBERT HORTON
Ames, Iowa

February 2004

1 Soil Solid Phase

Soil is an extraordinarily complex medium, made up of a heterogeneous mixture of solid, liquid, and gaseous material, as well as a diverse community of living organisms. Soil is the growth medium for plants, which form the base of the terrestrial food chain now supporting over 6 billion people worldwide. Thus, soils must not only hold the water needed for plants, but also provide nutrients and oxygen to the plant roots. These diverse tasks require that the soil solid phase have a unique set of properties and a very special composition.

The solid phase is made up of minerals and organic matter. The mineral fraction contains particles of widely varying sizes, shapes, and chemical composition. The organic fraction is also very heterogeneous, containing a diverse population of live, active organisms as well as plant and animal residues in various stages of decomposition. The solid phase may contain particles of vastly different sizes, spanning the lower limits of the colloidal state to the coarsest fractions of sand and gravel. The various particles, especially those of colloidal size, may be found in states ranging from almost complete dispersion to nearly perfect aggregation. In most soils, however, there is only a partial aggregation of the individual particles.

The spaces between the solid particles are fully or partially filled with water, which contains many chemicals that may have dissolved from the soil mineral phase or entered through the soil surface. Soil water is pulled down by gravity but is also attracted to the surface of the solid matrix. Its mobility varies significantly, depending on local conditions. The gaseous or vapor phase occupies the part of the pore space between the soil particles not filled with water; its composition can differ considerably from that of the air above the soil surface and may change dramatically in a short period of time.

The soil solid phase has a dominant influence on heat, water, and chemical transport and retention processes. Therefore, characterizing the physical and chemical properties of the soil solid phase is essential to an understanding of many of the practical agricultural and environmental problems within the purview of modern soil physics research.

1.1 CHARACTERISTICS OF THE PRIMARY PARTICLES

The solid mineral phase is made up of the products of weathering of the parent rocks that preceded the formation of soil. The most important characteristics of the soil solid phase are the sizes and shapes of the individual particles forming the matrix, their chemical and mineralogical composition, and the physical and chemical properties of their surfaces.

1.1.1   Characterization of Particle Size

Two procedures, sieving and sedimentation, are used to measure the size of individual soil particles. A soil sample whose mineral phase is to be characterized is first pretreated to remove organic material and to disperse aggregates, and then is passed through a series of coarse screens of specified opening. The smallest screen for which this operation is feasible has an aperture of about 0.05 mm (Gee and Or, 2002). Care must be taken to ensure that none of the small particles are excluded from passage through the screen at this stage.

The sizes of the remaining dispersed particles are characterized indirectly by sedimentation procedures using either the hydrometer method or the pipette method. Both methods are based on Stokes' law, which is used to establish a relationship between the settling velocity V (the speed with which a particle falls through a solution under the influence of gravity) and the particle size.

When a spherical particle of density ρ_s and radius R falls through a liquid of density ρ_l and viscosity η, it is subject to three forces: gravitation, buoyancy, and viscous drag. The force of gravity F_g is calculated from *Newton's law*:

$$F_g = m_s g = \rho_s V_s g = \rho_s \frac{4\pi R^3}{3} g \tag{1.1}$$

where m_s is the particle mass and V_s the particle volume. This force acts downward. The buoyancy force F_b is calculated by Archimedes' principle. It states that a solid object fully immersed in a fluid is buoyed upward by a force equal to the weight of the fluid displaced by the solid volume. Thus,

$$F_b = m_l g = \rho_l \frac{4\pi R^3}{3} g \tag{1.2}$$

where m_l is the liquid mass displaced by the solid volume. This force is upward. The viscous drag force F_d that the particle feels from the surrounding liquid as it falls at velocity V is calculated with *Stokes' law* (Bird et al., 2001), which states that

$$F_d = 6\pi R \eta V \tag{1.3}$$

This force is always opposite to the direction of the velocity V of the spherical particle with respect to the fluid. Thus, it is upward for a particle falling downward.

These three forces equilibrate rapidly, and the particle reaches a final velocity that may be calculated by setting the net force equal to zero:

$$\sum F_i = 0 = F_g - F_l - F_d \tag{1.4}$$

When (1.1)–(1.3) are inserted into (1.4) and the resulting equation is solved for V, we obtain

$$V = \frac{(\rho_s - \rho_l) D^2 g}{18\eta} \tag{1.5}$$

where $D = 2R$ is the particle diameter. Equation (1.5) gives the velocity of a spherical particle of diameter D and density ρ_s falling through a fluid of density ρ_l and viscosity η. Thus, in a solution containing a perfectly mixed suspension of particles of different diameters, the individual particles will settle to the bottom at different rates. For example, in a time $t = t'$ after stirring of the suspension stops, all particles of diameter D greater than D' will have fallen a distance greater than or equal to $X = V't$, where V' is the value of (1.5) when $D = D'$. Therefore, there will be no particles of diameter larger than D' between the water surface and depth X in the suspension beaker at time t'. In the hydrometer method of measuring particle size, a precalibrated floating object rests in the solution that is undergoing sedimentation. As the particles fall out of the upper part of the solution, the float sinks lower into the solution because the density of the displaced solution is decreasing. The position of the float at any given time can be used with the calibration curve for the device to calculate the density of the solution and hence the mass of the particles of a given size range that are still in the upper part of the suspension beaker.

The pipette method involves direct sampling of the solution in suspension at various times. Further details of this procedure are given in Gee and Or (2002).

Example 1.1   Use (1.5) to calculate the amount of time required for particles of density $\rho_s = 2.7$ g cm^{-3} and effective diameters of 0.2 mm, 0.05 mm, and 0.002 mm to fall 10 cm through water. The viscosity η of water is approximately 0.01 g cm^{-1} s^{-1}.

SOLUTION: Since $g = 980$ cm s^{-2}, (1.5) may be written as

$$V = 9256D^2 \quad \text{cm s}^{-1}$$

if D is in centimeters. Thus, V = 3.70, 0.231, and 0.00037 cm s^{-1}. The settling times for moving 10 cm are thus 2.7 s, 43 s, and 7.5 h. Clearly, sedimentation is practical only for particles smaller than 0.05 mm.

1.1.2   Classification of Textural Size Fractions

The different particles in soils are classified into groups of various sizes on the basis of their effective diameter, which is an experimental measurement of the size of the particles. Thus, the effective diameter of the larger particles that can be removed by mechanical sieving (i.e., > 0.5 mm) refers to the diameter of a sphere that cannot pass through a hole of a given size in a sieve. For the smaller particles, which must be separated by sedimentation techniques, the equivalent diameter of a particle refers to the diameter of a sphere that has the same density and settling velocity in the dispersing medium.

Several systems of soil texture classification have evolved over a period of years. The two classification systems, the U.S. Department of Agriculture (USDA) system and the International Society of Soil Science (ISSS) system, shown in Fig. 1.1, are used by agricultural scientists. The ISSS classification was suggested by Atterberg (1912), who established the upper limit of the clay fraction of soil on the basis that particles smaller than 0.002 mm (2 μm) exhibited Brownian movement in aqueous suspension and thus would not fall freely in solution under the influence of gravity. This definition for the clay particle size was given further justification by mineralogical studies that showed (see Section 1.1.3) that relatively few unweathered primary minerals existed in fractions smaller than 0.002 mm. Thus, the clay fraction may be regarded as a mixture of secondary minerals or weathering products, in contrast to the coarser fractions that are typically primary minerals. Prior to 1938, the USDA classification system used 0.005 mm (5 μm) as the upper limit of clay particle size; this limit had been established arbitrarily in 1896, because it represented the smallest size that could be detected by the microscopes used then to observe the size of the particles. In 1938, the 2-μm limit of clay particle size from the Atterberg classification was adopted by the USDA system without changing the size limits of the other separates. The two systems are not too dissimilar, except that the classification used by the USDA makes more separations in the sand fractions than the one used by the ISSS.

The clay fraction has a significant influence on many physical and chemical processes that occur in soil, primarily because the small particles have such a large and reactive surface area. In contrast, the sand and silt fractions typically do not have much influence on chemical processes, and their relatively small surface areas do not

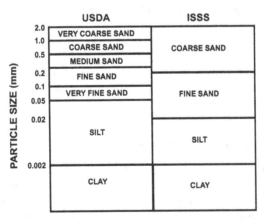

Figure 1.1   Particle-size classification systems of the U.S. Department of Agriculture and the International Soil Science Society.

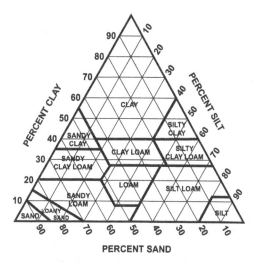

PERCENT SAND

Figure 1.2   Textural triangle, showing the ranges of sand, silt, and clay comprising a given soil textural name.

adsorb or retain water nearly to the degree that clays do. Consequently, the sand and silt portion of the soil matrix may be regarded as a passive entity whose influence on soil water is manifested primarily by the geometric arrangement of the particles.

Since clay has a distinctly different effect on soil water and chemicals than sand or silt, we will be able to make some generalizations about various characteristics of field soils that contain dominant amounts of one or the other of these particle size fractions. For example, sandy soils, because they do not retain water significantly, require more frequent water additions to avoid plant water stress than do soils in the same climate that contain significant amounts of clay. In this book we will make many such generalizations about "sandy" and "clayey" soils based on the dominant properties of the respective particle size fractions.

Soils are given different names based on the amount of sand, silt, and clay present in them. The textural triangle (Fig. 1.2) is a convenient way of showing the relationship between the soil textural name and its composition.

1.1.3   Particle-Size Distribution

Although the classification system is useful for describing the general features of a given soil type, many of the important soil physical properties discussed in this book are sensitive to the distribution of particle and pore sizes in a given soil. For this reason it is often useful to construct a continuous representation of the mass fraction of soil particles as a function of the effective particle diameter. This function, the *particle-size distribution* (Gee and Or, 2002), contains much more information about the soil than the textural classification. For example, Fig. 1.3 shows measured particle-size distributions for three silt loam soils and one clay soil according to the classification

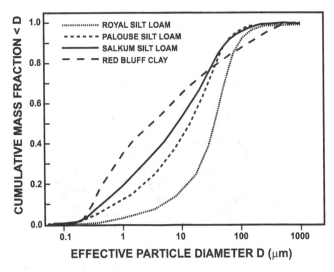

Figure 1.3   Measured particle-size distributions of four soils. (Adapted from Bittelli et al., 1999.)

scheme of Fig. 1.2. This graph reveals substantial differences in the abundance of particles of a given diameter within the single classification of silt loam.

1.1.4   Chemical and Mineralogical Properties

For the most part, in this book we discuss only chemical and mineralogical characteristics of soil properties that are relevant to the transport and retention of water, chemicals, and heat in soil. There are adequate references in the literature that provide comprehensive information about soil mineralogy (e.g., Dixon and Weed, 1988; Marshall, 1964). However, to understand the physical and chemical properties of the soil clay fraction, it is necessary to describe its mineralogical composition in some detail. For the more inert silt and sand fractions, it will suffice merely to indicate which minerals comprise the bulk of each particle-size group.

Sand and Silt Fraction   The sand and silt fractions of soil contain many primary minerals that are important in soil weathering and development processes and may exert an influence on certain chemical processes in the soil as well. Some of these primary minerals directly determine the mineralogical nature of the clay particles formed in the weathering process. For example, soils derived from schists that have a high content of mica in the sand fraction may contain a considerable amount of vermiculite along with mica in the clay fraction. This is in agreement with the recognized weathering sequence of mica → vermiculite → montmorillonite (Douglas, 1988). Because the larger particles that make up the sand and silt fractions have relatively small specific surface area (surface area per unit mass), they have only a minor influence on many soil chemical and physical properties.

Most of the primary minerals in the coarser soil fractions consist of quartz and aluminosilicates, primarily feldspars (Bohn et al., 1979). Mineralogical analyses have shown that the feldspars are not found in abundance in particle-size fractions smaller than 2 μm, although there are quartz particles as well as weathering products smaller than this size. Mineralogical analyses of New Zealand soils (Fieldes, 1962) indicated the presence of 15 to 21 mineral species in the coarser fractions and 8 to 12 different types in the clay fraction. Although quartz was the most abundant mineral found in the sand fraction of nonvolcanic soils in Fieldes's (1962) study, there were many other minerals present as well, most of which were aluminosilicates and oxides. Feldspars were a primary constituent of the sand fraction.

Clay Fraction   The clay fraction is composed primarily of Si, Al, Fe, H, and O along with varying amounts of Ti, Ca, Mg, Mn, K, Na, and P. Although the elemental chemical composition of clay was known for many years, mineralogical and structural units formed from these primary compounds were identified only after modern analytic techniques became available.

Prior to the development of quantitative means for the identification of crystalline structure, clay was thought to be made up of a mixture of hydrated oxides of silicon, aluminum, and iron, with the bases present in the adsorbed state. Only when x-ray diffraction techniques were applied to mineralogical studies of clays was this concept replaced with a quantitative model of clay structure (Hendricks, 1942; Hofmann et al., 1933; Kelley et al., 1931; Marshall, 1935a; Pauling, 1930; Ross and Shannon, 1926). X-ray analyses proved that clays are made up primarily of distinctly crystalline minerals, although there may be certain amounts of noncrystalline material present. The major constituents of the minerals are silicon, aluminum, ferrous and ferric iron, magnesium, and oxygen atoms plus hydroxyl groups.

Clay minerals are formed from two basic structural units. The first unit is the silicon tetrahedron, in which four oxygen atoms form a tetrahedron held together by a silicon atom in the center (see Fig. 1.4a). It is possible for the aluminum atom to substitute for silicon in the tetrahedron. In an idealized structure, the tetrahedra thus formed are linked together in a planar formation known as the *silica* or *tetrahedral sheet*. In this sheet, the three oxygen atoms that form the base of the tetrahedron are jointly shared by three adjacent tetrahedra. Figure 1.4b shows a basal view of the silica sheet with the oxygen atom attached to the apex of the tetrahedron pointed downward below the oxygens in the plane of the paper. Thus, the holes in this sheet are ringed by six oxygen atoms in a symmetrical hexagonal arrangement. When this structure repeats in a plane, the unit cell within the long sheet consists of four silicon atoms and six oxygen atoms, as indicated by the dashed rectangle in Fig. 1.4b. Two of the silicon atoms are located entirely within the cell (in tetrahedra 3 and 4), whereas four more (in tetrahedra 5, 6, 7, and 8) are on the cell boundary and hence are shared with the adjacent cell. The six oxygen atoms in the unit cell consist of three from tetrahedron 4 and one-half of each of the two atoms from tetrahedra 3, 7, and 8 that

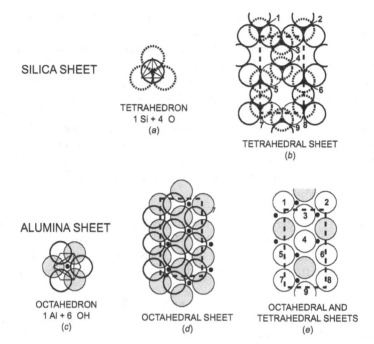

SILICA SHEET

TETRAHEDRON
1 Si + 4 O
(a)

TETRAHEDRAL SHEET
(b)

ALUMINA SHEET

OCTAHEDRON
1 Al + 6 OH
(c)

OCTAHEDRAL SHEET
(d)

OCTAHEDRAL AND
TETRAHEDRAL SHEETS
(e)

Figure 1.4 Structural arrangements in clay minerals, with open circles denoting O groups and shaded circles denoting OH groups. The dashed rectangles show a unit cell. (Adapted from van Olphen, 1963.)

are on the edge of the cell. Since the silicon atoms lie within the tetrahedra, the base of the unit cell is an oxygen surface.

The second basic structural unit of clay is the aluminum octahedron, in which six hydroxyl (OH) groups or oxygen atoms form an octahedron held together by an aluminum atom in the center (see Fig. 1.4c). It is possible for magnesium and ferrous and/or ferric iron atoms to substitute for aluminum in the center of the octahedron. The octahedra thus formed are linked together in a planar formation known as the *alumina* or *octahedral sheet*, in which the six OH groups that form the octahedron are jointly shared by three adjacent octahedra. This arrangement is illustrated in Fig. 1.4d, which shows a top view of the octahedral sheet, where six OH groups form a symmetrical hexagonal pattern that produces a closely packed arrangement with another OH in the center. In a long plane formed by repeated groups having this structure, the unit cell consists of four aluminum atoms and six OH groups, as shown in the dashed rectangle in Fig. 1.4d. Both the top and bottom of the alumina sheet are hydroxyl surfaces.

The silica and alumina sheets join together to form a layer silicate clay mineral. To visualize how this occurs, the two sheets in Fig. 1.4 are altered as follows. First, the OH groups in the upper plane of the alumina sheet shown in Fig. 1.4d, corresponding to positions 1 to 9 in Fig. 1.4e, are removed. Second, the silica sheet shown in Fig. 1.4b is placed on top of the alumina sheet, so that the oxygen atoms in apex positions 1 to 9 occupy the corresponding positions in Fig. 1.4e. Thus, the two sheets are held together by jointly shared oxygen atoms that are on the apexes of the silica tetrahedra. Figure 1.4e shows only the oxygen–hydroxyl boundary layer between the alumina and silica sheets along with the aluminum atoms of the accompanying octahedra. When so joined, the aluminum in each octahedron is now surrounded by four OH groups and two oxygen atoms. The OH groups at the boundary are found directly opposite the holes in the silica sheet. When the top silica–oxygen and the bottom alumina–hydroxyl sheets are added, the resulting structure is a single layer or 1:1 clay mineral.

A schematic drawing of the basic structural unit of a 1:1 clay mineral lattice is shown in Fig. 1.5. The oxygen surface of the tetrahedral sheet has six atoms, the hydroxyl surface of the octahedral sheet has six OH groups, and the boundary between the sheets has four oxygen atoms and two OH groups. This figure shows that the unit cell of the clay lattice has 28 positive and 28 negative charges to make it electrostatically neutral. A second silica sheet can be added to the bottom of the alumina sheet in the manner just described to create 2:1 clay minerals. A schematic drawing of such a mineral is shown in Fig. 1.5. There are now two boundaries and two oxygen surfaces. The unit cell has 44 positive and 44 negative charges to make it electrostatically neutral. The structures of the clay minerals shown in Fig. 1.5 are idealized. In reality, many clays have important irregularities that impart different properties to the mineral. First, isomorphic substitution of one atom for another of similar size in the crystal can change the properties of the clay mineral (Marshall, 1935b). The trivalent aluminum atom, which has a radius of 0.57 Å (1 Å = 10^{-10} m), can substitute in the silica sheet for the tetravalent silicon atom, which has a radius of 0.39 Å. When this occurs, the larger aluminum atom forces the four oxygen atoms of the tetrahedron farther apart than those that surround the silicon atoms, thus introducing a strain in the crystal. This substitution also decreases the positive charge

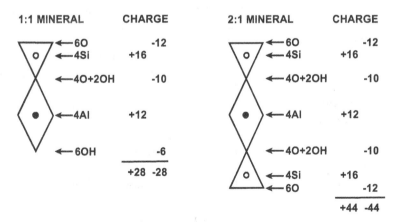

Figure 1.5   Diagrammatic sketches of sheets in clay minerals. (Courtesy of Larry P. Wilding.)

on that particular tetrahedron by one unit, thus creating a net negative charge that must be balanced either by a nearby positive charge or by adsorbing a cation from the surrounding solution.

Similarly, the divalent magnesium and ferrous iron atoms and trivalent ferric iron, which have radii of 0.78, 0.83, and 0.67 Å, respectively, can substitute for the aluminum atom in the alumina sheet. These larger atoms will push the hydroxyl groups farther apart than those that surround the aluminum atoms, thereby causing strains in the octahedral units. Also, the divalent atoms will increase the negative charge of the octahedron where the isomorphic substitution takes place, which is usually balanced by an extra cation adsorbed on the surface. Cations adsorbed on the surface of charged clay minerals can be replaced by other cations in the surrounding soil solution. Therefore, they are often called *exchangeable cations*. The total number of exchangeable cations that a surface will adsorb (expressed in SI units as cmol kg^{-1} of clay[1]), the *cation exchange capacity* (CEC), reflects the degree of isomorphic substitution within the crystals of the clay[2]. Cation exchange significantly influences the transport of positively charged ions through soil, slowing their movement relative to the velocity of the solution in which they are dissolved.

In addition to strains resulting from isomorphic substitution of larger, lower-valence atoms within the crystal, clay minerals can have other physical properties that differ from those of a mineral possessing the ideal structure shown in Fig. 1.5. It has been suggested that a few of the silica tetrahedra may be inverted instead of having their apex pointing into the octahedral sheet (Edelman and Favejee, 1940). Such a variation in arrangement would change the nature of the crystal surface, creating crystal strains and making the mineral surface rough rather than smooth. Moreover, electron microscope observations have indicated that the edges of minerals may have both beveled and frayed structure (Jackson, 1964).

Radoslovich (Radoslovich, 1960; Radoslovich and Norrish, 1962) has postulated that the silica tetrahedra can rotate freely within the crystal, thus distorting the hexagonal symmetry of the ideal structure and imparting more of a ditrigonal symmetry to the surface. These distortions and the strain they create in the crystal are partially relieved by a contraction of the tetrahedral layer caused by rotation of up to 300 of the basal tetrahedra. The silica tetrahedra can rotate until the six nearest oxygen atoms are in contact with the interlayer cation.

Because there are many different ways in which silica and alumina sheets can combine to form a lattice structure and because there are several different substituting atoms in the crystal, there are a host of different clay minerals, many of which have distinctly different properties. Layer silicate clay minerals may be divided into five major groups (Dixon and Weed, 1988): (1) kaolin group with a 1:1 crystal lattice; (2) mica (illite) group with 2:1 crystal lattice; (3) vermiculite group with a 2:1 expanding crystal lattice; (4) smectite group with a 2:1 expanding crystal lattice; and (5) chlorite group with a 2:2 crystal lattice. Regular and random interstratified combinations of these groups of clay minerals have also been identified (Sawhney,

---

[1] This unit has the same value as the traditional unit me/100 g used in earlier textbooks.
[2] The CEC is discussed in more detail in Section 1.1.6.

1988). A fibrous clay group, consisting of palygorskite and sepiolite with alternating 2:1 ribbons comprising the crystal lattice, is less abundant in soils.

The mineral structure of the kaolin group consists of one silica and one alumina sheet. The chief minerals belonging to this group are kaolinite, dickite, nacrite, and halloysite (Brindley, 1951). There is little, if any, isomorphic substitution in the crystal lattice of these clay minerals. Unit layers are held together by hydrogen bonding between the hydrogen atoms in the hydroxyl surface of one unit and the oxygen atoms in the oxygen surface of the adjacent unit. The unit layers are bonded together so tightly that ions or water molecules cannot enter the interlayer positions between adjacent kaolinite layers. Consequently, the physical and chemical properties of kaolin minerals are determined by the external mineral surfaces only. There is little swelling or shrinkage and low plasticity. The charge imbalance in this mineral group is principally caused by broken bonds on the edges of the clay mineral, but the CEC is generally less than 10 cmol kg^{-1} of clay. A diagrammatic edge view of two unit layers of kaolinite is shown in Fig 1.6a.

The mica group structure consists of one alumina sheet between two silica sheets. There is isomorphic substitution of aluminum for silicon in the silica sheets and magnesium or iron for aluminum in the alumina sheet. The two major types of minerals belonging to this group are biotite and muscovite (Fanning et al., 1988). The major isomorphic substitutions in muscovite occur in the silica sheets, whereas three Fe^{2+} and three magnesium ions replace aluminum in the octahedral sheet of biotite. The resulting negative charges are balanced by potassium ions, whose radius (1.33 Å) just fits into the holes within the hexagonal oxygen rings of the silica sheets.

If potassium ions fit perfectly in the space, they would hold the structure together while being surrounded by 12 oxygen atoms. However, the potassium ion apparently does not fit completely in the basal network because of some rotated tetrahedra (Radoslovich and Norrish, 1962) and consequently is in a six-fold rather than a 12-fold coordination with the oxygen atoms. Nevertheless, the unit layers of mica

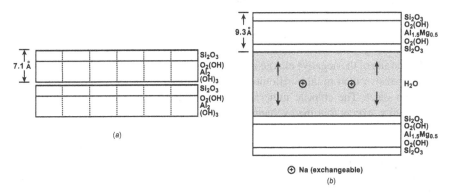

Figure 1.6   Diagrammatic edge views: (a) two layers of kaolinite six unit cells in width; (b) montmorillonite with interlayer water. (Courtesy of Marshall, 1964.)

minerals are held together by the potassium ions, even though their size does not permit as close a spacing as would be expected if the ions fitted completely in the holes. The CEC of mica minerals varies between 20 and 40 cmol kg^{-1}. Since the unit layers are held together so tightly by the potassium ions, the exchange capacity resides mainly on the external surfaces or on frayed or scrolled crystal edges, except where there may be an interstratification with an expanded, hydrated layer that would create some internal adsorbing surface area. Mica is one mineral in which the CEC does not reflect the degree of isomorphic substitution in the crystal.

Vermiculite minerals differ from the mica minerals primarily because magnesium, rather than potassium, is the interlayer exchangeable cation that helps to balance the negative charges on the lattice arising from a sizable substitution of aluminum for silicon in the silica sheet. The magnesium ions are highly hydrated. Consequently, the interlayer space contains both exchangeable cations and a double layer of water molecules that hold the unit layers together. The lattice has only limited expansion, depending primarily on the size of the exchangeable ions present in the interlayer. The CEC of vermiculites varies between 100 and 150 cmol kg^{-1}, which includes both external and internal surfaces.

The mineral structure characteristic of the smectite group (Borchardt, 1988) consists of one alumina and two silica sheets. The spacing between the 2:1 unit layers may expand or contract depending on the amount of water and interlayer cations present. The most important members of this group are montmorillonite, beidellite, and nontronite. In an ideal montmorillonite mineral, substitution consists of iron or magnesium for aluminum in the alumina sheet. Thus, negative charges originate primarily in the octahedral sheets and are balanced by exchangeable cations that reside between the unit layers. The hydration of these cations, along with the adsorption of water molecules on the oxygen surfaces of the silica sheets through hydrogen bonding, causes interlayer swelling and an expansion of the crystal lattice. The extent of expansion or contraction of the lattice varies with the nature of the exchangeable cation and the degree of hydration of the internal surfaces. A diagrammatic edge view of montmorillonite with the interlayer of water and cations between the basic units of the crystal is shown in Fig. 1.6b. The CEC of montmorillonite varies between 80 and 150 cmol kg^{-1}, including both internal and external surfaces.

Beidellite is an end member of the smectite group in which, ideally, most of the negative charge arises from substitution of aluminum for silicon in the silica sheet. Thus, the site of the negative charge is closer to the interlayer cation, making beidellite less expansive than montmorillonite. In nontronite, iron replaces aluminum in the alumina sheet. The chlorite mineral has a structure that is very similar to that of vermiculite, except that the exchangeable magnesium and water interlayer between the basic unit layers is replaced by an interlayer crystalline sheet in which magnesium instead of aluminum is in octahedral coordination with OH groups. In the chlorite mineral, substitution of aluminum for silicon in the silica sheets gives it a negative charge, while substitution of aluminum and Fe^{3+} for magnesium in the interlayer crystalline sheet gives it a positive charge. Thus, the three layers of this mineral are bound together electrostatically. Consequently, there are two silica, one alumina, and one magnesia sheet in the chlorite structure. Because there is an overall electrostatic balance within the crystal, the CEC ranges between 10 and 40 cmol kg^{-1}. Its physical properties are similar to those of mica. The palygorskite or fibrous clay group has a 2:1 lattice structure in which the silica tetrahedra are inverted in alternate strips on both sides of the basal oxygen plane. This produces a box or chainlike structure (Bradley, 1940). There are two members of this group, palygorskite and sepiolite.

The clay minerals discussed in the preceding, with the exception of members of the fibrous clay group, all have layered structures consisting of either tetrahedral or octahedral arrangements of oxygen atoms or OH groups with atoms of Al, Si, Mg, or Fe. There are also clay minerals made up of a mixed stacking of unit layers in the crystal instead of a uniform stack of the same units. When this occurs, the minerals are called *interstratified* or *mixed-layer clays*. Such interstratification can create surface properties that are different from those of the original clay mineral, particularly with respect to ion exchange and hydration. The reader is referred to the references on clay mineralogy cited previously for further information about the structure and chemistry of the crystalline materials that constitute the clay fraction of soils. It suffices for the purposes of this text to realize that differences in the crystal lattice structure of various clays will have an impact on their physical and chemical properties. Some of these are discussed further in this chapter.

There are also some noncrystalline materials within the clay size fraction that might be part of the composition of a soil depending on the nature of the soil and the conditions under which it was formed. For example, the volcanic ash soils often contain the noncrystalline material allophane. Allophane is a general name applied to hydrous aluminosilicates with a predominance of Si$-$O$-$Al bonds, variable composition, and short-range order (Wada, 1988). Since it lacks rigid structural characteristics, it possesses a very large specific surface, has high cation and anion exchange capacities, and contains tremendous quantities of water. Imogolite, a paracrystalline assemblage of a one-dimensional structural unit, has properties and abundance very similar to allophane.

The free oxides (including hydroxides) of aluminum, iron, and manganese are present in varying amounts in most soils. They are most abundant in Oxisols, where advanced weathering has depleted the soil of more soluble minerals. Aluminum hydroxide is typically present in soils as the mineral gibbsite [Al(OH)$_3$], whereas goethite (FeOOH) and hematite (F$_2$O$_3$) are the most common iron oxides in soils. Crystalline and, especially, noncrystalline aluminum and iron oxides play an important role in the stabilization of soil aggregates (Hsu, 1988; Schwertmann and Taylor, 1988). Their small particle size and positive charge at low pH[3] make them effective as binding agents of negatively charged clay minerals. Goethite and hematite are also responsible for the yellow-brown and red colors, respectively, commonly observed in soils. Opaline (amorphous) silica is a common constituent in soils with an abundant source of readily soluble silica, such as volcanic ash or basic igneous rocks, in subhumid Mediterranean and arid climates that do not completely leach silica from the

[3] pH is discussed in Section 2.1.1.

soil (Drees et al., 1988). Opaline silica can act as a cement to form extensive duripans and may be important in the weaker cementation of quartz-rich tillage pans.

1.1.5   Shape of Soil Particles

Although many conceptual models of soil assume that individual soil particles are spherical in shape, most of the experimental evidence indicates that the particles making up the smaller soil fractions are distinctly nonspherical. Modern spectroscopic techniques have made it possible to determine the shape of clay particles quite

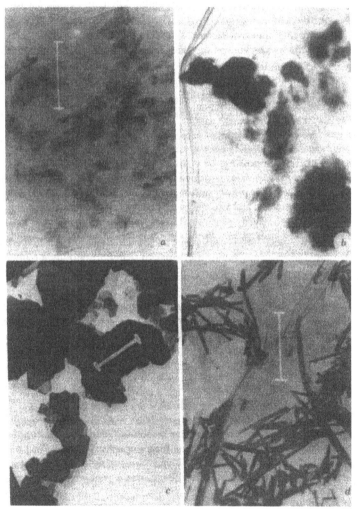

Figure 1.7   Electron micrographs of clay minerals: (a) beidellite from 100- to 500-nm Putnam clay; (b) montmorillonite; (c) kaolinite; (d) halloysite. Bars on each slide are 1 μm in length.

accurately by direct observation (Wilson, 1994). Photomicrographs of different clay particles such as those shown in Fig. 1.7 reveal the platelike attributes of the minerals. In Fig. 1.7a, showing beidellite in the 100- to 500-nm fraction of Putnam clay, the particles are very well defined and are obviously platelike, with a disk thickness of about 16 nm. Montmorillonite (Fig. 1.7b) shows a variety of structures, ranging from extremely thin plates to material that appears to be amorphous.

The roughly hexagonal shape of the plates of kaolinite, with sharp, well-defined crystal edges, is shown in Fig. 1.7c. In contrast, halloysite, a 1:1 lattice-type mineral like kaolinite, has definite rod-shaped particles that appear to be composed of twin sections (Fig. 1.7d). From the large body of information provided by electron micrographs, it has been established that the majority of the clay minerals found in soils are platelike or disk shaped.

1.1.6   Surface Area of Soil Particles

Relationship to Particle Size   Systems composed of dispersed particles of small size have a large amount of surface area per unit mass of material. Colloids, which possess very large surface area per unit mass, have enormously reactive properties. The amount of surface area of a dispersed system is usually expressed in terms of specific surface, which is defined as the surface area per gram or per unit volume of the dispersed phase. The relationship of specific surface to particle size is explored in Example 1.2.

Example 1.2   Calculate the specific area per unit mass s of a spherical particle of density ρ and radius R.

SOLUTION: The surface area a of a sphere may be expressed in terms of its radius by the formula $a = 4\pi R^2$. Similarly, the mass m of the sphere may be expressed as $m = \rho V = \rho 4\pi R^3/3$, where V is the volume of the sphere. Thus, the specific surface area s, or the area per unit mass, is given by

$$s = \frac{a}{m} = \frac{3}{\rho R} \tag{1.6}$$

Note that this is also equal to the specific surface area S of a total mass $M = Nm$ of identical spheres, since this mass has a total area $A = Na$.

As shown in Example 1.2, the specific surface area S of an arbitrary quantity of identical spheres (1.6) is inversely proportional to the radius of the sphere. This inverse relationship between particle size and specific surface area will also be true for other shapes of particles, as shown in Example 1.3.

Example 1.3   Calculate the specific surface area of a circular, disk-shaped platelet of density ρ, radius R, and thickness T. Assume that $T \ll R$.

SOLUTION: The total surface area of the disk is the sum of the two circular surface areas $2\pi R^2$ and the edge surface area $2\pi RT$. Thus, $a = 2\pi R^2 + 2\pi RT$. The mass m

of the platelet is the product of its density ρ and volume V, or $m = \rho V = \rho \pi R^2 T$. Consequently, the specific surface of the disk-shaped platelet is

$$s = \frac{a}{m} = \frac{2(R+T)}{\rho R T} \approx \frac{2}{\rho T} \tag{1.7}$$

since $T \ll R$.

Thus, the specific surface area of a thin disk-shaped platelet (1.7) is inversely proportional to the thickness of the platelet. The formulas in Examples 1.2 and 1.3 may be used to estimate specific surface areas for the soil particles associated with various textural classes. This procedure is illustrated in Example 1.4.

Example 1.4   Calculate the specific surface area in $m^2 \, g^{-1}$ of the following idealized soil particles: a 2-mm-diameter gravel sphere, a 0.05-mm-diameter sand sphere, a 0.002-mm-diameter silt sphere, and a circular clay platelet of diameter 0.002 mm and thickness 10^{-6} mm. For simplicity, assume that all particles have a mineral density $\rho = 2.7 \, g \, cm^{-3}$.

SOLUTION: Table 1.1 summarizes the calculated specific surface area of these prototype particles using (1.6) and (1.7) from Examples 1.2 and 1.3 and an effective diameter d_{eff} to characterize their size. Thus, the specific surface area of the clay platelet is almost 17,000 times larger than the specific surface area of the sand particle.

Disk-shaped particles also combine to create different bulk properties than spherical particles. For example, disk-shaped particles can be arranged in more intimate contact than spherical ones, creating more contact surface area when closely packed together and hence more cohesion. In addition, flat particles that are stacked together can slide over each other or shear under an applied force, whereas spheres that are in a closely packed hexagonal orientation will not shear until sufficient force has been applied to roll them farther apart.

The soil organic fraction often has a highly reactive surface that may have properties characteristic of material with an extremely high specific surface area (i.e., as high as $1000 \, m^2 \, g^{-1}$). Consequently, the apparent surface area of soil may be influenced significantly by the organic as well as the clay mineral fraction.

TABLE 1.1   **Specific Surface Area S of Idealized Shapes Representing Prototype Particles Found in Soil**

Particle	d_{eff} (cm)	Mass (g)	Area (m²)	S (m² g⁻¹)
Gravel	2×10^{-1}	1.13×10^{-2}	1.26×10^{-5}	0.0011
Sand	5×10^{-3}	1.77×10^{-7}	7.85×10^{-9}	0.044
Silt	2×10^{-4}	1.13×10^{-11}	1.26×10^{-11}	1.11
Clay[a]	2×10^{-4}	8.48×10^{-15}	6.28×10^{-12}	740

[a] $T \approx 10^{-7}$ cm.

TABLE 1.2   Specific Surface Area S^a, Cation Exchange Capacity CEC^b, and Density of Charge ζ^c of Various Clay Minerals

Clay Mineral	S (m^2 g^{-1})	CEC (cmol kg^{-1})	ζ (μmol m^{-2})
Kaolinites	5 – 20	3 – 15	6.0 – 7.5
Micas (illites)	100 – 200	10 – 40	1.0 – 2.0
Vermiculites	300 – 500	100 – 150	3.0 – 3.3
Smectites	700 – 800	80 – 150	1.1 – 1.9

[a] Data from Fripiat (1964).
[b] Data from Grim (1962).
[c] Calculated from S and CEC.

Relationship to Clay Mineralogy   The evidence obtained on the structure of clay minerals indicates that all groups have both external planar and external edge surfaces. Because their lattices expand, smectites and vermiculites also have internal planar surfaces. Table 1.2 shows the variations that occur in total surface area both between and within the clay mineral groups. Because they do not have interlayer surfaces and tend to form stacks containing many layers, kaolin minerals have a very low specific surface area compared to the other groups. The nonexpanding micas have a specific surface area about 10 times that of the kaolin group, whereas the specific surface area of the vermiculite minerals is about midway between the micas and the expanding-lattice smectites. Although there may be as much as 100% or more variation within the same group, the difference in surface area between groups is large enough to be significant.

The methods for the measurement of specific surface area usually make use of the adsorption properties of the mineral surfaces. A brief description of some of these methods follows the discussion of the surface properties of clay particles.

1.1.7   Surface Properties of Clay Particles

Density of Charge   The total charge on a mineral surface is called the *intrinsic surface charge density*, which is the sum of the structural surface charge density resulting from isomorphic substitutions in the soil mineral and the proton surface charge density resulting from proton-selective surface functional groups (Sposito, 1984). The structural charge density is a permanent charge determined by the degree of substitution in the crystal lattice and is independent of the conditions surrounding the mineral. In contrast, the proton surface charge density is due to the imbalance of complexed proton and hydroxyl charges on the surface, primarily on the exposed periphery of the mineral, and varies with the pH of the surrounding solution.

Most soils have a negative total surface charge density because of the negative charges on layer silicates and organic matter, but some highly weathered soils that contain substantial amounts of allophane and hydrous oxides can actually develop a net positive charge at sufficiently low pH. The pH dependence of colloids can be characterized experimentally by measuring the zero point of charge (ZPC), which is defined as the pH value at which the total surface charge vanishes.

The pH-dependent surface charge density will contribute significantly to mineral properties only for those clays that have a high surface area of positive edges of hydroxyaluminum. The vermiculite and smectite groups have a large permanent negative structural charge resulting from isomorphic substitution, which accounts for about 80% of the total charge density (Grim, 1962). In contrast, kaolinite has little isomorphic substitution and therefore a significant variation in its total charge density as pH is varied. Highly weathered soils, which are abundant in iron and aluminum oxides, also display significant pH dependence in their surface charge.

Because both the structural and proton-dependent surface charges depend strongly on mineral composition, the pH dependence of the charge density of different soils varies greatly. Figure 1.8 shows a graph of net surface charge versus pH for several clay mineral types. The negative charges of the 2:1 layer silicate clays montmorillonite and the Rothamsted subsoil (predominantly illite) vary little with pH, except in the high range. In these clays, particularly montmorillonite, the permanent negative charges overshadow the pH-dependent ones. There is a fairly sizable increase in net negative charge above pH 5.5 for Whatatiri soil clay (metahalloysite, gibbsite). Egmont soil clay, which contains allophane, shows the largest increase in net negative charge of any of the minerals shown, occurring principally between pH 6.0 and 7.6. Moreover, the Whatatiri and Egmont clays have a ZPC around pH 5.7 to 6.0, so that anion adsorption becomes the dominant exchange reaction at low pH.

Ionic Adsorption   A detailed discussion of the physical chemistry of ion exchange in soil is beyond the scope of this book and is reviewed elsewhere (e.g., McBride, 1994; Sposito, 1981). However, since many of the physical properties of soils are affected by adsorbed ions, it is important to characterize the amount of exchangeable

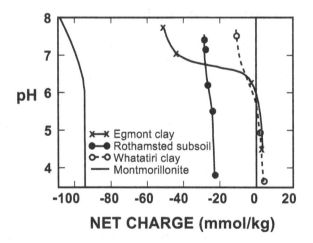

Figure 1.8   Net electric charges of clays in relation to pH. (After Fieldes and Schofield, 1960.)

ions per unit mass and the energy with which different ions are held on the surface of soil minerals.

Cation Adsorption   The negative charges present on soil minerals produce a positive electrostatic attraction for cations in soil solution. As a result, the fluid in the immediate vicinity of a negatively charged mineral surface has an excess of cations and a deficit of anions compared to the bulk solution farther away from the surface. Cations present in this surface fluid layer are retained by electrostatic attraction to the charged surface unless they exchange with other cations in solution. The total quantity of cations that may be retained in this manner is the cation exchange capacity (CEC) of the soil. The CEC values of different clay mineral groups are influenced principally by the lattice construction and the degree of isomorphic substitution. The order of CEC for the various mineral groups is vermiculite > smectite > mica > kaolinite (Table 1.2).

The CEC is an increasing function of the amount of internal and external surface and therefore should increase with decreasing particle size. The dependence on particle size is most pronounced for those clay minerals in which broken bonds are responsible for many of the exchange sites. One would not normally expect the expanding-lattice clays to exhibit increasing CEC with decreasing particle size, since 80% of the adsorption sites are on the planar surfaces (Grim, 1962). The relationship between CEC and particle size for the smectite mineral beidellite is illustrated in Fig. 1.9.

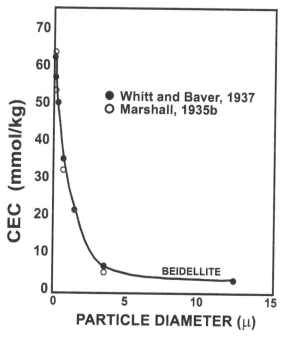

Figure 1.9   Relationship between CEC and particle size for the smectite mineral beidellite.

Diffuse Double Layer   Several models have been proposed to describe the distribution of cations and anions in the water layer adjacent to a clay mineral. The simplest representation, the *Helmholtz model*, assumes that all balancing cations are held in a single layer between the mineral surface and the bulk solution. However, such a picture is oversimplified, because cations possess thermal energy that will cause a dynamic concentration gradient to form away from the surface, creating a diffuse double layer (Bolt, 1982; van Olphen, 1963).

Standard but cumbersome calculations using potential theory may be performed to calculate cation and anion concentrations within the double layer between the mineral surface and the bulk solution. If the entire double layer is regarded as diffuse, the model is called a *Gouy–Chapman double layer* (Bohn et al., 1979), and the calculated concentration gradients change smoothly from the mineral surface to the bulk solution, as in Fig. 1.10*a*. If the double layer is treated as having a rigid region next to the mineral surface and a diffuse layer joining with the bulk solution, the model is called a *Stern double layer* (Bohn et al., 1979). Then the calculated concentration gradients are less steep in the diffuse layer since the rigid layer (commonly called the *Stern layer*) acts to lower the surface charge, which must be neutralized by the solution (Fig. 1.10*b*). Since anions are also attracted into the diffuse region by the cation excess, the total cation concentration must neutralize both the anion charge and the surface charge. The cations that neutralize the surface charge are exchangeable with other cations in solution.

Addition of more electrolytes to the system will produce a compression of the double layer and a decrease in the electric potential. This is especially true at high concentrations, as seen by comparing areas ABD and A'B'D' and diffuse layer thicknesses BD and B'D' in Fig. 1.10*a*.

The thickness of the liquid layer around the soil particle is assumed to be infinite in the double-layer model calculations. If the calculated double-layer thickness is

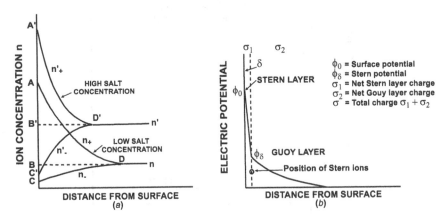

Figure 1.10   Diffuse double-layer relationships: (*a*) concentration of ions versus distance from surface; (*b*) electric potential of Stern and Gouy layers. (Adapted from van Olphen, 1963. Used with permission of John Wiley & Sons, Inc.)

greater than the actual thickness of the liquid layer, the theory is not applicable. The double layer can never extend beyond the surface of the liquid layer since the cations necessary to neutralize the surface charge must remain within the liquid layer. Therefore, when the liquid layer is small, the concentration never reaches a constant value (i.e., there is no equilibrium solution). If water is added to such a system, the double layer will expand and produce swelling in the medium unless the particle binding forces are strong enough to oppose the swelling pressure. A sodium–montmorillonite has few binding forces and swells freely as water is added. Conversely, a calcium–montmorillonite is strongly bound and tends to swell only to a very limited extent.

Corrections to the Gouy–Chapman and Stern theories, which take into account the influence of ionic interactions, polarization of ions, and dielectric saturation, are small for colloidal clay suspensions if the density of charge does not exceed 2 to 3 μmol m^{-2} (Bolt, 1955). However, the size of the ion and its extent of hydration have a considerable influence on the distribution of cations between the rigid Stern layer and the diffuse portion of the double layer (Shainberg and Kemper, 1966, 1967). This effect may be visualized by making the following assumptions:

1. The exchange sites in the surface are fixed charges.
2. Part of the surface charges are neutralized by ions in the Stern layer.
3. The remaining surface charge is balanced by cations in the diffusion portion of the double layer.
4. The balancing cations may be either fully hydrated or unhydrated with no water molecules between them and the adsorbing surface.

The total density of charge Q_t on the mineral surface is equal to the sum of the charge densities of the hydrated (Q_h) and unhydrated (Q_u) charge-balancing cations:

$$Q_t = Q_h + Q_u \tag{1.8}$$

The unhydrated ion is assumed to be directly on the surface in the Stern layer, whereas the hydrated ions reside in the diffuse phase because of their larger sizes. The hydration shell around a sodium ion, for example, would increase the distance from the center of the ion to the plane of negative charge in the octahedral sheet of the crystal lattice of montmorillonite by over 50% above that of the dehydrated cation. Computations of the distribution of the charge densities of monovalent cations in the double layer in relation to their hydration and polarization energies are shown in Table 1.3. These results confirm the well-known lyotropic series describing the ease of replacement of adsorbed cations: Li > Na > K > Cs. Only 17% of the highly hydrated lithium ion is in the Stern layer, compared with 76% of the unhydrated cesium ion. The difference between sodium and potassium ions is of the same order of magnitude as the variations in their maximum bonding energies.

Cation Exchange Equations   Many different equations have been proposed to model cation exchange reactions. Common features of these models are that they

TABLE 1.3   Distribution of Monovalent Cations between Stern and Diffuse Layers

Ion	Radius (Å)	Q_u (C m^{-2})	Q_h (C m^{-2})	Percent in Stern Layer
Li	0.60	0.017	0.080	17.3
Na	0.95	0.037	0.090	41.2
K	1.33	0.057	0.039	59.1
Cs	1.69	0.070	0.026	76.0

assume the mineral surface has a constant total CEC at a given pH and the exchange reaction is stoichiometric and reversible. The most general type of reaction model is the mass action equation, which describes $Ca^{2+} - Na^+$ exchange as follows (Bohn et al., 1979):

$$CaX + 2Na^+ \leftrightharpoons 2NaX + Ca^{2+} \tag{1.9}$$

where X represents the exchange phase of the cation. The equilibrium reaction coefficient for the reaction in (1.9) is therefore

$$K_K = \frac{(NaX)^2(Ca^{2+})}{(CaX)(Na^+)^2} \tag{1.10}$$

where $(\cdot)$ denotes activity.

All cation exchange models are limited by the fact that there are no models or measurements that characterize the activity of the adsorbed phases exactly. As an approximation, the activities in (1.9) are frequently replaced by concentrations. This approximate representation is called a *Kerr model* and may be reasonably valid over narrow concentration ranges. However, the reaction coefficient varies substantially over the entire range of cation concentration in the exchange phase (Marshall and Garcia, 1959).

A modified mass action equation was developed by Gapon in 1933, who expressed the exchange reaction in terms of chemically equivalent quantities in solution and adsorbed phases and used concentrations in place of activities. These assumptions produce the following reaction coefficient:

$$K_G = \frac{[NaX][Ca^{2+}]^{1/2}}{[Ca_{1/2}X][Na^+]} \tag{1.11}$$

where $[\cdot]$ denotes concentration. The Gapon equation (1.11) is inaccurate over large ranges of exchange concentrations, but it is reasonably accurate within the range of values of sodium exchange concentration found in most agricultural settings (U.S. Salinity Laboratory Staff, 1954).

Other exchange models have been proposed that differ primarily in the way they treat the activity of the exchange phases. For example, Vaneslow (1932) assumed that the activities of the exchangeable cations were proportional to their mole fractions.

This model works best when the ions are of equal charge, such as with $Ca^{2+} - Mg^{2+}$ exchange (McBride, 1994).

Anion Adsorption   Anions react with mineral surfaces in two different ways. The first reaction is the nonspecific electrostatic attraction to or repulsion from the positively or negatively charged mineral surface. Since most minerals are negatively charged at normal pH, the net anion adsorption is actually a repulsion from the surface, sometimes called *negative adsorption* or *anion exclusion*. This process has an influence on chemical transport through soil, as it embodies anions with a somewhat higher mobility than neutral species moving with water by excluding the anions from the slowest-moving portion of the water volume closest to the stationary solid surfaces.

The second anion reaction with soils is ion specific and involves positive adsorption to oxide surfaces. This reaction, known as *ligand exchange*, is postulated to occur when the oxygen ions on a hydrous oxide surface are replaced by anions that can coordinate with Al^{3+} or Fe^{3+} ions. The reaction can occur on surfaces with positive or negative charge and is selective for different anions in the order $SiO_4^{2-} > PO_4^{2-} \gg SO_4^{2-} > NO_3^- \sim Cl^-$ (Bolt and Bruggenwert, 1976). At normal pH, the adsorption of SO_4^{2-}, NO_3^-, or Cl^- is usually negligible, and these ions experience a negative net adsorption from electrostatic interactions.

Adsorption of Nonelectrolytes   Nonelectrolytes in solution may also be adsorbed by soil solids, either as a result of partial ionization in solution or by other mechanisms to be discussed in what follows. Some nonelectrolytes will accept protons in acid solutions, forming a cationic complex that will react with the negatively charged mineral surfaces. In contrast, organic acids such as 2,4-D, dinoseb, and picloram will dissociate into anionic complexes in solution (Green, 1974).

Nonionic species that do not ionize in solution can still react with mineral surfaces by hydrogen bonding and van der Waals reactions. The hydrogen bond reaction between a nonelectrolyte and a mineral surface occurs between organic functional groups and either siloxane oxygen atoms or surface hydroxyl group but does not appear to be a significant factor in the adsorption of dissolved organics to soil (Sposito, 1984). The reaction may be much more important for adsorption to organic matter surfaces (Burchill et al., 1978).

For nonpolar organic compounds such as DDT, van der Waals forces induced by instantaneous dipole moments in nearby nonpolar molecules are the principal mechanisms causing adsorption, which occurs predominantly on organic surfaces. The structure of soil organic matter is so complex, however, that the relative importance of specific reaction mechanisms is poorly understood.

1.1.8 Clay Flocculation and Swelling

Clay particles have complex charge surfaces, resulting potentially in attraction to or repulsion by other particles, depending on conditions. The repulsive forces are electrostatic, arising from the cations in the double layer surrounding the negatively charged particles surfaces. If clay platelets get close enough, however, the double-layer cations merge into a single positive layer that attracts the negatively charged surfaces of the platelets. The resulting stack of clay minerals, called a *tactoid*, can involve many individual platelets. The positively charged edges contribute to flocculation because of the electrostatic attraction between positive edges and negative planar faces of adjacent minerals (Schofield and Samson, 1954). In the presence of electrolytes, the thickness of both diffuse double layers is diminished and flocculation can be enhanced by face-to-face or edge-to-edge association of different minerals because of van der Waals forces. Flocculation is generally a desirable state because it stabilizes the soil and prevents clay migration and clogging.

The extent of hydration of exchangeable cations has long been known to be a dominant factor in the stability of clay suspensions (Jenny and Reitemeier, 1935; Tuorila, 1928; Wiegner, 1925). Divalent exchangeable cations result in flocculated clay systems; monovalent exchangeable cations produce dispersed systems.

As discussed in Section 1.1.3, the 2:1 layer clay mineral has a water layer between its sheets. In the case of smectites, addition of water can cause substantial lattice expansion and promote dispersion. When polyvalent cations such as Ca^{2+}, Mg^{2+}, or Al^{3+} are in the double layer, however, they form cation bridges that oppose the swelling and keep the clay flocculated (McBride, 1994). In contrast, monovalent Na^+ does not form a bridge, and the swelling process in a sodium-dominated exchange process can lead to dispersion and soil clogging. Soils high in exchangeable sodium have very poor structural characteristics, leading to swelling, crusting, poor drainage, and tendency to erosion. Thus, the exchangeable sodium percentage

$$ESP = 100 \times \frac{NaX}{CEC} \qquad (1.12)$$

where NaX is the exchangeable sodium concentration, is used to indicate the sodicity of the soil, and therefore its tendency to disperse (U.S. Salinity Laboratory Staff, 1954). The sodium adsorption ratio (SAR) is a characteristic of the soil solution, defined as (McBride, 1994)

$$SAR = \frac{[Na^+]}{\sqrt{([Ca^{2+}] + [Mg^{2+}])/2}} \qquad (1.13)$$

where all concentrations [·] are in mmol L^{-1}. SAR and ESP are highly correlated in soil, and can be related reasonably well by the equation (McBride, 1994)

$$\frac{ESP}{100 - ESP} = 0.015 \, SAR \qquad (1.14)$$

thereby allowing the exchange composition to be inferred from the solution composition.

Adsorption Isotherms   The adsorption of a substance in solution on soil particle surfaces is usually pictured as occurring over one molecular layer on the surface.

Beyond that distance, the substance is assumed to be in the bulk solution. The relation between the amount of substance adsorbed and the concentration of the substance in solution at any given temperature is known as the *adsorption isotherm*, one of the fundamental experimental characteristics of a dissolved chemical in soil.

Isotherms may have a variety of shapes depending on the characteristics of the adsorbent and adsorbing surface and sometimes on other constituents in solution. Although many isotherm shapes have no direct physical interpretation, the Langmuir isotherm (Langmuir, 1918), relating the adsorbed C_a and solution C_l concentrations

$$C_a = \frac{aQC_l}{1 + aC_l} \tag{1.15}$$

may be derived from a set of reasonable assumptions, as shown in Example 1.5.

Example 1.5   Derive the relationship between C_a and C_l for a surface that has a finite number Q of identical adsorption sites per unit mass, assuming that the adsorbing molecules do not interact with each other.

SOLUTION: At equilibrium the rate of adsorption r_A must equal the rate of desorption r_D. If the molecules do not interact with each other, the rate of desorption r_D should be proportional to the number of adsorbed molecules per unit mass of surface C_a, which we may write as

$$r_D = k_1 C_a$$

where k_1 is a rate constant. Similarly, the rate of adsorption r_A should be proportional to the concentration C_l in solution but also to the number $Q - C_a$ of unfilled sites on the surface. Thus,

$$r_A = k_2(Q - C_a)C_l$$

At equilibrium, $r_A = r_D$, so that we may equate

$$k_1 C_a = k_2(Q - C_a)C_l$$

This expression may be solved for C_a, with the result

$$C_a = \frac{aQC_l}{1 + aC_l}$$

where $a = k_2/k_1$.

The Langmuir isotherm is shown in Fig. 1.11a for various values of a. This type of isotherm, in which C_a increases linearly with C_l at low concentrations and approaches the constant Q at high concentrations, is most appropriate in soil for processes such as cation exchange, which have a finite adsorption capacity.

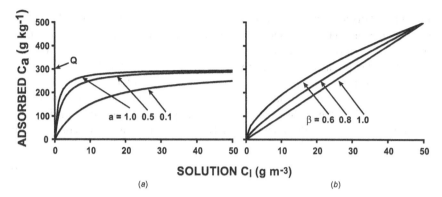

Figure 1.11   (*a*) Langmuir and (*b*) Freundlich adsorption isotherms.

Many compounds do not adsorb to soil according to the Langmuir isotherm but instead, act as though the surface contained different types of adsorption sites. For these compounds, the shape of the isotherm may often be described by the Freundlich isotherm

$$C_a = K_f C_l^\beta \tag{1.16}$$

where K_f and β are constants with $\beta \leq 1$. Plots of the Freundlich isotherm are shown in Fig. 1.11*b*. Sposito (1981) has shown that (1.16) may be derived by assuming that the soil surface is composed of a distribution of adsorption sites, each of which obeys a Langmuir isotherm. A special case of the Freundlich isotherm for $\beta = 1$,

$$C_a = K_d C_l \tag{1.17}$$

is called the *linear isotherm*, where K_d is the distribution coefficient.

The adsorption isotherm is extremely important in assessing the mobility of dissolved chemicals in soil. This subject is developed fully in Chapter 7.

Adsorption of Gases   The adsorption of gases on solid surfaces has been studied in a number of different research disciplines. Although many surfaces have only a finite number of sorption sites for gases to adsorb onto, the Langmuir adsorption isotherm does not describe the process well, because more than one monomolecular layer of gas can be condensed onto a surface. Brunauer et al. (1938) extended the Langmuir approach to multilayer adsorption by assuming that the heat of adsorption of the first layer of gas has a higher value than the heat of adsorption of all succeeding layers, the latter of which is set equal to the heat of vaporization of the liquid adsorbate. This leads to an equation of the form

$$\frac{x}{V(1-x)} = \frac{1}{V_m C^*} + \frac{(C^* - 1)x}{V_m C^*} \tag{1.18}$$

where $x = P/P_0$ is the ratio of the vapor pressure of the gas to the saturated vapor pressure of the liquid at the temperature specified, V is the volume of gas adsorbed, V_m is the volume adsorbed when the surface is covered with one monomolecular layer, and C^* is a constant related to the heat of adsorption. Equation (1.18) is known as the Brunauer–Emmett–Teller (BET) equation, whose parameters can be evaluated from adsorption data if the isotherm has a shape compatible with the model.

Commonly, the left side of (1.18) is plotted versus x to determine if the BET equation is valid. If the plot is linear, the model is regarded as correct, and the BET parameters V_m and C^* are calculated from the slope and intercept. The BET equation appears to work reasonably well for the adsorption of nitrogen on soil minerals at the temperature of liquid N_2.

Measurement of Specific Surface Area   Several methods involving the sorption of chemicals whose surface area coverage is known have been developed for measuring the external and internal surfaces of clays. Dinitrogen (N_2), a nonpolar gas, does not have access to the interlayer planes of expandable clay minerals, and therefore can be used to estimate the external areas. Polar molecules such as ethylene glycol (EG) or ethylene glycol monoethyl ether (EGME) do penetrate the interlayer surfaces and can estimate the total surface area (Pennell, 2002). The BET equation (1.20) has been used to interpret the adsorption of nitrogen. A common technique involves the use of a small glass bulb containing the solid sample to which N_2 is introduced in small increments and allowed to equilibrate. The mass of gas adsorbed may then be calculated from the change in pressure and an isotherm produced. The surface area is then calculated from the BET parameter V_m, describing the volume of a monomolecular layer of gas and the known surface area per unit volume of N_2 (Pennell, 2002).

The method for measuring total surface area by adsorption of polar molecules such as EG or EGME is to saturate a dry surface with the polar liquid in a vacuum desiccator. The excess is removed under vacuum over either anhydrous $CaCl_2$ or a $CaCl_2$–ethylene glycol mixture. When the weight of the clay mineral–glycol mixture reaches a constant value, the surface area is calculated from the equation

$$S = \frac{W_g}{\Omega W_a} \tag{1.19}$$

where S is the specific surface in $m^2\ kg^{-1}$, W_g the weight of the polar compound retained in the sample, W_a the weight of the oven-dried sample, and Ω a constant that represents the grams of polar liquid required to form a monomolecular layer per square meter of surface. For EG and EGME, $\Omega = 3.1 \times 10^{-7}$ and 2.86×10^{-7} kg m^{-2}, respectively [see Pennell (2002) for details]. Table 1.4 shows external and total specific surface areas measured by N_2 and EG/EGME sorption for a number of soils and clay minerals.

Measurement of Cation Exchange Capacity   There are a number of different methods used for the determination of the CEC of soil, but most of them use the same

TABLE 1.4   Comparison of Specific Surface Values Obtained by N_2 Gas Adsorption and EG/EGME Retention for a Range of Soils and Clay Minerals

Soil Sample	Organic C Content (g kg^{-1})	N_2 Surface Area (m^2 g^{-1})	EG/EGME Surface Area (m^2 g^{-1})
Wyoming bentonite[a]	0.0	65.0	372.0
Silica[b]	0.1	4.9	8.7
Kaolinite[b]	0.1	8.5	21.3
Lula aquifer material[c]	0.1	7.7	10.5
Montmorillonite[b]	0.2	97.4	733.0
Boston silt[a]	26.6	28.6	46.0
Webster soil[d]	33.2	8.2	168.4
Ashurst soil[a]	45.5	6.3	25.8
Houghton muck[d]	445.7	0.8	162.9

Source: After Pennell (2002).

[a] Call (1957).
[b] Ong and Lyon (1991).
[c] Rhue et al. (1988).
[d] Pennell et al. (1995).

principle, which consists of adding a sufficient amount of new cation to saturate the exchange sites completely and measuring the total quantity of cations displaced from the surfaces by the replacing cation. Two common reagents used for this purpose are ammonium acetate (pH 7.0) and sodium acetate (pH 8.2) (Chapman, 1965). Various sources of error in this determination are discussed in Bohn et al. (1979).

1.2   COMPOSITE SOIL PROPERTIES

Many of the important transport and retention processes in the soil are influenced strongly by the macroscopic composite or bulk properties of the soil matrix. These properties are commonly characterized with samples that contain many individual soil particles, void spaces, and water films. Hence, composite soil properties are said to be volume averaged.

1.2.1   Volume Fractions

A soil volume V containing a total mass M of solid, liquid, and gaseous material may be partitioned into contributions from the various constituents within it. For example, the volume may be subdivided into

$$V = V_g + V_l + V_s \tag{1.20}$$

where g, l, and s refer to gaseous, liquid, and solid phases, respectively. If the total volume V is divided into each side of (1.20), the result may be written as

$$\phi = 1 - \frac{V_s}{V} = a + \theta \tag{1.21}$$

where ϕ is the volume of void space per total volume and is called the *porosity*, a is the volume of gas space per total volume and is called the *volumetric air content*, and θ is the volume of liquid per total volume and is called the *volumetric water content*.

1.2.2 Bulk and Particle Densities

It is often useful to characterize the density of the solid soil matrix. The bulk density ρ_b is defined as the mass of dry soil per volume of soil. It is equal to the density of the composite solid matrix in its natural configuration. The particle density ρ_s is defined as the mass of dry soil per volume of soil solids. It is equal to the density of the solid material comprising the soil matrix. The relationship between ρ_b and ρ_s is derived in Example 1.6.

Exampel 1.6   Derive a relationship between ρ_b and ρ_s using the previous definitions.

SOLUTION: By definition,

$$\rho_b = \frac{m_s}{V} = \frac{m_s}{V_s}\frac{V_s}{V}$$

where m_s is the mass of dry soil in V. Thus, since $\rho_s = m_s/V_s$ and $1 - \phi = V_s/V$ by (1.21), then

$$\rho_b = \rho_s(1 - \phi) \tag{1.22}$$

Measurement of Bulk Density   There are several recognized field methods for measuring bulk density, depending on the state of the soil (Grossman and Reinsch, 2002). If the soil is cohesive enough, a soil core sample may be taken that does not appreciably deform the sample from its preexisting state. The soil is dried for 24 h at 105°C and its dry mass determined by weighing. The volume is assumed to be that of the sampling tube. If the soil is too dry or of insufficient structure to remain in a sampling tube, a sample may be removed from the soil and the volume of the remaining hole measured by filling it with a known quantity of sand or other substance. If the sidewalls of the hole are sealed, water may also be used as fill. If the soil is highly structured, an intact clod may be removed and coated with paraffin or a resin, after which it is weighed in water and in air to determine the volume of water displaced (Blake and Hartge, 1986). Bulk densities may also be determined on intact clods of soil by coating their surfaces with a thin layer of paraffin and measuring the volume of displaced water when the clod is lowered into a full beaker.

Measurement of Particle Density   The most common way to determine soil particle density is to weigh a dry soil sample in air and in water. From the weight in water the volume of the soil solids can be determined. A constant-volume container called a pycnometer bottle is often used for this purpose (Flint and Flint, 2002).

The particle density of a soil can be estimated from the weighted average of the solid components. Although different minerals in soil have different densities, many commonly occurring minerals have a density ρ_m of about 2.65 g cm^{-3}. In contrast, soil organic matter has a much lower density ρ_{om} of about 1.3 g cm^{-3}. The following equation can be used to estimate soil particle density:

$$\rho_s = \rho_m X_m + \rho_{om} X_{om} \tag{1.23}$$

where X_m and X_{om} are mineral volume fraction and organic matter volume fraction, respectively, expressed as a fraction of the total solid volume.

Example 1.7 Estimate the particle density ρ_s of a soil that has 95% of the solid fraction as minerals.

SOLUTION: If we assume that $\rho_m = 2.65$ g cm^{-3} and $\rho_{om} = 1.3$ g cm^{-3}, then by (1.23) we have

$$\rho_s = 2.65 \times 0.95 + 1.3 \times 0.05 = 2.58 \ \text{g cm}^{-3}$$

1.2.3   Soil Horizons

Soils develop distinct layers below the soil surface, in response to the action of climate and chemical weathering processes. The upper layer, the *A horizon*, comprises the topsoil. It is usually higher in organic matter than the layers beneath it, and often darker in color. Below this layer is the *B horizon*, the subsoil layer. It is usually higher in clay than the surface layer, and brighter in color. The *C horizon* lying underneath the B horizon consists of the parent material from which the soil was formed and extends downward into bedrock. It can be of any thickness, or even missing from a profile. The A and B layers together with the upper part of the C horizon are considered to be the soil profile. There are numerous other subclassifications within these horizons. For a full description of soil taxonomy, see Soil Survey Staff (1999).

1.2.4   Soil Structure

Soil structure and its stability play an important role in a variety of processes in the soil, such as erosion, infiltration, root penetration, aeration, or mechanical strength. Since these processes all have observable characteristics, soil structure is often evaluated by methods that correlate it to the properties of the process of interest. An agricultural scientist may be interested in *soil tilth*, which is a general term that signifies the ability of the granules or aggregates to withstand destruction by the impact of implements, raindrops, or running water. Tilth also involves water retention and the workability of soils, both of which depend to a large extent on the basic textural nature of the soil. In contrast, a plant scientist may be interested primarily in the effect of soil structure on root penetration and may therefore seek to correlate penetrometer resistance or root growth with aggregate size or some other index of structure. Soil structure can be evaluated by determining the extent of aggregation, the stability of the aggregates, and the nature of the pore space (Dullien, 1979). These characteristics, which significantly influence plant response to water management practices, will change with tillage practices and cropping systems. Generally, soil structure refers to the arrangement and organization of the soil solid phase; however, soil pore structure is more and more being considered to be the most relevant index of soil orientation.

Soil structure may be classified into three broad structural groups: single-grained, massive, and aggregated. The *single-grained* designation refers to the geometric arrangement of soil separates into a porous formation, with little or no cementation of particles. Coarse-textured sands frequently orient in this manner, producing a rigid solid framework made up of primary minerals of roughly spherical shape. These soils are often called *structureless*, to indicate the absence of cementation or flocculation among its constituents. A *massive* structure represents the opposite extreme of a complete consolidation of soil with no apparent lines of weakness. This arrangement is typical of parent material that has a high clay content (Troeh and Thompson, 1993).

In between these two extremes are *aggregated* soils, in which individual particles stick together to form larger units of various shapes and strength, called *peds*. These structural units form by weathering out of the parent material under the influence of wetting and drying cycles, freezing and thawing, plant roots, animal burrows, or other disruptive forces. Plants, animals, and the microbial community produce cementing agents that help bond the peds together. Ped structure has been subclassified into four groups: granular, platy, blocky, and prismatic, to reflect the shapes formed (Troeh and Thompson, 1993). A schematic of the six structural categories is given in Fig. 1.12.

Aggregation is a highly desirable state for soils, because it promotes aeration and water transmission and prevents erosion. Soil aggregation is characterized experimentally by the size distribution, quantity, and stability of the aggregates. These parameters of aggregation are important in determining both the amount and distribution of the pore spaces associated with the aggregates and the susceptibility of the soil to water and wind erosion. Testing procedures to determine aggregate strength and stability properties must be standardized because they represent artificial measures of the behavior of the aggregates under natural conditions.

The *stability* of structure refers to the resistance that the soil aggregates offer to the disintegrating influences of water and mechanical manipulation. Water content is important in structural stability and is almost always a factor in determining the degree to which particular mechanical forces will cause structural breakdown. For example, rather compact and coherent aggregates may be found in the dry state, but if these secondary particles disintegrate in water, the aggregation is not very stable. Water may cause the deterioration of aggregation in two ways. First, hydration causes a disruption of the aggregate through the processes of swelling and the exploding of entrapped air. Second, the impact of falling raindrops on exposed soil can break up the aggregates. The dispersed particles are then carried into the soil pores, causing increased compaction and decreased porosity. Intense rains destroy the granulation and open structure of the top few centimeters or more of soil to form a dense,

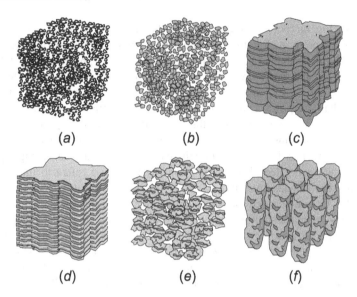

Figure 1.12 Six categories of soil structure: (*a*) single-grained; (*b*) granular; (*c*) massive; (*d*) platy; (*e*) blocky; (*f*) prismatic. (Adapted from USDA Agricultural Information Bulletin 199.)

impervious surface known as a *crust*. This type of structure degradation is least common with aggregates that are stabilized with humus or iron compounds. In many instances, raindrops are the major cause of the dispersion of soil aggregates. Their immediate influence is confined to a shallow layer in the surface, but the structure of this layer may be broken down to limit the air and moisture relations of the entire profile.

Cultivation and other tillage operations generally cause a continued decrease in the stability of aggregates unless the organic matter level of the soil is kept relatively high and mechanical manipulation of the soil is performed at optimum moisture contents. Tisdall and Oades (1982) studied soil organic matter and water stability of soil aggregates. They classified soil organic binding agents based on turnover rates in soil. They also looked at occurrence of the agents in different aggregate size fractions and found that large aggregates had temporary binding agents, while microaggregates had more persistent ones. Tillage management seems to influence the temporary and transient binding agents more than it does the persistent binding agents. Thus, tillage management will more likely influence the larger aggregates far more than soil microaggregates.

Aggregate Analysis   In their review of the principal methods for characterizing aggregates, Nimmo and Perkins (2002) identify four properties: dry-aggregate size distribution, wet-aggregate size distribution, dry-aggregate stability, and wet-aggregate stability. Aggregates should be dried under humidity and temperature conditions representative of the field regime rather than in an oven, after which they are sieved to determine their distribution. Air-dry sifting is considered to give a better picture of aggregation of some arid-zone soils than wet sieving, since the aggregates are so weakly held together when moist that the mechanical action of sieving is sufficient to destroy them. Clogging of the flat sieves and breaking up of weak aggregates by the mechanical action required in the sieving operation are major problems with this technique. These difficulties have been overcome by the use of a rotary sieve (Chepil, 1962; Lyles et al., 1970). Dry sieving of aggregates provides an important index for characterizing the susceptibility of soils to wind erosion. No special preparation of the sample is required for consistent results.

The greatest problem in wet sieving is achieving a consistent method of wetting the sample for analysis. Aggregate wetting procedures are quite varied, including rapid wetting of dry aggregates under high vacuum, slow wetting by vapor or wicking from moist filter paper, or by organic solvents (Nimmo and Perkins, 2002). Elliot (1986) compared several aggregate wetting techniques and found a significant correlation between size distribution and the initial water content of the samples. The wet-aggregate size distribution is determined by a specialized sieving procedure. In this method, the soil is spread on top of a nest of sieves (4.76, 2.00, 1.00, and 0.21 mm) immersed in water with the finest sieve on the bottom. The sieves are slowly raised and lowered in the water 30 times by standard mechanical procedures (Kemper, 1965; Yoder, 1937). The weight of soil on each sieve is then determined.

Dry aggregate stability was originally characterized by the percentage of sample weight remaining after a prescribed period of dry sieving (Chepil, 1953). Another method is the drop shatter procedure of Marshall and Quirk (1950), which measures the height of drop required to shatter an aggregate on a hard plate. More recently, Perfect and Kay (1994) have introduced the rupture threshold method, which directly measures the force required to rupture an aggregate between parallel plates.

Wet aggregate stability is determined by standardized action on wet samples placed on a sieve of 0.26-mm aperture for a specified time period while being subjected to oscillation to promote water flow across the aggregates. Those aggregates that do not pass through the sieve during the test are deemed to be stable (Kemper and Rosenau, 1986).

1.2.5 Soil Strength

The resistance that soil offers to penetration or shearing may be informally thought of as the strength of the soil. Soil strength is clearly important in seedling establishment and subsequent root growth, and also in tillage. The resistance to penetration is measured with a soil penetrometer, which consists of a rod terminating in a cone or flat end. The flat end is most useful in near-surface evaluations, while the cone with sharpened tip is used for deeper penetration into the soil profile. Various devices are used to record the force required to push the device into the soil (Lowery and Morrison, 2002).

Another index of soil strength used by soil engineers is the *modulus of rupture*, which is the force required per unit area to break an intact soil block apart. It has been modified to simulate soil crust formation by conducting the test on a prepared sample which is wetted and dried in a rectangular mold before the test is conducted (Richards, 1953). More recently, soil resistance to breaking and deformation has been generalized into a test of shear strength. A thorough discussion of this concept and the methods used to characterize it is given in Fredlund and Vanapalli (2002).

1.2.6   Crust Formation

Formation Mechanisms   Soil crusts usually are formed as a result of compaction at the immediate surface due to an externally applied force. This force is supplied primarily by the impact of raindrops as the soil is wetted and the radiant energy of the sun as the soil dries. When the raindrops fall on dry soils, there is almost instantaneous slaking of the soil aggregates, followed by dispersion and orientation of the finer particles, and the clogging of the pores as these particles are carried into the soil. A compacted zone of higher bulk density is formed at the surface. The soil crust formed by raindrop impact consists of two distinct parts (McIntyre, 1958). First, there is a thin skin approximately 0.1 mm thick formed on the surface due to the compaction that results from the impact. Second, the dispersed particles that arise from impact are washed into the soil with the infiltrating water and clog the pores immediately beneath the surface to give a layer of decreased porosity. The water permeability of the "washed-in" zone is reduced about 200 times below that of the undisturbed soil beneath; that of the surface skin is reduced 2000 times. Aggregates from virgin soils with rather stable structure slake but do not disperse under raindrop impact. Consequently, there is no washed-in zone and the water permeability of the crust is a function of the permeability of the thin surface skin. The zone immediately below the thin surface skin consists of oriented clay particles with very few isolated air pores (Evans and Buol, 1968). It seems, therefore, that dispersion of aggregates rather than slaking is responsible for soil crusting. Upon drying, surface tension forces cause particle interaction and orientation as shrinkage takes place.

Particle rearrangement at the immediate surface can also be brought about by intense slaking and dispersion of aggregates when the soil is wetted to saturation, as is often the case in surface irrigation. A compacted zone is formed during infiltration of the turbid water, which becomes a hard crust upon drying.

Evaluating Crust Strength   Evaluation of the strength of soil crusts can be made in at least three ways. In addition to the modulus of rupture and soil penetrometer methods discussed previously, the determination of crust strength can be approached from the point of view of the emerging seedling (Arndt, 1965a,b; Morton and Buchele, 1960). Morton and Buchele developed a penetrometer to simulate a mechanical seedling and measured the force required to push it upward through a 3-in. layer of soil compacted to different bulk densities. Using various sizes of probes to represent different seedling diameters, they found that the emergence energy increased directly with seedling diameter, degree of compaction, initial soil-moisture content, and depth of planting.

Arndt (1965a,b) designed an instrument that records the force required to bring about emergence of a mechanical probe that is buried prior to formation of a surface crust. This type of probe does not penetrate the crust as it is forced upward. In the case of a wet crust, there is a conical rupture as a cone-shaped mass of soil is forced out of the crust ahead of the probe. In a dry soil, there is first a cracking of the crust, followed by formation of a dome-shaped structure of tilted, broken pieces of crust. The modulus of rupture values, which reflect tensile strength, were only about 20 percent of the force required for emergence of the in situ probe through the crust. This probe produces both compression and shear stresses.

Crust strength, as measured by the modulus of rupture technique, has been observed to increase as the rate of drying decreases (Hillel, 1960; Lemos and Lutz, 1957). These results are not in agreement with the observations of Arndt, who found that larger impedance values were obtained with his method when the evaporation rates were high. Crust strength is greater when the rain falls on air-dry rather than wet soil (Hanks, 1960). There is a direct relationship between clay content and crust strength when the soil does not contain organic matter.

The mechanical composition of the surface soil plays an important role in crust formation, determining the strength of the crust as well as the frequency and width of cracking on drying (Arndt, 1965a). One must also consider the tremendous effect of organic matter on increasing the resistance of soil aggregates to the destructive impact of raindrops.

Seedling Emergence and Crust Characteristics   Restriction of seedling emergence can take place in two ways. First and foremost is the effect of mechanical impedance due to crust strength. Resistance at the surface may be so great that the seedlings buckle, grow in a horizontal direction, and fail to emerge. The emergence of grain sorghum seedlings decreased if the soil strength, as measured with a penetrometer, exceeded 3 bar and ceased when the strength reached 13 to 18 bar; this depended on the type of soil (Parker and Taylor, 1965). A crust strength of 273 mbar prevented the emergence of bean seedlings in Pachappa fine sandy loam (Richards, 1953). On the other hand, critical modulus of rupture values ranged from 1200 to 2500 mbar for the emergence of sweet corn on Pachappa loam that contained exchangeable sodium (Allison, 1956). It is necessary to specify the moisture content in establishing a critical crust strength for restricting seedling emergence (Hanks, 1960). For example, modulus of rupture values for a silt loam soil vary for both crop and moisture content of the crust (Table 1.5). These data emphasize the difficulties of establishing critical soil strengths because of the variations encountered due to the nature of the plant, the moisture content of the crust at time of emergence, and the experimental technique itself.

Arndt (1965a,b) made a study of the morphology of soil crusts in relation to cracking and seedling emergence. He observed that the natural pattern of cracking and the size of the seedlings may be more important factors in emergence than the strength of the soil crust. The location of the seedling in relation to the natural cracks in the crust is of considerable significance. In the case of thin seedlings, the cracks must be frequent enough to enable the seedlings to emerge freely. For coarse seedlings, both the size and frequency of the cracks are important for emergence. The cracks must be wide enough so that there is no jamming of the plates between cracks as the coarse seedlings lift plates of soil during emergence.

TABLE 1.5 Effect of Crop and Water Content on Critical Crust Strength for Seedling Emergence

	Critical Modulus of Rupture Range (bar)	
Crop	$\theta = 0.25$	$\theta = 0.145$
Wheat	3.21–6.41	0.8–1.6
Grain Sorghum	1.61–3.21	0.0–0.8
Soybeans	6.41	0.0–0.8

Source: After Hanks (1960).

Strong crusts with high bulk densities can impede aeration under moist conditions if the pores in the crusted layer contain sufficient water to prevent effective diffusion of oxygen into the soil. Diffusion is not restricted significantly in the case of dry crusts. If lack of aeration does become a problem, germination of the seeds would be impaired, which would also account for decreases in emergence.

Crusting of soils can be controlled by surface mulches, which protect the soil from the impact of raindrops. As stated previously, organic matter promotes the formation of stable aggregates that resist dispersion. Certain artificial soil conditioners also reduce soil crusting by producing stable aggregation.

PROBLEMS

1.1 Use (1.5) for the settling velocity to calculate the amount of time required for particles of diameter $D = 2.0, 0.05, 0.002$, and 0.001 mm to fall 10 cm in water solution. Assume that $\rho_s = 2.7$ g cm^{-3}.

1.2 Fifty grams of soil containing 10% clay by weight is added to a beaker containing 1000 mL of water and mixed thoroughly. Calculate the density of the suspension before and after the sand and silt settle out. Ignore any volume change in the solution as the soil is added and removed.

1.3 Since most of the surface area of a disk lies on the circular faces, why does the formula (1.7) for specific surface not depend on the circular radius R?

1.4 If the distribution coefficient K_d defined in (1.17) for a linear equilibrium adsorption process is regarded as the slope of the $C_a - C_l$ curve, calculate the effective K_d of the Freundlich isotherm (1.16) and show that it is a function of the dissolved concentration. For a compound with a Freundlich coefficient of $\beta = 0.8$, calculate the ratio of the effective K_d at $C_l = 1$ mg L^{-1} and $C_l = 1000$ mg L^{-1}.

1.5 A soil is made up of identical spherical aggregates of bulk density $\rho_b = 1.3$ g cm^{-3} whose solid phase has a particle density of $\rho_m = 2.65$ g cm^{-3}. The aggregates are packed into an arrangement with an interaggregate porosity of 0.4. Calculate the bulk density of the soil, the aggregate porosity, and the porosity of the soil. What is the wet bulk density of the soil if the aggregates are water saturated and the interaggregate space is filled with air?

2   Water Retention in Soil

Soil has an amazing property without which life as we know it would never have survived on the land—it can retain water for substantial periods of time. Despite the incessant pull of gravity, water entering the soil surface by rainfall or irrigation stays in the upper zone long enough for plant roots to extract what they need to survive. Moreover, water is held in such a manner that gases can also move through air spaces, allowing oxygen to reach the roots. Soil is thus much more than a water storage reservoir. It is an ideal growth medium. To understand how soil manages this complex task, we must examine in detail how water is retained by the solid framework of the porous medium.

The two most important characteristics of the soil water phase are the amount of water in a given amount of soil and the forces holding water in the soil matrix. Many processes are influenced by the amount of water in soil, including gas exchange with the atmosphere, diffusion of nutrients to plant roots, soil temperature, and the speed with which dissolved chemicals move through the root zone during irrigation or rainfall. The forces exerted on water by the solid framework also affect many processes, including the efficiency of water absorption by plant roots, the amount of drainage occurring due to gravity, and the extent of upward movement of water and solutes against gravity. In this chapter we introduce a formalism for quantifying water retention in soil and describe how the parameters and functions that describe soil water properties are measured.

2.1   PROPERTIES OF WATER

Many of the bulk properties of soil water are a consequence of the molecular characteristics of water. Therefore, prior to discussing the fluid properties of water, it is worthwhile to describe the attributes of the water molecule that are most relevant to understanding the behavior of the bulk fluid in soil.

2.1.1   Molecular Properties of Water

The water molecule H_2O consists of two hydrogen atoms that are covalently bonded to an oxygen atom by sharing electrons. The oxygen and hydrogen protons in the water molecule are separated by about 0.97 Å, whereas the hydrogen protons are 1.54 Å apart, and the angle formed by the $H-O-H$ bond is about 105° (Pauling, 1948). Since the water molecule contains the same number of positively charged protons as negatively charged electrons, it is electrically neutral. However, because the center of positive charge is displaced from the center of negative charge in the water molecule, it possesses a dipole moment that produces an electric field. As a result, water molecules interact with each other, with dissolved ions in solution, and with the electric field of solid minerals and organic material in soil. The electric fields of the dipole moments of adjacent water molecules create an attractive force, forming a relatively weak intermolecular hydrogen bond between the proton of the hydrogen atom of one molecule and the oxygen atom of the other. The effect of the weak attraction is to form an effective particle made up of many bonded molecules.

The hydrogen bonds linking water molecules together are significantly weaker than the covalent bonds joining a given molecule. Hence, liquid water does not form a stable crystalline structure, because the energy of agitation of individual molecules is occasionally high enough to break the intermolecular bonds. The crystalline structure is more complete in ice, where each water molecule bonds to neighbors in a hexagonal structure like that shown in Fig. 2.1.

When ice crystals melt, the hydrogen bond length increases to about 2.9 Å. In contrast to most liquids, the density of liquid water actually increases with increasing temperature between 0 and 4°C, above which the density decreases monotonically. The reason for this density increase above the melting point is that some hydrogen bonds break as the bonds are stretched upon melting and warming, so that a crystalline structure develops that is less ordered than the ice structure and is of higher density. As the thermal agitation of the liquid water increases at higher temperature, the lattice of this structure expands. However, some crystalline properties are maintained in the liquid state at temperatures well above freezing. In the liquid state, each water molecule can be surrounded at any one time by five or more adjacent molecules joined by hydrogen bonding. Despite fluctuations in the positions of the molecules because

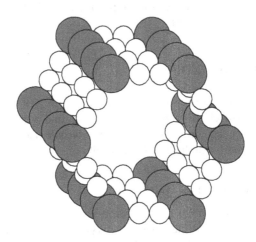

Figure 2.1   Postulated structure of ice, showing a hexagonal arrangement of water molecules and their orientation. The dark circles are oxygen nuclei and the open circles are hydrogen nuclei.

of thermal motion, the structure is sufficiently stable that clusters of water molecules behave in some respects like single large molecules, causing water to exhibit polymerlike characteristics. Only in the vapor state do water molecules completely lose their intermolecular hydrogen bonds.

Because the covalent electron bond between the hydrogen and oxygen molecules of an individual molecule is so strong, the hydrogen nucleus has a finite probability of acquiring sufficient thermal energy to break the attraction to its own electrons and to ionize, forming a hydrated hydronium ion (H_3O^+) with the water molecule to which it is hydrogen bonded. The remnant hydroxyl ion (OH^-) from the nucleus, having lost the positively charged proton from one of the hydrogen atoms forming the neutral water molecule, is negatively charged. This reversible reaction,

$$2H_2O \rightleftharpoons H_3O^+ + OH^- \tag{2.1}$$

ionizes a small fraction of the water molecules. The charged hydronium and hydroxyl ions have electric fields that differ considerably from the dipole field of the undissociated molecules. Consequently, the dissociated fraction of water significantly influences the behavior of charged particles (e.g., dissolved ions, mineral surfaces) in contact with the solution and has a critical effect on many chemical reactions occurring within the solution.

The dissociation constant K_w for the ionization reaction given in (2.1) is equal to

$$K_w = [H^+][OH^-] = 10^{-14} \tag{2.2}$$

where $[H^+]$ and $[OH^-]$ are concentrations in moles per liter (M). Equation (2.2) has been derived from the more general formula appropriate for the reaction in (2.1) by assuming that the concentration of neutral water molecules is constant and that the dissociated species are dilute (Robinson and Stokes, 1959).

Because the dissociation is so slight, there is a negligible difference between the activity and concentration of the hydrogen and hydroxyl ions. The measure of the degree of ionization of water is called pH and is equal to the negative of the base 10 logarithm of the molar concentration of the hydrogen ion $[H^+]$:

$$pH = -\log_{10}[H^+] \tag{2.3}$$

The pH of pure water at 25°C is 7, implying that

$$[H^+] = [OH^-] = 10^{-7} \, M \tag{2.4}$$

This ionization state is called *neutral*. Whenever there are more $[H^+]$ ions than $[OH^-]$ ions, the solution has a pH less than 7 and is called *acidic*. Conversely, a solution with less $[H^+]$ ions than $[OH^-]$ ions has a pH greater than 7 and is called *basic* or *alkaline*. The pH of soil water strongly affects chemical reactions in solution and mineral weathering rates, as well as influencing the biological environment of the subsurface.

2.1.2 Fluid Properties of Water

Thermal and Mechanical Properties Water is an unusual fluid. It has a very high boiling point, a high melting point, a low fluid density, and a liquid phase that is denser than the solid phase. It requires a substantial quantity of heat (the heat of fusion H_F) to melt a unit mass of ice and an even larger quantity of heat (the heat of vaporization H_V) to evaporate a unit quantity of water. It has a large dielectric constant, which makes it a good electrical insulator, and a large specific heat, causing it to experience a smaller temperature rise as it absorbs heat than other fluids. Hence water has a strong moderating influence on climate. Water is also an excellent solvent, allowing chemicals to be dissolved, to undergo a host of reactions, and to be transported through the subsurface.

The temperature dependence of the density of water has an enormous influence on aquatic ecosystems that freeze in winter. Since ice is less dense, the surface of lakes and rivers freeze first, leaving the underlying regions in a liquid state and allowing inhabitants to survive the winter. When the ice melts in spring and the surface begins to warm, water reaches its maximum density at 4°C and sinks, pushing the deeper water up. This process mixes oxygen and minerals between the water layers and enriches the aquatic ecosystem significantly. The reverse process happens in the fall, and the surface layer cools (Raven and Berg, 2002).

The important fluid properties of water are summarized in Table 2.1. These bulk characteristics are all a consequence of the molecular properties discussed in the preceding. Several fluid properties of water that strongly affect its behavior in soil are discussed further in the sections that follow.

Surface Tension and Interfacial Curvature When a fluid such as water forms an interface with another fluid or with a solid, molecules near the interface are exposed to different forces than molecules within the fluid. For example, at an air–water

TABLE 2.1 Some Physical Properties of Pure Water

Property	Value	SI Unit	T (°C)
Density			
Liquid	9.98×10^2	kg m^{-3}	20
	1.00×10^3	kg m^{-3}	4
Solid	9.10×10^2	kg m^{-3}	0
Vapor	1.73×10^{-2}	kg m^{-3}	20
Heat of fusion	3.34×10^4	J kg^{-1}	0
Heat of vaporization	2.45×10^5	J kg^{-1}	20
Specific heat	1.00×10^{-3}	J kg^{-1} °C^{-1}	20
Dielectric constant	8.0×10^2	—	20
Thermal conductivity	6.03×10^{-2}	J m^{-1} s^{-1} °C^{-1}	20
Viscosity	1.00×10^{-3}	kg m^{-1} s^{-1}	20
Surface tension	7.27×10^{-2}	J m^{-2}	20

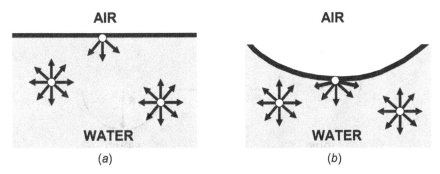

Figure 2.2 Forces on a water molecule (*a*) in bulk solution and (*b*) at the air–water interface. When the interface curves toward the liquid (*b*), molecules at the boundary feel an extra attraction into the bulk fluid.

interface, water molecules within the bulk fluid away from the interface are hydrogen-bonded to adjacent molecules and experience no net attraction from the water in any direction. However, molecules at the air–water interface feel a net attraction into the liquid because the density of molecules on the air side of the interface is lower than that on the liquid side (Fig. 2.2). This unequal attraction deforms the hydrogen bonds of the molecules at the interface and imparts some "membranelike" properties to the surface, which stretches over the water volume like a skin. As a consequence, water molecules require extra energy to remain at the interface. The extra energy per unit surface area possessed by molecules at the interface is called the *surface tension* σ. It may be defined alternatively as the energy per unit area required to increase the surface area of the interface or as the force per unit length holding the surface together (Batchelor, 2000).

The curvature of an air–water interface at equilibrium is related to the pressure difference across the interface. If the water is pure and the interface flat, the pressure is the same above and below the interface. However, if the interface is curved, the pressure is greater on the concave side of the interface by an amount that depends on the radius of curvature and the surface tension of the fluid. The right side of Fig. 2.2 shows that the effect of curvature into the liquid phase on a water molecule at the air–water interface is to expose the molecule to additional forces drawing it away from the surface compared to the case of a flat interface. For a hemispherical interface of curvature R between pure water and air, the pressure difference ΔP between the air and liquid sides of the interface is given by

$$\Delta P = \frac{2\sigma}{R} \tag{2.5}$$

where $\Delta P = P_a - P_l$ when the interface curves into the liquid (i.e., an air bubble in water) and $\Delta P = P_l - P_a$ when the interface curves into the gas (i.e., a water droplet in air). Figure 2.3 illustrates the relationship between curvature and pressure difference for various interfaces.

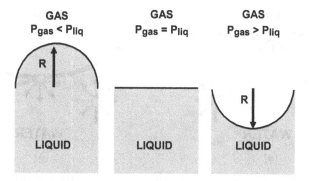

Figure 2.3 Relationship between interfacial curvature and pressure difference between the liquid (shaded) and gas (unshaded) phases. The air–water interface is shown as a solid line.

Equation (2.5) may be derived from a force balance for certain types of ideal interface geometry. The derivation is illustrated for a spherical water droplet in Example 2.1.

Example 2.1 Derive (2.5) for a spherical water droplet of radius R in air.

SOLUTION: To derive the relationship between the air pressure P_a pushing on the droplet from outside, the water pressure P_l pushing outward, and the surface tension σ holding the surface together, imagine that the droplet in Fig. 2.4a is cut in half and the right hemisphere is replaced by a network of forces that exactly equal those exerted by the right hemisphere when it was in place (Fig. 2.4b). Since the forces balance exactly, the left hemisphere remains unchanged and in static equilibrium with zero net force on it. The force pushing to the left,

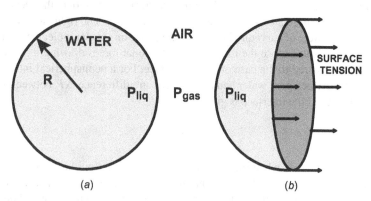

Figure 2.4 (a) Spherical droplet of water in air; (b) force balance on the left hemisphere of the droplet.

$$F_1 = \pi R^2 P_l \qquad \text{(to left)}$$

is that which was exerted by the water from the right hemisphere on the cross-sectional area $A = \pi R^2$ of the circle where the cut was made. There are two forces acting to the right. The first is the net force exerted from the air pressure P_a on the outside area of the hemisphere. The components pushing up and down cancel by symmetry, leaving a net force

$$F_2 = \pi R^2 P_a \qquad \text{(to right)}$$

(This may be derived exactly using calculus.) The third force, acting to the right, is the force required to keep the surface in place. Since the surface tension σ is a force per unit length of surface and the perimeter of the circle along the cut is $2\pi R$,

$$F_3 = 2\pi R\sigma \qquad \text{(to right)}$$

Thus, at equilibrium,

$$F_1 = F_2 + F_3 \rightarrow P_l - P_a = \frac{2\sigma}{R}$$

Note that $P_l > P_a$, in accordance with Fig. 2.3.

Contact Angle   When liquid is present in a three-phase system containing air and solids, the angle measured from the liquid–solid interface to the liquid–air interface is called the *contact angle* γ. When the liquid is preferentially attracted to the solid phase compared to its cohesive attraction to other liquid molecules, this angle is small and the liquid is said to "wet" the solid (Fig. 2.5*a*). Conversely, when the cohesive force of the liquid is much stronger than the attractive force to the solid, the liquid is said to *repel* the solid (Fig. 2.5*b*). Bachmann et al. (2000) describe a rapid procedure for measuring contact angle by microscopy.

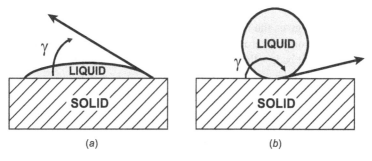

Figure 2.5   Contact angle γ for the case where (*a*) the liquid wets the solid and (*b*) the liquid is repelled by the solid.

Capillary Rise   When a small cylindrical glass capillary of radius R is placed in contact with a water reservoir open to the atmosphere, water (which has a small contact angle with glass) will spread over the inside walls of the capillary, curving the air–water interface within the tube. This curvature lowers the water pressure at the interface below atmospheric pressure [see (2.5)], causing water to rush into the capillary and to rise upward into the tube. A new equilibrium is reached when the force exerted on the column of water across the air–water interface is balanced by the weight of water in the tube.

Example 2.2   Calculate the height of rise of water in a clean glass capillary of radius R ($\gamma = 0$) at equilibrium.

SOLUTION: Since the water wets the glass surface completely, the radius of curvature of the air–water interface will be equal to the radius R of the capillary, and the pressure difference will be, using (2.5),

$$\Delta P = P_l - P_a = \frac{2\sigma}{R}$$

The column of water will be approximately cylindrical, with volume $V = AH$, where $A = \pi R^2$ is the cross-sectional area and H is the height of rise in the capillary tube. Thus, the upward force will be

$$F_{up} = A\,\Delta P = \pi R^2\,\Delta P = 2\pi R\sigma$$

The downward force is the weight of the column Mg, where $M = \rho_w V$ is the mass of water. Thus,

$$F_{\text{down}} = \rho_w V g = \pi R^2 \rho_w H g$$

At equilibrium, these two forces must be equal, from which we obtain

$$H = \frac{2\sigma}{\rho_w g R} \tag{2.6}$$

Equation (2.6) gives the height that water will rise in a capillary tube of radius R, with which it forms a zero contact angle. When the contact angle is nonzero, the height reached will be less than this value, as shown in Example 2.3.

Example 2.3   Repeat Example 2.2 for the case of water rising in a capillary of radius R made of a substance with which water forms a contact angle γ.

SOLUTION: As shown in Fig. 2.6, the radius of curvature of the interface is

$$r = \frac{R}{\cos \gamma}$$

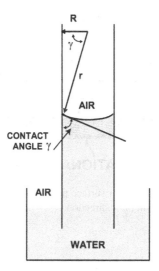

Figure 2.6   Water held in a capillary whose wall forms a contact angle γ with the water.

Thus, the pressure difference is, from (2.5),

$$\Delta P = \frac{2\sigma}{r} = \frac{2\sigma \cos \gamma}{R}$$

and the force balance this time produces

$$H = \frac{2\sigma \cos \gamma}{\rho_w g R} \tag{2.7}$$

In this case, the water rises less than in Example 2.2 since $\cos \gamma < 1$. In fact, for capillaries made of substances that repel water, γ is greater than 90°, $\cos \gamma < 0$, and water in the capillary will be pushed below the height of water in the reservoir.

Viscosity   Since adjacent water molecules are attracted to each other, they will resist any attempt to accelerate a quantity of water within the fluid that has a force exerted on it. This resistive force is called a *drag* or *shear force*. A hypothetical example will illustrate the effect of this interaction. Assume that a large volume of water is in a basin filled to a height L (Fig. 2.7). A large massless plate is laid over the top of the water and is pushed to the right with a force F, which because of the resistive forces, will attain a terminal velocity V_{max} when the applied force is balanced. The water attached to the plate will move at a velocity V_{max} and the water attached to the bottom of the basin will not move, setting up a linear change of water velocity in the vertical direction. The ratio between the force per unit area of plate (called the *tangential force per unit area* or *shear force* τ) and the velocity gradient perpendicular to the motion (equal to V_{max}/L in this example) is called the *coefficient of viscosity* v (units of mass per length per time):

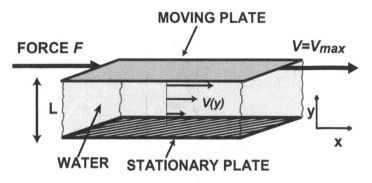

Figure 2.7   Massless plate on a water surface pushed by a force F to a final velocity V_{max}. The velocity increases linearly in the y direction.

$$v = \frac{F/A}{V_{max}/L} = \frac{FL}{V_{max}A} \tag{2.8}$$

The general expression of this relation between drag force and velocity is called *Newton's law of viscosity*:

$$\tau = \frac{F}{A} = -v\frac{dV}{dy} \tag{2.9}$$

where y is the direction perpendicular to the fluid flow.

Osmotic Pressure   The water molecule has a dipole moment, which produces an electric field that attracts it to charged ions in solution. As shown later in this chapter, the effect of this attraction is to lower the energy state of the water. If a membrane permeable to water but impermeable to solutes is used to separate pure water from a solution containing ions, water from the pure side of the membrane will cross over into the solution side. This mass transfer will continue indefinitely unless stopped by an opposing force. If the solution is sealed inside a flexible volume such as a rubber diaphragm, the pure water entering the volume will expand it and cause the hydrostatic pressure to rise, eventually stopping the water flow. The final hydrostatic pressure of the solution, which balances the ionic attraction of the water at equilibrium, is called the *osmotic pressure* π of the solution. For dilute solutions, this pressure is given by the approximate formula (McBride, 1994)

$$\pi \approx C_s RT \tag{2.10}$$

where π has units of $J\,m^{-3}$, C_s $(M\,m^{-3})$ is the concentration, T $(°K)$ is the temperature, and $R = 8.32\,J\,M^{-1}\,K^{-1}$ is the universal gas constant. Equation (2.10) describes the contribution made by a single ion species.

Example 2.4   Calculate the osmotic pressure of a 0.01 M solution of HCl at $T = 300$ K.

SOLUTION: Using (2.10), we obtain

$$\pi = 10 \ M \ \text{m}^{-3} \times 300 \ \text{K} \times 8.32 \ \text{J} \ M^{-1} \ \text{K}^{-1} = 2.5 \times 10^4 \ \text{Pa} \approx 0.25 \ \text{atm}$$

for each ion species. Thus, the total osmotic potential is approximately 0.5 atm.[1]

2.1.3   Water near Particle Surfaces

Charged mineral surfaces create an electrical double layer where ions congregate. Because of its dipole moment, water is attracted to charged solid surfaces and to ions in the electrical double layer. The layer acts in some ways like a semipermeable membrane, imparting unusual properties to water in its vicinity. Detailed discussions of this phenomenon may be found in Bolt (1955) and Gast (1977).

As discussed in Chapter 1, colloidal clays are characterized by the presence of oxygen atoms and hydroxyl radicals on their surfaces, particularly at broken edges. Hydrogen bonds readily form an association with the outer-shell electrons of oxygen and with the hydrogen protons. The strength of this hydrogen bond to oxygen in the lattice approaches the strength of covalent bonding, because lone-pair electrons of the oxygen atom are somewhat repelled by excess electrons in the negatively charged crystal lattice; this makes possible a closer tie with a hydrogen proton of the water molecule. Bonding of water molecules to surfaces leads to "structured" water close to such surfaces. Evidence exists that water close to particle surfaces is less dense than normal water because of the increased structural organization (Low, 1961). It has also been shown that the viscosity of water in the first few molecular layers may be several times greater than that in water unaffected by such surfaces (Kemper, 1961a,b; Low, 1959).

Evidence of strong attractive forces at particle surfaces has been obtained through experimental measurements such as those performed by Goates and Bennett (1957), who measured the heat necessary to absorb the first molecular layer of water vapor to the particle surface. They found the energy of adsorption of water on a kaolinite clay to be about -60.7 kJ M^{-1} with 50% of the surface covered and about -54.4 kJ M^{-1} with the entire surface covered by a monolayer. (The negative sign indicates that work must be done to remove the adsorbed water.) This is about 13 to 17 kJ M^{-1} more energy than was required to condense vapor on a free water surface. The energy of adsorption or desorption of successive layers of water after one monolayer has covered the surface rapidly approaches values characteristic of the bulk solution. The energy of adsorption is also influenced to some extent by the nature of the particle surface and by the type of exchangeable cations present on the surface.

Montmorillonite, which has a high cation exchange capacity and a large specific surface area, has a high capacity to adsorb water. In contrast, kaolinite has the lowest

---

[1] 1 Pa = 1 N m^{-2} = 10^1 erg cm^{-2} = 10^{-5} bar = 10^{-2} cm = 9.9×10^{-6} atm.

adsorption capacity of the common soil minerals, and lliite has adsorption properties intermediate between the two. At 50% relative humidity (at which the soil is very dry), the mass of water adsorbed per mass of soil is on the order of 21, 4.5, and 0.5% for montmorillonite illite, and kaolinite, respectively. These differences arise from the makeup of the crystal lattices and the sites of the negative charges. As shown by Thomas (1928), Baver and Homer (1933), and numerous investigators more recently, the nature of the exchangeable cation has a sizable impact on water adsorption. Generally, clays saturated with divalent cations exhibit greater water adsorption than do those saturated with monovalent cations. An exception is the Li^+ ion, which behaves more like a divalent ion. Adsorption of water vapor decreases with increasing size of the ion within both the monovalent and divalent groups.

Inasmuch as the freezing temperature of water is affected by its structure, the presence of ions and the proximity of solid surfaces that affect water structure will also depress its freezing-point temperature. Water at a considerable distance from soil particles freezes first as the temperature is reduced, and the highly structured water adjacent to particle surfaces freezes last. Freezing-point depression has been used as a measure of the energy state of soil water (Babcock and Overstreet, 1957; Day, 1942). However, water can be supercooled appreciably before freezing. The degree of supercooling depends heavily on the presence or absence of freezing nuclei. Consequently, the complex nature of the soil solid and solution phases make the degree of supercooling difficult to predict, and therefore measurements are sometimes difficult to interpret (Low, 1961).

As a consequence of water structuring near particle surfaces and near hydrated ions, water or ion movement close to particle surfaces is considerably more difficult to describe than is flow farther from such surfaces. Water film thicknesses, ranging from a few molecular layers (ca. 8.0 Å in kaolinite) to many layers (up to 68.0 Å in montmorillonite), have been reported to be affected by clay surfaces. The extent of this effect depends on the nature of exchangeable cations on the surfaces. As porous materials dry out, increasingly greater proportions of the remaining water are found in thin films strongly attached to particle surfaces. Thus, water in dry soils has a very low energy state.

2.2 SOIL WATER CONTENT

2.2.1 Definitions

The quantity of water in soil is expressed in two different units, as the volumetric water content θ_v and the gravimetric water content θ_g. The volumetric water content is the volume of liquid water per volume of soil, and the gravimetric water content is the mass of water per mass of dry soil.

Example 2.5 Calculate a relation between θ_v and θ_g.

SOLUTION: The mass of water per volume of soil is obtained by multiplying θ_v by ρ_w, the density of water. This is also equal to the product of θ_g and ρ_b, the dry soil bulk density. Thus,

$$\rho_w \theta_v = \rho_b \theta_g$$

or

$$\theta_v = \frac{\rho_b \theta_g}{\rho_w} \tag{2.11}$$

2.2.2 Direct Measurement of Soil Water Content

Measurement of θ_g   The gravimetric water content θ_g of a soil sample is measured by weighing it, drying it to remove the water, and then reweighing it. The customary method of drying is to place the sample in a convection oven at 105°C for 24 h. This removes the interparticle water but not the water molecules trapped between clay layers, and can also volatilize some components of organic matter. Thus, the drying time and temperature are somewhat arbitrary standards used to define oven-dry soil. As an alternative, microwave drying of the sample may be used. Although rapid drying may be achieved by this method, it suffers from numerous problems. Temperatures of the interior of the sample may be uneven and quite high in some locations. As a result, this method is not considered to be as accurate as convective oven drying. The merits and limitations of both methods are discussed in Topp and Ferré (2002a).

Example 2.6   Calculate θ_g in the following experiment. A can of moist soil is brought to the lab and weighed, dried, and reweighed. The following data are taken:

Mass of can with moist soil	140 g
Mass of can with dry soil	120 g
Mass of empty can	20 g

SOLUTION:

$$\theta_g = \frac{\text{mass of water}}{\text{mass of dry soil}} = \frac{140 - 120}{120 - 20} = 0.2$$

Measurement of θ_v by Mass and Volume Estimation   The volumetric water content may be estimated from θ_g with (2.11) if the bulk density ρ_b is known. Methods for measuring bulk density are discussed in Chapter 1.

2.2.3 Indirect Measurement of θ_v by Electromagnetic Radiation Methods

Because water has so many unusual molecular properties, it has a pronounced influence on transmission of various forms of electromagnetic radiation. This means that a volume of soil will have a water content–dependent transmission of radiation, allowing various devices to be used to measure θ_v. A thorough discussion of the theoretical foundation of these devices and of how they are used to measure water content is given in Topp and Ferré (2002b). In the discussion that follows we describe briefly the principal methods in use today.

Measurement of θ_v by Time-Domain Reflectometry   Time-domain reflectometry (TDR) is a method by which volumetric water content is estimated indirectly by measurement of the permittivity[2] ϵ of the soil and subsequent calibration of this property with θ_v (Dalton and van Genuchten, 1986; Topp et al., 1980). The permittivity is measured by placing a prong with two arms (usually about 30 cm long or less) forming two parallel waveguides into the soil and sending a step pulse of electromagnetic radiation along the guides. The pulse is reflected at the end of the prong and returned to the source, where its travel time and velocity can be estimated with an oscilloscope. The permittivity of the material (the soil) between waveguides causes the velocity of the pulse to deviate in a known manner from the velocity of light in vacuum. Hence, the permittivity can be estimated from the travel time, and the water content can be calculated by a regression equation from the permittivity.

Various regression methods have been proposed for this purpose. Topp et al. (1980) used the following empirical third-order polynomial to calculate θ_v based on measurements from several soils:

$$\theta_v = -5.3 \times 10^{-2} + 2.92 \times 10^{-2}\epsilon_T - 5.5 \times 10^{-4}\epsilon_T^2 + 4.3 \times 10^{-6}\epsilon_T^3 \quad (2.12)$$

where ϵ_T is the apparent permittivity measured by the probe.

Dobsen et al. (1986) used a theoretically based expression for the permittivity of a composite medium, from which θ_v could be calculated. The latter model was found to estimate the relation between ϵ and θ_v accurately for a variety of different soils (Roth et al., 1990).

The volume of soil that is sampled by a TDR probe has been analyzed theoretically by Knight (1992). He concluded that for the commonly used parallel probes, the ratio of the prong spacing to the thickness of the prongs is a critical parameter and should not exceed 10. The TDR probe can also be used to measure solute concentration under certain conditions (see Chapter 7).

Measurement of θ_v by Neutron Attenuation   The neutron attenuation method (Gardner, 1986; Gardner and Kirkham, 1952) for measuring volumetric water content is used exclusively in the field. The device consists of a compact radiation source and detector that is small enough to move inside a hollow access tube in the ground. The radiation source, usually radium–beryllium or americium–beryllium, emits high-energy neutrons in the range of 5 MeV, which collide with the nuclei of atoms in the surrounding soil. Since the nuclei of most atoms are substantially heavier than the neutrons, most collisions will not slow neutrons down from their initial energy state (see Problem 2.1). However, when neutrons collide with hydrogen nuclei, they are slowed substantially and reach velocities characteristic of the thermal motion of the hydrogen atoms in the soil after a few collisions. The detector, which is located alongside the radiation source, is sensitive only to neutrons moving at *thermal velocities*, which are the velocities characteristic of the hydrogen atoms in the soil. Since it requires an enormous number of collisions to slow down neutrons when any atom other

---

[2] *Permittivity* is another name for the dielectric constant.

than hydrogen is struck, the thermalized neutron counts essentially are proportional to the density of hydrogen atoms in the vicinity of the source. Thus, a calibration curve may be obtained, giving the number of counts per unit time received by the detector versus the amount of hydrogen present, which is predominantly in the form of liquid water. Background hydrogen present in, for example, organic matter or kaolinite will just register as a constant factor in the intercept of the calibration curve. When simultaneous measurements are made of water content (by soil coring) and neutron count rate, the calibration curve may be converted to a relationship between volumetric water content and thermal neutron count rate. The sphere of influence surrounding the radiation source varies between about 15 cm (wet soil) and about 70 cm (very dry soil) (Van Bavel et al., 1956). To work correctly, this method obviously requires that the background counts do not change with time. Therefore, it would be unsuitable in swelling soil unless there was a unique relationship between the soil bulk density and water content. In this case, the count rate could still be calibrated against θ_v, although the curve would probably be nonlinear.

Measurement of θ_v by Gamma-Ray Attenuation   The volumetric water content may be measured nondestructively for enclosed soil samples by gamma-ray attenuation (Gardner, 1986). In this method a narrow beam of gamma radiation is sent through a soil sample of known thickness and is collected beyond its exit from the sample by a detector. Because the detector is beyond a narrow slot that is aligned with the incident beam, it records only those gamma rays that pass through the soil without scattering off of an atom along the way. The gamma radiation has a narrow range of wavelengths and has a characteristic probability of interacting with any obstacle in its path, which depends on the type of substance and its density. In a soil that does not swell, all of the solid-phase material (soil, column walls, etc.) influence the absorption identically during each measurement, and any change in the reading of transmitted gamma radiation from one time to the next is attributed to a change in water content. The gamma-ray equation may be written as

$$n(L) = n_0 \exp(-\nu_m \rho_b L - \nu_w \rho_w \theta_v L) \tag{2.13}$$

where $n(L)$ is the number of counts of gamma radiation per unit time recorded at the detector, n_0 the background count rate when soil is removed, ν_m the soil mineral gamma-ray absorption coefficient, ν_w the water gamma-ray absorption coefficient, and L the thickness of the soil sample. The absorption coefficients may be measured for a given system, as shown in the next example.

Example 2.7   Devise an experiment to measure ν_w.

SOLUTION: Place an empty tray of inner width L in the path of the beam and measure the number n_0 of counts in 30 s. Fill the tray with water and measure the number of counts n_1 in 30 s. Equation (2.13) may then be written as

$$n_1 = n_0 \exp(-\nu_w \rho_w L)$$

since $\rho_b = 0$ and $\theta_v = 1$ when no soil is present. Thus,

$$v_w = -\frac{1}{\rho_w L} \ln \frac{n_1}{n_0}$$

The mineral absorption coefficient may be measured in a similar way, as illustrated in Problem 2.8.

It is also possible to use gamma-ray scanning with two beams of different energies to measure both ρ_b and θ_v in swelling systems in which bulk density changes over time (Reginato, 1974) (see Problem 2.9).

2.3   ENERGY STATE OF WATER IN SOIL

It is not sufficient to know simply the amount of water in the soil, because depending on conditions, a given amount of water might be held so tightly by the force fields of a soil that it is essentially immobile. The energy state characterizes the effects of the forces exerted on soil water by its surroundings and hence expresses the water's availability. As we will see, two soils with identical water content may be at strikingly different energy states, such that one will easily allow plants to extract water, while the other may completely resist attempts to remove water from its location in the soil matrix. Another reason for characterizing the energy state of water will be that it will allow us to determine whether water will flow from one point to another.

Water in soil moves slowly enough that its kinetic energy may be neglected in most applications. Therefore, our characterization of the energy state of soil water will consist of defining its potential energy. In the discussions to follow, we assume that the soil water is isothermal, so that energy changes associated with temperature changes do not have to be taken into account.

2.3.1   Potential Energy of Water in Soil

Potential energy is the energy that a body has by virtue of its position in a force field. For example, a mass possesses greater potential energy in a gravitational field than an identical mass lying below it, because the latter would have to have work performed on it to move it up to the former's position. Water molecules experience many different forces in a porous material such as soil. The gravitational field of Earth pulls vertically downward on the water. The weight of water and sometimes the additional weight of soil particles that are not constrained by the soil matrix can also exert downward forces on underlying water. Ions in solution attract water molecules. Attractive force fields from solid surfaces hold water in various configurations, both adsorbing it to the surface and curving air–water interfaces into the liquid phase in unsaturated soil, thereby lowering the water pressure (see Fig. 2.3).

The variety of forces and the directions in which they act make it all but impossible to characterize the force field in a real soil. However, it is possible to calculate the potential energy of a unit quantity of water as a consequence of the forces acting upon it. Potential energy differences from point to point in isothermal systems determine both the direction of water flow and the net force driving it. We use this fact in Chapter 3 to develop equations describing water movement in soil.

2.3.2 Reference or Standard State

The potential energy of water in soil must be defined relative to a reference or standard state, since there is no absolute scale of energy. The standard state is customarily defined to be the state of pure (no solutes), free (no external forces other than gravity) water at a reference pressure P_0, reference temperature T_0, and reference elevation z_0 and is arbitrarily given the value zero (Bolt, 1976). The soil water potential energy is defined as the difference in energy per unit quantity of water compared to the reference state. This statement may be made more rigorous through the definition given in the next section.

2.3.3 Total Soil Water Potential

The total soil water potential of the constituent water in soil at temperature T_0 is the amount of useful work per unit quantity of pure water that must be done by means of externally applied forces to transfer reversibly and isothermally an infinitesimal amount of water from the standard state to the soil liquid phase at the point under consideration (Bolt, 1976).

There are several systems of units in which the total potential and its components may be described, depending on whether the quantity of pure water mentioned in the preceding definition is expressed as a mass, a volume, or a weight. Table 2.2 summarizes these systems and their units.

The first system of units is used extensively in chemical thermodynamics. We use the latter two systems in this book.

Example 2.8 Derive relations among μ_T, ψ_T, and h_T.

SOLUTION: Since pure water is the standard quantity in all cases, the relationship between the mass m_w and volume V_w of pure water is given by $m_w = \rho_w V_w$, where ρ_w is the density of water. Thus, $\mu_T = \rho_w \psi_T$. The relationship between the mass m_w and weight W_w of water is $W_w = g m_w$, where g is the acceleration of gravity. Thus, $\mu_T = g h_T$ and $\psi_T = \rho_w g h_T$.

2.3.4 Components of the Soil Water Potential

The transformation of water from the reference state to the soil water state may be broken up into a series of steps, each of which partially converts water to the final state. As long as each step is performed reversibly and isothermally, the total change in potential energy of the water may be set equal to the sum of the potential energy changes corresponding to each of the steps. These sequential potential energy changes

TABLE 2.2   Systems of Units of Total Soil Water Potential

Units	Symbol	Name	Dimensions	CGS Unit	SI Unit
Energy/mass[a]	μ_T	Chemical potential	L^2/T^2	g^{-1}	$J\,kg^{-1}$
Energy/volume	ψ_T	Soil water potential	M/LT^2	cm^{-3}	$N\,m^{-2}$
Energy/weight	h_T	Soil water potential head	L	cm	m

Source: Adapted from Sposito (1981).

[a] Or energy per mole.

will be called *components* of the total water potential. In accordance with the recommendations of the 1976 soil physics terminology committee of the ISSS (Bolt, 1976), we define the transition from the reference pool to the soil-water state in three steps.

Gravitational Potential ψ_z (z in Head Units)   The gravitational potential ψ_z is the energy per unit volume of water required to move an infinitesimal amount of pure, free water from the reference elevation z_0 to the soil water elevation z_{soil}. It has the value

$$\psi_z = \rho_w g\,(z_{soil} - z_0) \tag{2.14}$$

Solute Potential ψ_s (s in Head Units)   The solute or osmotic potential ψ_s is the change in energy per unit volume of water when solutes identical in composition to the soil solution at the point of interest in the soil are added to pure, free water at the elevation of the soil.

The other components of the water potential have been defined somewhat differently by various authors, in part because the remaining effects on water caused by the presence of the soil matrix are difficult to isolate from each other. In 1976, the ISSS terminology committee completed the transition from the reference state to the soil water with the following definition.

Tensiometer Pressure Potential ψ_{tp}   The tensiometer pressure potential ψ_{tp} is the energy per unit volume required to transfer reversibly and isothermally an infinitesimal amount of solution containing solutes (which are identical in composition to the soil water) from a reservoir of solution (which is at reference pressure and is located at the elevation of the soil) to the point of interest in the soil.

This component describes the effects of all the forces on soil water other than gravity and solutes. Its influence may include the effects of binding to soil solids, interfacial curvature, air pressure, weight of overlying soil solid material, and hydrostatic water pressure in saturated soil. This definition is not in common use in the literature because soil scientists have traditionally divided the energy change associated with the transfer of a unit quantity of water from a pool of solution to soil water into several other components, which account separately for the effects of air pressure, weight of soil solids, hydrostatic water pressure, and binding by the soil matrix (matric effects).

As pointed out by the 1976 commission of the ISSS, however, these divisions must be made carefully because the various forces on a soil water element can interact with each other. This interaction will become apparent from the discussion that follows. A potential component that has been used since the formulation of soil water potential relations began is the matric potential, which describes the potential energy change associated with the addition of the soil matrix.

Matric Potential ψ_m (h in Head Units)   The matric potential ψ_m is the energy per unit volume of water required to transfer an infinitesimal quantity of water from a reference pool of soil water at the elevation of the soil to the point of interest in the soil at reference air pressure.

Thus, the matric potential differs from the tensiometer pressure potential in that the soil air pressure is maintained at reference pressure. The transition to the final state is achieved with the following definitions, depending on whether the soil water is unsaturated and has an air phase that can exert pressure on the water or the soil water is saturated and experiences hydrostatic pressure from an overlying water phase.

Air Pressure Potential ψ_a (a in Head Units)   The air pressure potential ψ_a is defined as the change in potential energy per unit volume of water when the soil air pressure is changed from the pressure P_0 of the reference state to the pressure P_{soil} of the soil. If this change does not alter the geometry of the soil liquid phase,

$$\psi_a \approx P_{\text{soil}} - P_0 = \Delta P \tag{2.15}$$

which is the gauge pressure of the soil air relative to the standard-state air pressure.

Hydrostatic Pressure Potential ψ_p (p in Head Units)   The hydrostatic pressure potential ψ_p is defined as the water pressure exerted by overlying unsupported (saturated) water on the point of interest in the soil. By definition, it is

$$\psi_p = \rho_w g(z_{\text{soil}} - z_{\text{wt}}) \tag{2.16}$$

which is the water pressure exerted by the height of water between the point of interest z_{soil} and the water table (saturated–unsaturated soil interface) z_{wt}.

The potential components discussed thus far are sufficient to describe soil water energy states whenever the soil solid matrix is rigid and self-supporting. For these soils, called *nonswelling soils*, the weight of soil solid material above the point of interest is borne entirely by the rigid matrix. No external forces exerted on the solid boundary may influence the energy state of the soil water. In this case, the matric potential described earlier may be characterized as a function of the amount of water in the soil, since the soild framework remains fixed during changes in wetting or drying.

The situation is far more complex in swelling soils. Since the solid particles are not in complete contact with each other, part or all of the weight of overlying solid material may be exerted on an element of soil water. Furthermore, changes in water content cause reorientation of the solid particles, which can create large changes in interfacial curvature. In swelling soils, air pressure variations may cause the soil geometry to change, thereby altering the interfacial curvature of water in soil. For these reasons, the matric potential is a far less useful concept in swelling soils than in rigid soils, and it may not be possible to characterize it adequately from a knowledge of water content alone. For such applications, it is customary to break the matric potential up into two additional components, one that describes the effect of the soil geometry (i.e., adsorptive forces, interfacial curvature) and one that describes the effect of the mechanical forces exerted by the pressure of the solid material.

Overburden Pressure Potential ψ_b (b in Head Units)   The overburden potential ψ_b is defined as the change in energy per unit volume of soil water when the envelope pressure P_e (mechanical pressure exerted by the unsupported solid material on the soil water) is changed from zero to P_e. When the swelling soil is sufficiently saturated that there is no contact between the solid particles and there is no external load on the system, the envelope pressure is equal to the pressure due to the weight of the overlying solids. When air is in the system, however, the overburden is more complicated to determine. To see why this is so, we can imagine an experiment in which a soil is brought in contact with a water reservoir at atmospheric pressure and allowed to expand freely until a final volume is reached, while contained in a closed environment where the air pressure is equal to $P_{atm} + \Delta P$. Thus in this case $P_e = \Delta P$. Then the water is slowly removed by lowering the water pressure of the reservoir while keeping the air pressure in the vicinity of the soil system at $P_{atm} + \Delta P$. Figure 2.8 shows a plot of the volume of the soil system as a function of

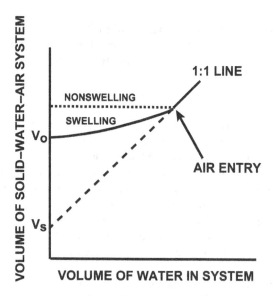

Figure 2.8   The shrinkage curve for a swelling and nonswelling soil, where water is removed while the air pressure exerts a constant envelope pressure.

TABLE 2.3 Components of Total Soil Water Potential for Various Applications

Application	Components
Unsaturated swelling soil	$\psi_T = \psi_z + \psi_s + \psi_{tp}$
	$\psi_T = \psi_z + \psi_s + \psi_a + \psi_m$
	$\psi_T = \psi_z + \psi_s + \psi_a + \psi_b + \psi_w$
Unsaturated rigid soil	$\psi_T = \psi_z + \psi_s + \psi_{tp}$
	$\psi_T = \psi_z + \psi_s + \psi_a + \psi_m$
Saturated swelling soil	$\psi_T = \psi_z + \psi_s + \psi_{tp}$
	$\psi_T = \psi_z + \psi_s + \psi_p + \psi_b$
Saturated rigid soil	$\psi_T = \psi_z + \psi_s + \psi_{tp}$
	$\psi_T = \psi_z + \psi_s + \psi_p$

the volume of liquid water removed from the system by this procedure. This graph is called a shrinkage curve (Sposito, 1972). A nonswelling soil would never change its volume after air entered the system, and so would follow the horizontal dotted line as water was removed. In contrast, the swelling soil (solid line) would move down the 1:1 line as long as the particles were separated until approximately the point where air entered the system, and then would embark on a curved pathway to the residual volume V_0 which it would approach horizontally. The slope of this line at different envelope pressures is needed to estimate the overburden potential, which is given approximately by (Philip, 1972; Sposito, 1972)

$$\psi_b = \int_0^{P_e} \left(\frac{\partial V}{\partial V_w} \right)_{P_e = P'} dP' \qquad (2.17)$$

Wetness Potential ψ_w (w in Head Units)   The wetness potential ψ_w is the value of the matric potential at zero external air pressure and zero envelope pressure. Table 2.3 summarizes the various ways in which the total potential is divided for different applications.

2.4 ANALYSIS OF SYSTEMS AT EQUILIBRIUM

When two systems at the same temperature containing water at different energy states are brought into contact with each other, water will flow from the region of higher total potential to the region at lower total potential. This flow will occur until the total potential energy of the two systems is the same. At this time, flow will cease and the systems are said to be *at equilibrium.* Equivalently, all points of a water system at equilibrium have the same value of total water potential energy.

The *equilibrium principle* may be used to estimate components of the total water potential using various measurement devices. The analysis of a system at equilibrium by this procedure should proceed in the orderly sequence of steps given in Table 2.4.

TABLE 2.4   Analysis of Components of the Soil Water Potential at a Point of Interest Using the Equilibrium Principle

Step 1	Verify that the system is at equilibrium. Water flow must be zero or negligible everywhere and the temperature must be constant.
Step 2	Define a reference elevation and reference air pressure.
Step 3	Find a location in the system where the total water potential ψ_T may be estimated. Call this value ψ_{T0}.
Step 4	Set $\psi_T = \psi_{T0}$ at the point of interest.
Step 5	Divide ψ_T into appropriate components using Table 2.3.
Step 6	Evaluate directly those components for which appropriate information at the point of interest is available.
Step 7	Determine the remaining component or sum of components using the equation in step 4.

Example 2.9   Calculate the height of rise of water in a clean glass cylindrical capillary tube of radius R placed in an open vessel of pure water. Assume that the contact angle of the water on glass is zero and that the air above the interface is at the same pressure as the air above the open vessel. Assume that evaporation of water into the air has been suppressed or is negligible.

SOLUTION: Following the steps in Table 2.4:

Step 1: The system shown in Fig. 2.9 will reach equilibrium (if water evaporation is suppressed) as soon as the water in the capillary stops rising.

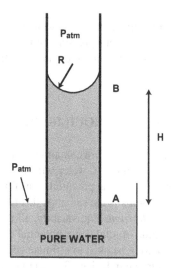

Figure 2.9   Capillary tube containing water at equilibrium everywhere.

Step 2: Define $z = 0$ at point A and $P_0 = P_{atm}$.

Step 3: At point A all components of ψ_T may be evaluated.

$$
\begin{aligned}
\psi_z &= 0 && \text{since } z = 0 \\
\psi_s &= 0 && \text{pure water} \\
\psi_a &= 0 && P = P_0 = P_{atm} \\
\psi_m &= 0 && \text{no soil} \\
\psi_p &= 0 && \text{no overlying hydrostatic pressure}
\end{aligned}
$$

Thus, relative to the reference state, $\psi_T = 0$ at A.

Step 4: At point B, $\psi_T = 0$ since point B is at equilibrium with point A where $\psi_T = 0$.

Step 5: Regarding point B as a location in a rigid unsaturated "soil," we may use the relation $\psi_T = \psi_z + \psi_s + \psi_a + \psi_m$. The matric potential ψ_m accounts for the decrease in liquid water pressure caused by interfacial curvature resulting from adsorption of water to the interior capillary walls.

Step 6: By definition, at point B the components are:

$$
\begin{aligned}
\psi_s &= 0 && \text{pure water} \\
\psi_a &= 0 && P = P_0 = P_{atm} \\
\psi_z &= \rho_w g (z - z_0) = \rho_w g H \\
\psi_m &= P_l - P_a = -2\sigma/R && \text{by (2.5)}
\end{aligned}
$$

Step 7: $\psi_T = 0 = \rho_w g H - 2\sigma/R \rightarrow H = 2\sigma/\rho_w g R$.

Notice that solute potential differences could not be present at equilibrium in this system, even if there were solutes in the water, since diffusion from one point of the system to another would create a uniform distribution at equilibrium (except for a small density stratification because of gravity), provided that no solute barriers were present. Therefore, the solute potential would be the same everywhere and could not affect the value of any of the other components. Thus, the only time that solute potential needs to be included in the analysis is when such barriers are present within the soil water system. This point is illustrated in the next example.

Example 2.10 Repeat Example 2.9 for the case where the open vessel contains a solution of osmotic pressure π (energy per volume) and the capillary has a perfect semipermeable membrane at the bottom that restricts the passage of solutes but allows water to flow into the capillary tube (Fig. 2.10). Assume that the surface tension and contact angle are unaffected by the addition of solutes.

SOLUTION: Repeating the steps of Table 2.4, we may write:

Step 1: Equilibrium is reached as soon as the water in the capillary stops rising.

Step 2: Define $z = 0$ at point A and $P_0 = P_{atm}$.

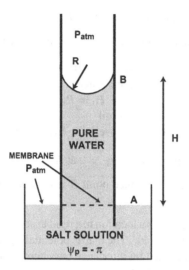

Figure 2.10 Capillary tube with an internal solute membrane, containing water at equilibrium with a solution of osmotic pressure π.

Step 3: At point A all components of ψ_T may be evaluated.

$$
\begin{array}{lll}
\psi_z & = & 0 & \text{since } z = 0 \\
\psi_a & = & 0 & P = P_0 = P_{atm} \\
\psi_m & = & 0 & \text{no soil} \\
\psi_p & = & 0 & \text{no positive hydrostatic pressure} \\
\psi_s & = & -\pi & \text{by definition}
\end{array}
$$

Thus, relative to the reference state, $\psi_T = \psi_{T0} = -\pi$ at A.

Step 4: At point B, $\psi_T = -\pi$.

Step 5: $\psi_T = -\pi = \psi_z + \psi_s + \psi_a + \psi_m$ at point B.

Step 6: By definition, at point B the components are:

$$
\begin{array}{lll}
\psi_s & = & 0 & \text{pure water in capillary} \\
\psi_a & = & 0 & P = P_0 = P_{atm} \\
\psi_z & = & \rho_w g(z - z_0) = \rho_w g H \\
\psi_m & = & P_l - P_a = -2\sigma/R & \text{assuming that } \gamma = 0
\end{array}
$$

Step 7: $\psi_T = -\pi = \rho_w g H - 2\sigma/R$. Thus,

$$
H = \frac{2\sigma}{\rho_w g R} - \frac{\pi}{\rho_w g} \tag{2.18}
$$

In this case, the water will not rise as high as it did in the system without a membrane, since the solutes in the vessel attract the water molecules on the capillary side of the membrane. In fact, if $\pi > 2\sigma/R$, no water will enter the capillary through the membrane.

Example 2.10 illustrates the important effects that solute membranes may have on a water system. The most important solute membranes in soil are found in plant roots. However, the air–water interface also acts as a solute membrane because salts in the liquid phase are left behind as water evaporates.

2.5 MEASUREMENT OF COMPONENTS OF WATER POTENTIAL

Many of the devices that measure components of the water potential operate on the equilibrium principle, in that they are placed in the soil water system and exchange water until they come to equilibrium with the point in the soil where the measurement is made. These devices are used to measure those components of the system that are not completely specified in terms of parameters that are directly measurable.

2.5.1 Direct Measurement of Potential Components

Several of the components discussed earlier in the chapter may be measured individually by standard devices:

1. *Gravitational potential:* Measure the vertical distance from the reference elevation to the point of interest with a ruler.
2. *Solute or osmotic potential:* Extract soil solution and measure its osmotic pressure or analyze for the solute concentration and use (2.10). The osmotic pressure of a solution may be measured by standard methods such as equilibrium dialysis or vapor pressure depression (Rawlins and Campbell, 1986).
3. *Air pressure potential:* Measure the soil air pressure with a barometer and subtract the reference air pressure.
4. *Hydrostatic pressure potential:* Measure the vertical height of saturated water above the point of interest. This requires finding the water table location, which may not be easy to detect in swelling soil systems (see Fig. 2.11).

2.5.2 Measurement Devices

The remaining components (matric potential, overburden potential, wetness potential) may not be evaluated directly from their definitions but must be inferred from the readings of devices that equilibrate with soil water. The principal devices used for such purposes are described in the following.

Piezometer Tube   The piezometer tube (Fig. 2.11) is a hollow tube that is placed into the soil with a water entry point at the place where a measurement is desired. The other end of the tube extends vertically from the measurement point and is open

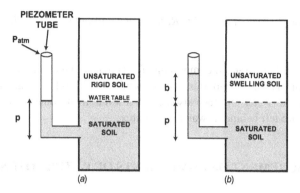

Figure 2.11 Piezometer tube used to measure components of the water potential in (*a*) non-swelling and (*b*) swelling soil.

to the atmosphere. If water in the soil is under positive pressure, it will enter the tube and rise to a height whose gravitational potential relative to the soil at the point of interest is equal to the tensiometer pressure potential of the soil water. (For practice, the reader may verify this using equilibrium analysis.) Excluding unsaturated soil, in which water would not enter the piezometer, the tensiometer pressure potential in a saturated system is equal to the sum of the overburden and the hydrostatic pressure potentials. Thus, in a rigid soil, the water in the piezometer will rise to a height equal to the water table height. In a swelling soil, the height reached will exceed the height of the water table by an amount equal to the overburden potential head (see Fig. 2.11).

The piezometer is a standard device for measuring water pressure head in groundwater systems. Young (2002) discusses the features of commercial piezometers and their applications.

Tensiometer A tensiometer consists of a water-saturated porous ceramic cup connected to a manometer through a water-filled tube (see Fig. 2.12). The ceramic consists of very fine pores that remain water saturated even when placed in contact with soil at relatively low water potentials. Upon contact, water moves from the tube to the soil, creating a suction at the manometer end until equilibrium is reached and the total potential of the water is equal everywhere in the system consisting of the soil and the device.

The equilibrium principle may be applied to the mercury manometer tensiometer shown in Fig. 2.12*b* to determine which components of the soil are measured and the relationship between these components and the height of rise X of the mercury in the manometer.

Assuming that equilibrium is reached rapidly, we may regard points A and B as being at the same total water potential. The reference height is taken as $z = 0$ at the soil surface, and the reference pressure is taken as the air pressure around the mercury reservoir. Since solutes will pass freely through the ceramic and through the tube, we may ignore ψ_s in the subsequent analysis.

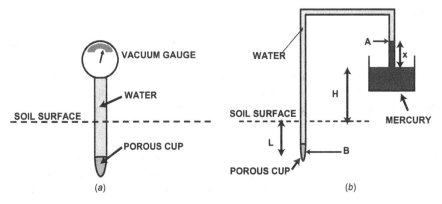

Figure 2.12 (a) Tensiometer for measuring soil water potential with a vacuum gauge; (b) mercury manometer.

At point A:

$$
\begin{aligned}
\psi_z &= \rho_w g(H + X) & &\text{relative to } z = 0 \text{ at the soil surface} \\
\psi_a &= 0 & &\text{no air inside the tube} \\
\psi_m &= 0 & &\text{no soil inside the tube} \\
\psi_p &= -\rho_m g X & &\text{where } \rho_m \text{ is the density of mercury}
\end{aligned}
$$

The hydrostatic pressure of the water in the tube must be less than the reference pressure, or mercury will not rise in the tube. The weight of suspended mercury per unit area $\rho_m g X$ must be equal to the pressure difference across the ends of the column. Thus the mercury pressure at the top, which is equal to the water pressure at the interface, is $-\rho_m g X$.

Therefore, at point A, $\psi_T = \rho_w g(H + X) - \rho_m g X$, which must be equal to the total potential at point B in the soil. Assuming that the soil is unsaturated, we may write

$$\psi_T = \rho_w g(H + X) - \rho_m g X = \psi_{zB} + \psi_{aB} + \psi_{mB}$$

where $\psi_{zB} = -\rho_w g L$. Thus,

$$\psi_{mB} + \psi_{aB} = \rho_w g(H + L) - (\rho_m - \rho_w)gX$$

This relationship shows that the manometer reading is a function of the distance $H+L$ between the ceramic and the mercury reservoir and that the tensiometer measures the sum of the matric potential (including overburden influences if any) and air pressure potential. However, the soil air pressure will usually not be different from atmospheric ($\psi_a \approx 0$) and

$$\psi_m \approx \rho_w g(H + L) - (\rho_m - \rho_w)gX \tag{2.19}$$

or in head units,

$$h \approx H + L - 12.6X \tag{2.20}$$

Often, the vacuum gauge and mercury manometer tensiometers will be offset to read matric potential directly. For the mercury manometer, this is accomplished by placing the $h = 0$ point of the scale reading at $X = (H + L)/12.6$. Young and Sisson (2002) present a discussion of various types of tensiometers and their use in soil systems.

Example 2.11   Calculate the matric potential measured by the tensiometer in the following application. A mercury manometer tensiometer is placed so that the ceramic is 100 cm below the mercury reservoir. The mercury rises to a height of 20 cm above the reservoir. Calculate the matric potential of the soil in the vicinity of the ceramic. Assume that $\psi_a = 0$.

SOLUTION: In this example, $H + L = 100$ and $X = 20$. Therefore, by (2.20), we obtain

$$h = 100 - 12.6 \times 20 = -152 \text{ cm}$$

or in units of energy per volume,

$$\psi_w = \rho_w g h = 1.49 \times 10^5 \text{ erg cm}^{-1} = -14.9 \text{ kPa}$$

The tensiometer has a practical range of about -80 kPa (-800 cm or -0.8 bar), below which gases dissolved in the water will begin to form bubbles and the liquid column will break up (Young and Sisson, 2002).

Soil Psychrometer   The soil psychrometer measures the relative humidity (RH) of the water vapor in the soil, from which the water potential of the vapor phase may be calculated using (Andraski and Scanlon, 2002)

$$\text{RH} = \frac{P_v}{P_v^*} = \exp\left(\frac{M_w \psi_w}{\rho_w RT}\right) \tag{2.21}$$

where P_v is the vapor pressure, P_v^* the saturated vapor pressure, and M_w the molecular weight of water. At equilibrium, the vapor water potential is equal to the liquid water potential. Since the vapor and liquid are at essentially the same elevation, the components of the soil water potential measured by the psychrometer are the sum of the matric and osmotic potentials, assuming that the air pressure is atmospheric.

The thermocouple psychrometer is the most widely used psychometric device in soil. A small (0.025-mm-diameter) thermocouple made of metals such as chromel–constantan or copper–constantan inside a thin-walled ceramic cup is buried in the soil so that the vapor in the atmosphere that surrounds the thermocouple is at equilibrium with the soil solution. By using a small direct current, the temperature of the thermo-couple is reduced to the dew point by Peltier cooling so that water is condensed on the junction. Cooling is then stopped and the temperature of the junction is measured with a microvoltmeter while the junction is cooled by evaporation. The temperature of the junction depends on the evaporation rate, which in turn depends on the relative humidity of the atmosphere. The device, when calibrated over osmotic solutions of known humidities, permits measurement of water potentials in a range from about -10 kPa (-100 cm) down to the order of -7 mPa (-70 bar), although it is least accurate in wet soil (Bruce and Luxmoore, 1986). Even lower values of water potential may be measured with special techniques (Rawlins and Campbell, 1986). The psychrometer also may be used in the laboratory on plant tissues and on soil and plant samples.

The psychrometer has many sources of error in soil, under wet conditions when the RH is not appreciably less than 1 and in dry conditions where modest temperature differences in the device can result in large errors in the estimate of water potential (Andraski and Scanlon, 2002).

2.6 WATER CHARACTERISTIC FUNCTION

2.6.1 Measurement

In rigid porous media, the matric potential as defined earlier represents the effect of adsorptive soil solid forces and interfacial curvature on water potential energy. The functional relationship between the matric potential and the gravimetric or volumetric water content is called the *water characteristic function* or *matric potential–water content function* $\psi_m(\theta)$. This function may be evaluated by measuring matric potential and water content simultaneously with the methods already discussed during a succession of water content changes. In the laboratory, $\psi_m(\theta)$ may be measured on replicated prepared samples over a large range of water contents. Virtually the entire range from water-saturated soil to very dry soil may be covered by using a hanging water column, a pressure membrane, and equilibration over salt solutions. These devices are illustrated below in terms of the equilibrium principle.

Hanging Water Column (Range -100 cm $< h < 0$)   A hanging water column consists of a water-saturated highly permeable porous ceramic plate connected on its underside to a water column, terminating in a reservoir open to the atmosphere. Water-saturated samples of soil held in rings are placed in contact with the flat plate when the water reservoir height is even with the top of the plate. Then the reservoir is lowered to a new height a distance H below the top of the plate (Fig. 2.13).

By the equilibrium principle, water will flow from the soil samples through the ceramic to the reservoir until the total water potential of the system is constant. At this time the potential of the free reservoir may be set equal to zero, and at the soil sample height H we may write (assuming that $z = 0$; $P = P_{atm}$; neglect solutes)

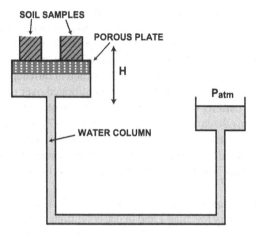

Figure 2.13   Desorption of soil water samples to a desired energy state with a hanging water column.

$$\psi_m + \psi_z = 0 = \psi_m + \rho_w g H \rightarrow \psi_m = -\rho_w g H$$

When equilibrium has been restored, some of the samples may be removed and their gravimetric or volumetric water content measured. The tube may then be lowered further and a new set of samples measured. If there is good contact between the soil and the ceramic, equilibrium will be reached rapidly (i.e., several hours) since the samples are quite moist. The range of the device is limited chiefly by the space available for lowering the water column.

Pressure Plate (Range $-15,000$ cm $< h < -100$ cm)   The pressure plate consists of an airtight chamber enclosing a porous ceramic plate connected on its underside to a tube that passes through the chamber to the open air. Saturated soil samples are packed into rings and placed in contact with the ceramic on the top side. The chamber is then pressurized, which causes water to flow from the soil pores through the ceramic and out the tube (Fig. 2.14).

At equilibrium, flow through the tube will cease. We may set the total potential equal to zero at the point where the water exits the tube. Inside the plate we may write (assuming $z = 0$; neglect solutes)

$$\psi_m + \psi_a = 0 = \psi_m + \Delta P \rightarrow \psi_m = -\Delta P$$

When equilibrium is reached, the chamber may be depressurized and the water measured. In this method an assumption is made that the matric potential of the sample does not change as the air pressure is lowered to atmospheric. This method may be used up to air gauge pressures of about 15 atm if special fine-pore ceramic plates are used. Since these devices have a very high flow resistance, it may require

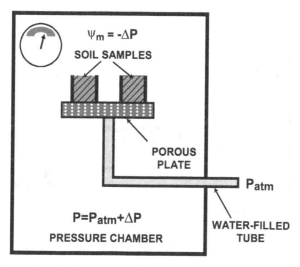

Figure 2.14   Desorption of soil water samples to a desired energy state with a pressure plate.

a substantial amount of time to remove the last small amount of water from the soil. Thus, the time of equilibrium is difficult to estimate.

Equilibration over Saturated Salt Solutions (h < −15,000 cm)   By adding precalibrated amounts of certain salts, the energy level of a reservoir of pure water may be lowered to any level specified. If this reservoir is brought into contact with a moist soil sample, water will flow from the sample to the reservoir. If the sample and the reservoir are placed adjacent to each other in a closed chamber at constant temperature, water will be exchanged through the vapor phase by evaporation from the soil sample and condensation in the reservoir until equilibrium is reached. Since the reservoir is a pool of saturated salt solution, at equilibrium the total potential will be $\psi_T = \psi_{s0}$ of the solution. In the soil, $\psi_T = \psi_m + \psi_s$, since the air–water interface acts as a solute membrane. Thus, $\psi_T = \psi_{s0} - \psi_s$, the difference between the solute potentials of the reservoir and the soil. In practice, the soil will usually not be saline enough for its solute potential to be significant compared to ψ_{s0} in the range where these measurements are made.

The equilibration time for this method can be shortened by creating a partial vacuum in the chamber (Campbell and Gee, 1986). Care should be taken that the sample and the reservoir are at the same temperature, because even small temperature differences will cause the soil and salt solution to equilibrate at very different potentials (Campbell and Gee, 1986).

Figure 2.15 shows typical matric potential–volumetric water content curves for a sandy soil and a finer-textured soil high in clay measured from soil initially at water saturation. The water characteristic function for a soil desorbed from saturation may be divided roughly into three regions, as shown in the figure. The air-entry region corresponds to the region at saturation, where the matric potential changes but the

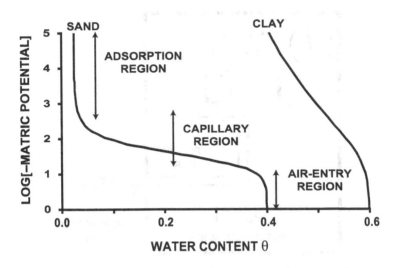

Figure 2.15   Matric potential–water content function for a sand and clay soil.

water content does not. The minimum suction that must be applied to a saturated soil to remove water from the largest pores, the *air-entry suction*, varies from about 5 to 10 cm for coarse sands to much higher values in unaggregated, fine-textured soils.

After air begins to enter the system, incremental increases in suction on the soil sample will drain progressively smaller pores, and the water content will drop. This intermediate part of the curve is called the *capillary region*, because the energy state of the water is determined primarily by the curvature of the interfaces between the air and water phases.

When essentially all of the water held in pores has been drained, only the tightly bound water adsorbed to particle surfaces remains. Large changes in matric potential in this region, called the *adsorption region*, are associated with small changes in water content.

The differences in the shapes of the water characteristic function for the prototype sandy and clay soils in Fig. 2.15 may be explained by considering the properties of the bulk solid phases. The clay soil generally has a lower bulk density and hence a higher water content at saturation. The clay soil has very few large pores and a broad distribution of particle sizes. Hence, it decreases gradually in water content with decreases in matric potential. The sandy soil, on the other hand, has much of its water held in large pores that drain at modest suctions. Hence it will have a very rapid decrease in water content in the capillary region. Finally, the clay soil has a very large surface area compared to the sand and will have a large amount of water adsorbed to the surfaces.

Example 2.12   Calculate the gravimetric water content of a sand and a clay containing only one monolayer of adsorbed water using the following information:

Montmorillonite, surface area/mass	S =	$800 \text{ m}^2 \text{ g}^{-1}$
Quartz sand, surface area/mass	S =	$5 \times 10^{-2} \text{ m}^2 \text{ g}^{-1}$
Area occupied by one adsorbed water molecule	A_m =	$1.05 \times 10^{-16} \text{ cm}^2$
Mass of one water molecule	m_w =	$3 \times 10^{-23} \text{ g}$

SOLUTION: The gravimetric water content is the mass of water per mass of soil. In a total mass M_s of soil there is:

Total area of soil	A_s =	$M_s S$
Number of water molecules adsorbed	N =	$A_s/A_m = M_s S/A_m$
Mass of water adsorbed	M_w =	$N m_w = m_w M_s S/A_m$
Mass of water/mass of soil	θ_g =	$M_w/M_s = m_w S/A_m$

Thus, for the montmorillonite, $\theta_g = 0.23$, and for the quartz sand, $\theta_g = 1.4 \times 10^{-5}$.

2.6.2 Hysteresis in Water Content–Energy Relationships

Water content and the potential energy of soil water are not uniquely related because the potential energy state is determined by conditions at the air–water interfaces and the nature of surface films rather than by the quantity of water present in pores. Soil pores are highly variable in size and shape and interconnect with each other in a variety of ways. Common to porous media are *bottleneck pores*, which have large cavities but narrow points of connection to adjacent pores. Water is held most tenaciously in small pores, which fill first when water is admitted to a system. But they do not always empty again during drying in the same order as they were filled.

We may illustrate the phenomenon of hysteresis by first imagining a soil that is completely free of water. As a small amount of water is added, small pores fill first, followed by successively larger and larger pores until all pores are filled and the matric potential is zero. At intermediate values of saturation, with enough water in the system so that air–water interfaces can exist between particles and in small pores, the curvature of such interfaces is given by (2.5). Water content and water potential in such a system will follow the wetting curve in Fig. 2.16.

If the system is dried either by evaporating water or by bringing the soil into contact with a dry, porous material that pulls water away from the system, pores will begin to empty, generally from large to small. However, liquid water may now be trapped in large pores in such a way that they will not empty in the order that they filled. Water will be held in large pores until conditions are reached where at least one interconnecting smaller pore can empty; at this time the larger pore empties quickly. As a result, the water content for a given matric potential is higher than for the wetting system, as shown by the drying curve in Fig. 2.16. Thus, the relationship between matric potential and water content is not unique in a given soil. Rather, it also depends on the wetting or drying history of the medium.

Figure 2.17 shows three different examples of how hysteresis can occur in soil water. Part (*a*) features a single pore of radius R, which has a narrow neck of radius $r < R$. When the pore is wet and the drying cycle begins, the pore will not drain until

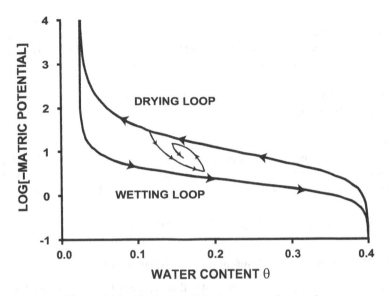

Figure 2.16 Matric potential–water content hysteresis.

the liquid pressure drops below atmospheric by an amount $\Delta P_{liq} = -2\sigma/r$, where r is the radius of the small pore at the neck. During the wetting cycle, however, the large pore will fill when the pressure is raised to $\Delta P_{liq} = -2\sigma/R$, at which point the curvature becomes less than the pore radius R. Thus, the filled pore is at equilibrium at two different liquid pressures, depending on whether it was wetted or dried.

It is also possible for a single pore to contain the same amount of water at two different water potentials, as shown in Fig. 2.17b. Here the single pore is represented as a hollow cylinder that has water condensing on its inner wall. Just before the water film grows large enough to connect at the center of the cylinder, the liquid bulges outward into the gas phase, indicating positive pressure.[3] However, when the water coalesces, the cylinder fills and the pressure drops to a value characteristic of the radius of curvature of the ends of the cylinder.

Surface wetting can also induce hysteresis. Unless particle surfaces are meticulously clean, they will form a nonzero contact angle with water when wetted (Fig. 2.17c). This results in thicker films than would be present in the drying phase, where water films are drawn tightly over the surface by adsorptive forces.

If the wetting or drying process is reversed at any intermediate water content, curves like that in the interior of the hysteretic loop in Fig. 2.16 are produced. These interior curves are called *scanning* or *secondary curves*. The outer curves produced by drying from saturation or wetting from complete dryness are called the *primary drying* or *wetting curves*, respectively.

---

[3] This statement assumes that many monolayers are present, so that the bulk water does not feel adsorptive forces.

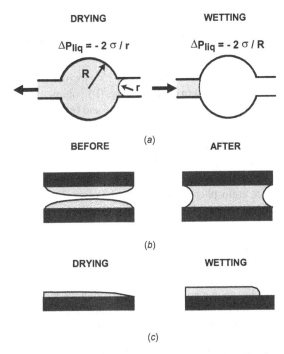

Figure 2.17   Three mechanisms for causing hysteresis in soil: (*a*) bottleneck pore, restricted by the pore radius R and pore neck radius r; (*b*) cylindrical pore condensing water on its inside before and after connection of the water film in the center; (*c*) water film during wetting and drying of a hydrophobic soil surface.

PROBLEMS

2.1   A sphere of mass m_1 moving at velocity v_1 strikes a stationary sphere of mass m_2 directly, so that all subsequent motion will occur along the line of flight. Using conservation of energy,

$$\sum_{j=1}^{2} \frac{1}{2} m_i v_i^2 = \text{const}$$

and conservation of momentum,

$$\sum_{j=1}^{2} m_i v_i = \text{const}$$

calculate the final velocities of m_1 and m_2. Evaluate the special cases $m_2 \gg m_1$ and $m_2 = m_1$. Discuss the relevance of your answers to the neutron moderation method of measuring water content in soil.

2.2 A 10-cm^3 sample volume of soil weighs 15 g before drying and 13 g after oven drying. Calculate the (**a**) bulk density, (**b**) volumetric water content, (**c**) gravimetric water content, (**d**) total porosity, and (**e**) volumetric air content of the sample. Assume that the soil has a particle density of 2.6 g cm^{-3}.

2.3 Calculate the height of rise of water in a clean glass capillary tube of radius $R = 0.001$ cm and height $L = 200$ cm that is sealed at the top before placing it in contact with the free water reservoir. Assume that the gas pressure is initially at the reference atmospheric pressure P_0 and obeys Boyle's law $PV = P_0V_0$ (V_0 is the volume of air in the capillary tube before water begins to rise) thereafter.

2.4 Trees can raise water to great heights by osmotic pressure. Calculate the osmotic pressure head s required for a redwood tree to absorb water from the soil where it is at a matric potential head of $h = -1000$ cm and raise it to the uppermost leaves at a height of $z = 100$ m above the roots. Assume that a solute membrane is present at the roots and that water is continuous between root and leaf. Neglect the solute potential in the soil.

2.5 A tensiometer is buried in an unsaturated soil sample at a depth of 15 cm below the soil surface. The surface of the mercury reservoir is 10 cm above the soil surface, and the mercury–water interface in the tube connecting the tensiometer to the mercury is 20 cm above the soil surface. What is the matric potential of the soil sample in head units (centimeters of water)?

2.6 A long capillary tube of radius $R = 0.0005$ cm has a solute membrane attached to its bottom end. It is placed in a salt solution and water rises to a final height of 100 cm above the top of the salt solution. What is the solute potential head s of the solution?

2.7 A curious student has constructed a manometer tensiometer that uses a liquid alloy of density 8.0 g cm^{-3} instead of mercury. A tensiometer is buried 50 cm below the soil surface where the matric potential head of the soil water is $h = -150$ cm. The surface of the reservoir containing the alloy is 10 cm above the soil surface. How high above the reservoir surface will the alloy rise in the manometer tube when equilibrium is reached in the system?

2.8 In a gamma-ray scanning experiment, an empty tray of width L is placed in the path of the radiation beam, and a number N_0 of counts are recorded in 15 s. The tray is then repacked with oven-dry soil to a known bulk density of ρ_b, and a number N_1 of counts are recorded in 15 s. Calculate the mineral mass absorption coefficient v_m of the soil as a function of L, N_0, N_1, and ρ_b.

2.9 A dual gamma-radiation beam uses two radiation sources of different energy that have different mass absorption coefficients for both water and soil.

Assuming that each beam is attenuated according to (2.13), demonstrate that changes in bulk density and of water content may be estimated.

2.10 Calculate the surface area per mass of a sphere of radius R and a circular disk of radius R and thickness $\tau \ll R$. Assume that each has a density $\rho = 1.7$ g cm^{-3}. Calculate the radius R of a spherical sand particle and the thickness τ of a montmorillonite plate that have surface/mass areas as given in Example 2.12.

3 Water Movement in Soil

Water is in continuous motion all over the planet. Rain falls from the clouds and strikes the land surface, where it may evaporate, be absorbed by plants, collect and move sideways, or seep deep into the soil to recharge subsurface water formations called *aquifers*. Once in an aquifer, water begins to flow sideways, or may be drawn up by a pumping well. Managing water efficiently is important for optimum crop production and to minimize pollution from agricultural chemicals. Design of pumping wells, flood control procedures, water conservation strategies, and a host of other water-use procedures all require an understanding of how water moves through soil or aquifer material. In this chapter we develop a theoretical foundation for describing water flow in porous media and use the theory to make flow calculations in a number of practical applications.

The soil water systems analyzed in Chapter 2 were in a state of thermodynamic equilibrium. In a system at equilibrium, the total water potential is constant everywhere and no water flow occurs. In this chapter, nonequilibrium problems are examined in which the total potential of water varies from point to point in the system. In such cases, water will flow from regions of higher potential to regions of lower potential at a rate that depends on the hydraulic resistance of the medium.

When two points of a porous medium at different potentials are brought into contact[1] with each other, water will flow from high to low potential until the two points are at equilibrium. However, if water is supplied or withdrawn from the boundaries of the system by an external control mechanism, it may not be possible for the system to reach equilibrium, and flow will continue. The control exerted on a system from its boundary is called a *boundary condition*.

A nonequilibrium flow system in which the flux or water potential depends on time is called *transient* or *time dependent*. A special class of nonequilibrium flows in which all variables have fixed values in time is called *steady state*. A steady-state system is characterized by water flows that do not cause storage changes within the soil.

3.1 WATER FLOW IN CAPILLARY TUBES

Before discussing water flow in soil, it is worthwhile for us to examine flow through a system with an ideal geometry. One of the simplest systems, which nonetheless has

---

[1] For movement of water in the liquid phase, bringing two points into contact means connecting them hydraulically so that water is able to move from one point to another.

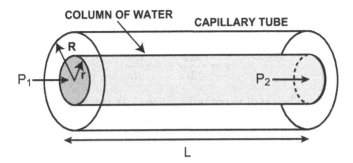

Figure 3.1   Section of capillary tube of radius R and length L filled with water flowing in response to a pressure difference $P_2 - P_1$. A force balance is conducted on a cylindrical volume of water of radius $r < R$ and length L.

many properties relevant to understanding flow in porous media, is a water-saturated capillary tube (Fig. 3.1). The horizontal cylindrical capillary tube of radius R in Fig. 3.1 has water flowing through it as a result of an imposed hydrostatic pressure difference $\Delta P = P_2 - P_1$ across the length L of the tube. At low flow rates, it is possible to neglect fluid acceleration and assume that the external applied force on the fluid (the pressure difference) is balanced by the viscous or frictional force described by Newton's law of viscosity (2.9).

Since the fluid is not accelerating, the net force on any water volume within the tube must be equal to zero. We may therefore derive an expression for the shear stress τ in (2.9) by conducting a force balance on the cylindrical water volume of radius $r < R$ and length L shown in the figure within the capillary tube. Each end of the tube has a force exerted on it by the imposed hydrostatic pressure ($F = PA$), so that the net force F_P caused by the unequal pressure across the two ends is equal to

$$F_P = \pi r^2 P_2 - \pi r^2 P_1 = \pi r^2 \, \Delta P \tag{3.1}$$

This pressure-induced force is exactly balanced by the fluid resistance or shear force[2] F_s exerted on the water volume by the water molecules in contact with the external surface area $2\pi r L$ of the cylindrical water volume:

$$F_s = (2\pi r L)\tau \tag{3.2}$$

where τ is the shear stress, or tangential force per unit area. When fluid acceleration is neglected, $F_s = F_P$, from which we obtain, using (3.1) and (3.2),

$$\tau = r \frac{\Delta P}{2L} \tag{3.3}$$

---

[2] The shear force is also called a *drag force* because it is the opposing frictional force that resists an attempt to drag one object across another.

Newton's law of viscosity (2.9) relates τ to the rate of change of velocity V perpendicular to the motion of the fluid. Thus, equating (2.9) and (3.3), we obtain

$$\tau = r\frac{\Delta P}{2L} = -\nu\frac{dV}{dr} \qquad (3.4)$$

where ν is the coefficient of viscosity. Equation (3.4) may be integrated to produce an expression for the velocity $V(r)$ as a function of r. The velocity must be specified at one value of r to eliminate the constant of integration. We will assume that the water molecules at the walls of the capillary tube adhere perfectly to the surface, so that $V(R) = 0$. In fluid dynamics this is called a *no-slip condition* (Bird et al., 2001). Here it amounts to assuming that the water molecules are attracted more strongly to the solid capillary walls than to other water molecules. To integrate (3.4), we place all factors that depend explicitly on r on the same side of the equation:

$$r\,dr = -\frac{2L\nu}{\Delta P}dV \qquad (3.5)$$

where now the left-hand side depends only on r and the right-hand side only on V. Integrating (3.5) from r to R, we obtain

$$\int_r^R r\,dr = -\frac{2L\nu}{\Delta P}\int_{V(r)}^0 dV = \frac{R^2 - r^2}{2} = \frac{2L\nu V(r)}{\Delta P} \qquad (3.6)$$

or

$$V(r) = \frac{\Delta P}{4L\nu}\left(R^2 - r^2\right) \qquad (3.7)$$

Equation (3.7) describes a parabolic velocity profile, which is plotted in Fig. 3.2, where the velocity changes from a value of zero at $r = R$ to a maximum

$$V_{\max} = V(0) = \frac{\Delta P R^2}{4L\nu} \qquad (3.8)$$

at $r = 0$.

3.1.1   Poiseuille's Law

To calculate the volume of water Q flowing per unit time through the capillary, we integrate (3.7) over the entire cross-sectional area of the tube. Because the cross-sectional area is circular, it is most convenient to use cylindrical coordinates to integrate over the area. The area element $dA = dx\,dy$ is equal to $r\,dr\,d\phi$ in cylindrical coordinates. For a circular area, r is integrated from 0 to R and ϕ from 0 to 2π radians. Since the velocity $V(r)$ is the same for all ϕ at any value of r, integration over $d\phi$ yields 2π. Thus, the volume flow rate is

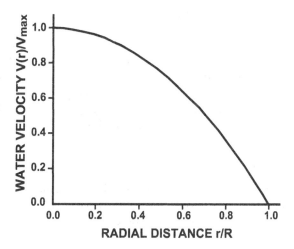

Figure 3.2   Parabolic velocity distribution in the capillary tube of Fig. 3.1.

$$Q = \int \int V(r)\, dA = \int_0^R \int_0^{2\pi} V(r) r\, dr\, d\phi = 2\pi \int_0^R V(r) r\, dr$$

$$= \frac{\pi\, \Delta P}{2Lv} \int_0^R (R^2 - r^2) r\, dr = \frac{\pi\, \Delta P}{2Lv} \left(\frac{R^4}{2} - \frac{R^4}{4} \right) \tag{3.9}$$

or finally,

$$Q = \frac{\pi R^4\, \Delta P}{8Lv} \tag{3.10}$$

Equation (3.10) is called *Poiseuille's law*. It says that for a given hydrostatic pressure difference ΔP across a length L of cylindrical capillary, the volume of water flowing per unit time Q will be proportional to the fourth power of the radius. Thus, a tube of radius $2R$ will have 16 times as much water flowing through it per unit time as a tube of radius R if each tube has the same pressure gradient $\Delta P/L$ acting on the water. By analogy, we can expect a significant nonlinear dependence of flow on the aperture of the pore space through which flow occurs in a porous medium.

As we shall see later in this chapter, water flow equations are commonly expressed in terms of the volume flow rate per unit area, or water flux J_w. The average flux through the preceding capillary tube is $Q/\pi R^2$, or

$$J_w = \frac{R^2\, \Delta P}{8Lv} \tag{3.11}$$

3.2   WATER FLOW IN SATURATED SOIL

Soil contains a large distribution of pore sizes and channels through which water may flow. The exact geometry of these openings is unknown, so that Newton's law of viscosity (2.9) may not be used directly to calculate flow rates in response to gradients of water potential. Instead, averages are taken over many pores to define macroscopic flow equations to describe movement of water through porous media. The first person to employ this method was Henry Darcy in 1856 (Darcy, 1856).

3.2.1   Darcy's Law

Darcy, an engineer working for the city of Dijon in France, measured the volume of water Q flowing per unit time through water-saturated packed sand columns of length L and area A when a hydrostatic pressure difference $\Delta P = P_2 - P_1$ was placed across them (Fig. 3.3).

After conducting a number of experiments, he developed the following relationship among the variables:

$$Q = \frac{K_s A \, \Delta P}{L} \tag{3.12}$$

where K_s, the *saturated hydraulic conductivity*, is a constant for rigid, saturated soil in a given geometric configuration. Darcy's law may be generalized to apply between any two points of a saturated porous medium provided that the total potential difference of the water between the two points is known. We assume that the soil is rigid and saturated and that no solute membranes exist within the water flow paths. Under these restrictions, the total water potential in saturated soil consists of the sum of the hydrostatic pressure and gravitational potential components. When head units are used, this combination is called the *hydraulic head*,

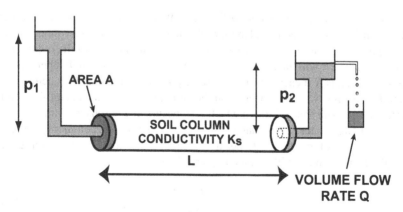

Figure 3.3   Soil column experiment illustrating Darcy's law in the horizontal direction.

$$H = p + z \tag{3.13}$$

Note that when water flow is in the horizontal direction, H reduces to p, since z is the same at all points (see Fig. 3.3). In this case, the distance must also be changed to the x coordinate. As in the case of equilibrium calculations, it is useful to employ a sequence of steps in applying Darcy's law to a flow problem. First, we must develop a sign convention for the flow direction. In this book we use the z coordinate to indicate the vertical direction and assume that the upward direction is positive. Similarly, we use the x coordinate for one-dimensional horizontal flow and assume that the direction pointing to the right is positive. Thus, upward water flow (i.e., evaporation) is positive and downward water flow (i.e., drainage) is negative. With this convention, Darcy's law may be written in the flux form $J_w = Q/A$ between two points 1 and 2 as

$$J_w = -K_s \frac{H_2 - H_1}{z_2 - z_1} \qquad \text{vertical flow} \tag{3.14}$$

$$J_w = -K_s \frac{p_2 - p_1}{x_2 - x_1} \qquad \text{horizontal flow} \tag{3.15}$$

where H_1 is the hydraulic head at point z_1 and H_2 that at z_2. Note that these equations have a minus sign, which must be included in all calculations. Also, K_s represents an average over the entire region between points 1 and 2. When the head form [(3.14) and (3.15)] of Darcy's law is used, K_s has units of length per time, as does J_w.

Table 3.1 lists the steps to be followed in using Darcy's law to calculate vertical water flow. A similar procedure is used in horizontal flow. The first example illustrates the application of this procedure.

Example 3.1 A 100-cm soil column (Fig. 3.4) containing packed sand with a saturated hydraulic conductivity of 100 cm day^{-1} is placed vertically with the bottom open to the atmosphere ($p = 0$). A constant 10-cm height of water is ponded continuously on the top surface. Calculate the steady water flux J_w through the soil.

SOLUTION: Following the steps in Table 3.1, we obtain:

Step 1: Define $z = 0$ at the bottom of the column.

Step 2: Let point A be the bottom of the column where $z_1 = 0$ and $p_1 = 0$. Thus, $H_1 = 0$. Let point B be the top of the column where $z_2 = 100$ cm and $p_2 = 10$ cm (since 10 cm of unsupported water lies above z_2). Thus, $H_2 = 110$ cm.

TABLE 3.1   Procedure for Using Darcy's Law in Vertical Flow Problems

Step 1	Define a reference elevation.
Step 2	Determine two points 1 and 2 where the hydraulic head H is known.
Step 3	Calculate the gradient $(H_2 - H_1)/(z_2 - z_1)$.
Step 4	Plug into (3.14).

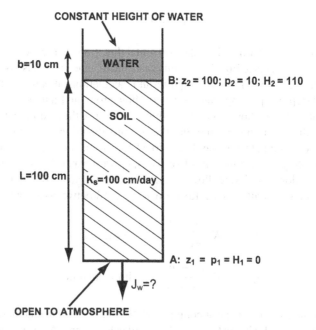

Figure 3.4 Calculation of water flow through vertical saturated soil.

Step 3: The gradient is $(H_2 - H_1)/(z_2 - z_1) = (110 - 0)/(100 - 0) = 1.1$.

Step 4: The flux is $J_w = -K_s(H_2 - H_1)/(z_2 - z_1) = -110 \, \text{cm day}^{-1}$. The negative sign means that the water is flowing downward.

This example illustrates a fundamental difference between equilibrium problems and flow problems. If the bottom of the column was sealed, then at equilibrium the hydrostatic pressure potential head at $z = 0$ would be 110 cm, since the weight of all the water above $z = 0$ is exerted on that point.[3] When the bottom is open to the atmosphere, however, water will leave the pores at the bottom of the column as soon as any positive pressure develops. Thus, $p = 0$ there. The difference between this case and the equilibrium case is that the weight of the water in the column is opposed by the viscous resistive forces on the water. An important difference between vertical and horizontal flow is illustrated in the next example.

Example 3.2 Assume that the soil column in Example 3.1 is placed horizontally[4] with 10 cm of water ponded over the left side while the right side is open to the atmosphere. Calculate the flow through the column.

---

[3] See the definition of ψ_p in Section 2.3.4.
[4] Ignore any elevation differences within the column.

SOLUTION: Following the steps in Table 3.1, we obtain:

Step 1: Define the left side of the column to be $x = 0$.

Step 2: Let point 1 be the left (inlet) end of the column. Here $x_1 = 0$ and $p_1 = 10$ cm. Let point 2 be the right (outlet) end of the column. Here $x_2 = 100$ cm and $p_2 = 0$ (open to the atmosphere).

Step 3: The gradient for horizontal flow is $(p_2 - p_1)/(x_2 - x_1) = (0 - 10)/(100 - 0) = -0.1$.

Step 4: By (3.15), the flux is $J_w = -K_s(p_2 - p_1)/(x_2 - x_1) = 100 \times 0.1 = 10$ cm day^{-1}. The flux is positive, meaning that water is flowing to the right.

Notice that the same externally applied 10 cm water pressure produced only 1/11 as much water flow in the horizontal column as in the vertical column. This occurred because gravitational potential was constant everywhere in the horizontal column, so that the hydraulic head difference was only 10 cm, whereas the hydraulic head difference across the vertical column was 110 cm.

3.2.2 Measurement of Saturated Hydraulic Conductivity

In Examples 3.1 and 3.2 we assumed that K_s is known and used Darcy's law to calculate J_w. However, since it is easy to measure J_w in a laboratory column experiment, Darcy's law may be used to measure K_s.

Assume that soil is packed uniformly in a vertical soil column of length L. A constant height b of water is maintained over the upper end by an external manometer, and the bottom end is open to the atmosphere so that $p = 0$. The water volume flow is collected at the bottom and is used to calculate J_w. For this system, if z is set equal to zero at the bottom, $H_1 = 0 + 0$ (bottom), $H_2 = b + L$ (top), the gradient is $(b + L)/L$, and by (3.14),

$$K_s = -\frac{J_w(b + L)}{L} \tag{3.16}$$

Since J_w is downward, it is negative, so that $K_s > 0$. This procedure is called the *constant-head method* of measuring K_s (Klute and Dirksen, 1986).

There is an even simpler method of measuring K_s in the laboratory using the preceding apparatus. Water is initially ponded to a height b_0 above the saturated column and is allowed to fall with time as water flows through the column and out the bottom where $p = 0$. This device is called a *falling-head permeameter* (Fig. 3.5). In contrast to the controlled pressure head device, here the hydrostatic pressure $p_2(t)$ at the surface is time dependent, as is the gradient $[L + b(t)]/L$. Consequently, the flux J_w is also time dependent and is equal to the time rate of change of the water height $J_w = db/dt$, provided that the water ponded on the surface is contained in a volume of the same cross-sectional area as the soil column. Each small decrease in the height of the water column per unit time represents a flow of the same amount of water per unit area per unit time through the column. Hence, we may write Darcy's law (3.14) as

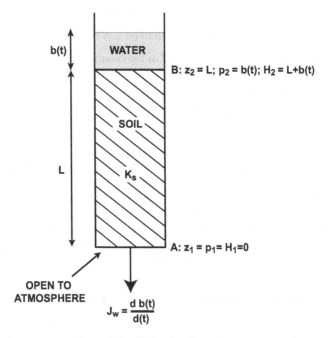

Figure 3.5 Falling-head permeameter.

$$J_w = \frac{db}{dt} = -\frac{K_s}{L}(b + L) \tag{3.17}$$

Equation (3.17) may be rearranged by placing all factors that depend on b on the left side:

$$\frac{db}{b + L} = -\frac{K_s}{L}dt \tag{3.18}$$

where $b = b_0$ at $t = 0$.

Equation (3.18) may be integrated from $t = 0$ to some time $t = t_1$ when b has fallen to $b_1 < b_0$. The left side of (3.18) is equal to

$$\int_{b_0}^{b_1} \frac{db}{b + L} = \ln(b + L)\Big|_{b_0}^{b_1} = \ln\frac{b_1 + L}{b_0 + L} \tag{3.19}$$

where ln is the natural (base e) logarithm. The right side of (3.18) may be integrated simply, producing

$$-\int_0^{t_1} \frac{K_s}{L}dt = -\frac{K_s}{L}\int_0^{t_1} dt = -\frac{K_s t_1}{L} \tag{3.20}$$

Thus, equating (3.19) and (3.20) and recalling that $\ln(A/B) = -\ln(B/A)$, we can solve for K_s:

$$K_s = \frac{L}{t_1} \ln \frac{b_0 + L}{b_1 + L} \tag{3.21}$$

Equation (3.21) demonstrates that K_s may be measured in the falling-head permeameter by measuring b_0, b_1, L, and t_1 only. This procedure is called the *falling-head method* of measuring K_s (Klute and Dirksen, 1986).

Example 3.3   A 100-cm-long soil column is saturated and 10 cm of water is ponded over the top at $t = 0$ in a vessel that has the same cross-sectional area as the column. At $t = 1$ h, the height of overlying water has fallen to 5 cm. Calculate K_s in cm day^{-1}.

SOLUTION: In (3.21),

$$L = 100 \text{ cm}$$
$$t_1 = 1 \text{ h} = 0.042 \text{ day}$$
$$b_0 = 10 \text{ cm}$$
$$b_1 = 5 \text{ cm}$$

Thus,

$$K_s = \frac{100}{0.042} \ln \frac{110}{105} = 111.65 \text{ cm day}^{-1}$$

Extreme care must be taken with all laboratory methods for determining K_s to saturate the soil thoroughly, because any air entrapped in the system will decrease the conductivity of the medium. Also, the value of K_s may change over time due to plugging of pores with suspended material or biological growth. In swelling soils the geometry of the medium will change continuously unless constrained (Reynolds et al., 2002).

3.2.3   Calculation of Hydrostatic Pressure in Soil Columns

Thus far, Darcy's law has been used only to calculate K_s or to estimate J_w when K_s is known. It is also possible to use Darcy's law to calculate pressure profiles within the soil if K_s is known everywhere. This procedure is illustrated in the next example.

Example 3.4   Assume that the hydraulic conductivity K_s is constant everywhere in the soil column of Fig. 3.4 and Example 3.1. Calculate the hydrostatic pressure at an arbitrary point z in the column.

SOLUTION: Since K_s is constant everywhere, Darcy's law may be written between the new point z, where p is unknown, and the bottom of the column at $z = 0$, where $p = 0$. Thus, using (3.14) with $H_2 = z + p$, we obtain

$$J_w = -K_s \frac{p + z}{z}$$

However, by using Darcy's law over the entire column, we already know the flux J_w in terms of the known pressure $p = b$ at $z = L$:

$$J_w = -K_s \frac{b + L}{L}$$

Since the flux is constant, these two expressions are equal. Solving for p, we obtain

$$p = \frac{b}{L} z$$

Thus, the pressure decreases linearly from $p = b$ at $z = L$ to $p = 0$ at $z = 0$.

This linear decrease will occur only if the soil column is homogeneous and K_s is constant everywhere. In layered soils that have a nonuniform K_s, the results are more complex.

3.2.4 Water Flow in Saturated Layered Soil

Figure 3.6 illustrates steady water flow through a layered soil column containing N layers of thickness L_J and saturated hydraulic conductivity K_J ($J = 1, \ldots, N$). We would like to calculate the water flux and hydrostatic pressure distribution as a function of known values of K_J, L_J, and b.

An easy way to approach this problem is to make use of the well-known law of resistances in electrical circuits. This law states that when current i is flowing from one point to another because of a voltage difference ΔV through a series of resistors in accordance with Ohm's law ($\Delta V = iR$), the total resistance between the two points is equal to the sum of the individual resistances. We may establish an analog with this law by rewriting Darcy's law in the form

$$\Delta H = H_1 - H_2 = J_w \frac{L}{K_s} \qquad \text{(Darcy's law)} \qquad (3.22)$$

$$\Delta V = V_1 - V_2 = iR \qquad \text{(Ohm's law)} \qquad (3.23)$$

where we have set $z_2 = L$ and $z_1 = 0$ in (3.14). Thus, since a current is a flux of electrons and a voltage is an electrical potential difference, Darcy's law and Ohm's law are mathematically identical. The electrical resistance R is by definition a potential difference divided by a flux. By analogy, we may define a hydraulic resistance R_H as

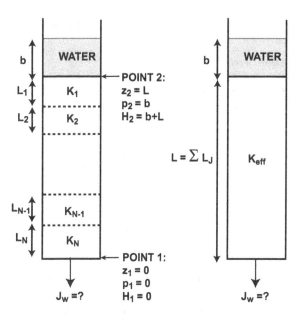

Figure 3.6   Heterogeneous soil column with N layers of thickness L_J and hydraulic conductivity K_J, $J = 1, \ldots, N$. The right side is a homogeneous soil of the same total thickness and equivalent total resistance to flow, which results in the same flux J_w when a height of water b is ponded on top of it.

$$R_H = \frac{\Delta H}{J_w} = \frac{L}{K_s} \tag{3.24}$$

Now we may return to the layered soil column in Fig. 3.6. The effective flow resistance R_{eff} of the entire column is, by the law of resistances,

$$R_{\text{eff}} = \sum_{J=1}^{N} R_{HJ} \equiv \frac{\sum_{J=1}^{N} L_J}{K_{\text{eff}}} \tag{3.25}$$

where K_{eff} is the effective hydraulic conductivity of the entire column. Since by (3.24), the hydraulic resistance R_{HJ} of each layer is L_J / K_J, then (3.25) may be written as

$$\frac{\sum_{J=1}^{N} L_J}{K_{\text{eff}}} = \sum_{J=1}^{N} \frac{L_J}{K_J} \tag{3.26}$$

or

$$K_{\text{eff}} = \frac{\sum_{J=1}^{N} L_J}{\sum_{J=1}^{N} L_J / K_J} \tag{3.27}$$

TABLE 3.2   Procedure for Calculating Water Flow through Saturated Layered Soil

Step 1	Define a reference elevation.
Step 2	Calculate K_{eff} using (3.27).
Step 3	Calculate J_w using the procedure given in Table 3.1.
Step 4	Write Darcy's law across each layer to determine p at the interfaces.

Once we have calculated K_{eff} with (3.27), we may replace the layered column with an equivalent column having a conductivity K_{eff} for the purpose of calculating J_w. Then, once J_w is known, the pressure drop across any homogeneous layer within the column may be calculated using Darcy's law with the actual saturated hydraulic conductivity of the layer being analyzed. The procedure to be used in calculating flow through layered columns is given in Table 3.2. This procedure is illustrated in the next example.

Example 3.5   A layered vertical soil column (Fig. 3.7) consists of a loamy textured soil ($K_1 = 5$ cm h^{-1}, $L_1 = 25$ cm) with a sandy textured soil ($K_2 = 25$ cm h^{-1}, $L_2 = 75$ cm) layer over it. The top of the column has water ponded at a constant height of 10 cm above it, and the bottom is open to the atmosphere. Calculate J_w and the hydrostatic pressure distribution $p(z)$ in the column.

SOLUTION: Following the steps in Table 3.2, we obtain:

Step 1: Let $z = 0$ at the bottom of the column.

Step 2: We use (3.27) to calculate

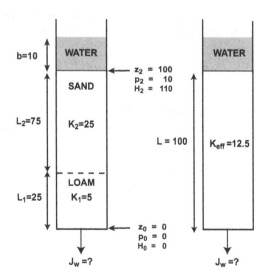

Figure 3.7   Layered soil column analyzed in Example 3.5.

$$K_{eff} = \frac{25 + 75}{25/5 + 75/25} = \frac{100}{5 + 3} = \frac{100}{8} = 12.5 \text{ cm h}^{-1}$$

Step 3: Let point 2 be located at the top of the column, where $z_2 = 100$, $p_2 = 10$, and $H_2 = 110$ cm. Place point 1 at the bottom of the column, where $z_1 = 0$, $p_1 = 0$, and $H_1 = 0$. Therefore, by Darcy's law, the flux across the column is equal to

$$J_w = -K_{eff}\frac{H_2 - H_1}{z_2 - z_1} = -12.5\left(\frac{110 - 0}{100 - 0}\right) = -\frac{110}{8} = -13.75 \text{ cm h}^{-1}$$

This is also equal to the flux across each homogeneous layer within the column.

Step 4: Now that J_w is known, we may apply Darcy's law to each layer within the column to calculate p at the interface, labeled point 3 in the figure. We then write Darcy's law across the bottom layer, whose upper point 3 has $z_3 = 25$, $p_3 = p_3$ (unknown), and $H_3 = p_3 + 25$. The bottom of the column has $H_1 = 0$. Thus,

$$J_w = -K_1\frac{H_3 - H_1}{z_3 - z_1} = -5\left(\frac{p_3 + 25 - 0}{25 - 0}\right) = -13.75 \rightarrow p_3 = 43.75 \text{ cm}$$

We may check our answer for consistency by writing Darcy's law across the upper layer:

$$J_w = -K_2\frac{H_2 - H_3}{z_2 - z_3} = -25\left(\frac{110 - p_3 - 25}{100 - 25}\right) = -13.75 \rightarrow p_3 = 43.75 \text{ cm}$$

From the analysis of flow through homogeneous soil, we know that the pressure change within a given layer will be linear (see Example 3.4). Hence, the pressure and hydraulic head distribution within the layered column are as shown in Fig. 3.8. The

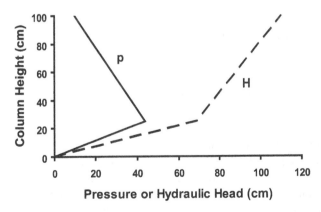

Figure 3.8 Pressure and hydraulic head distribution in the layered soil column analyzed in Example 3.5.

pressure distribution increases linearly within the sand layer and then drops sharply to zero within the loam layer. The hydraulic head decreases continually (which it must, since flow is downward), but more gradually within the sand than the loam.

The reason that the hydrostatic pressure builds up in the sand is that the loam offers substantially more resistance to flow than the sand and hence produces most of the force opposing the externally applied hydraulic head difference acting on the water in the column. This pressure must build up in order to "push" the water through the loam at the same rate as it is moving through the sand. We obtain a rather surprising result when we turn the column upside down and repeat the experiment.

Example 3.6   Repeat Example 3.5 with the column rotated 180° so that the loam lies over the sand.

SOLUTION: Following the steps in Table 3.2, we obtain:

Step 1: As before, $z = 0$ at the bottom.

Step 2: Since the resistance law does not depend on the order of the resistances, $K_{eff} = 12.5 \, cm \, h^{-1}$, as before. [Calculate it yourself using (3.27) if you do not agree.]

Step 3: Since K_{eff}, H_2, and H_1 are all the same as in Example 3.5, $J_w = -13.75$ cm h^{-1} as before.

Step 4: Writing Darcy's law across the sand layer, where $H_3 = 75 + p_3$, $H_1 = 0$, we obtain

$$J_w = -K_1 \frac{H_3 - H_1}{z_3 - z_1} = -25 \left(\frac{p_3 + 75 - 0}{75 - 0} \right) = -13.75 \rightarrow p_3 = -33.75 \, cm$$

In this case a negative hydrostatic pressure will develop at the interface in order to decrease the water flow through the low-resistance sand until it matches the flow through the high-resistance loam. In practice, this negative pressure will probably exceed the air-entry suction of the saturated sand, causing part of the soil in the column to unsaturate. As soon as the column unsaturates, the resistance will change, thereby invalidating the assumptions of the calculation.

3.3   WATER FLOW IN UNSATURATED SOIL

When soil is partially unsaturated, an air phase is present and the water flow channels are drastically modified from those in saturated soil. In unsaturated soil, the water phase is bounded partially by solid surfaces and partially by an interface with the air phase. In contrast to the positive water pressure found in saturated soil, the water pressure within the liquid phase is caused by water elevation, attraction to solid surfaces, and the surface tension of the air–water interface. Because of the effects of surface tension and solid surfaces, the liquid pressure is lower than the reference liquid pressure at the same elevation. As the water content decreases, the liquid pressure decreases and the water phase is constrained to narrower and more tortuous channels.

3.3.1 Buckingham–Darcy Flux Law

In 1907, Edgar Buckingham proposed a modification of Darcy's law (3.14) to describe the flux of water through unsaturated soil. This modification rested primarily on two assumptions:

1. The driving force for water flow in isothermal, rigid, unsaturated soil containing no solute membranes and zero air pressure potential is the sum of the matric and gravitational potentials.
2. The hydraulic conductivity of unsaturated soil is a function of the water content or matric potential.

In head units, the Buckingham–Darcy flux law may be expressed for vertical flow as

$$J_w = -K(h)\frac{\partial H}{\partial z} = -K(h)\frac{\partial(h+z)}{\partial z} = -K(h)\left(\frac{\partial h}{\partial z}+1\right) \qquad (3.28)$$

where $H = h+z$ (in units of length) is the hydraulic head in unsaturated soil and $K(h)$ (in units of length/time) is the unsaturated hydraulic conductivity. As in saturated flow, the flux J_w (length/time) is the water flow per unit cross-sectional area per unit time.

Several points should be stressed about (3.28). First, it is a differential equation that is written across an infinitesimally thin layer of soil over which h is constant and $K(h)$ is constant. It may not be written across a finite layer of soil unless the water content and matric potential of the layer are uniform, which occurs only under special conditions to be covered later. Second, the derivative in (3.28) is a partial derivative, because in unsaturated soil h may be a function of both z and t. The partial derivative $\partial h/\partial z$ implies that the derivative with respect to z is taken at constant t; it is the instantaneous value of the slope of $h(z, t)$:

$$\frac{\partial h}{\partial z} \equiv \left(\frac{\partial h}{\partial z}\right)_t = \lim_{\Delta z \to 0}\frac{h(z+\Delta z, t) - h(z, t)}{\Delta z} \qquad (3.29)$$

where $(\cdot)_t$ means that the derivative is evaluated at constant t. Partial derivatives are required for the mathematical description of transient (time-dependent) flow. If the system is at steady state, the partial derivative reduces to an ordinary derivative, since in steady state, h depends only on z.

3.3.2 Unsaturated Hydraulic Conductivity

The unsaturated hydraulic conductivity is a nonlinear function of water content or matric potential. Figure 3.9 shows typical $K(h)$ curves for a coarse-textured sandy

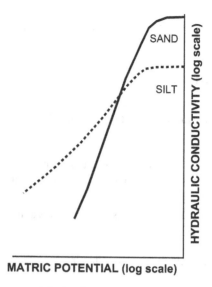

Figure 3.9 Typical unsaturated hydraulic conductivity–matric potential functions for a coarse-textured sandy soil and a finer-textured silty soil.

soil and a fine-textured silty soil. At saturation, the coarse-textured soil has a higher conductivity than the fine-textured soil, because it contains larger pore spaces, which are filled and conduct water. However, these pores drain at modest suctions, producing a dramatic decrease in hydraulic conductivity in the sandy soil. Eventually, the curves will cross and the sandy soil will actually have a lower hydraulic conductivity than the silty soil at the same matric potential, because the latter will retain considerably more water and will contain a greater number of filled pores.

We may gain insight into the relation between the pore sizes of a porous medium and its unsaturated hydraulic conductivity by using a simple capillary tube model of soil. This model is derived in the next section.

3.3.3   Capillary Tube Model of Unsaturated Hydraulic Conductivity

The value of the unsaturated hydraulic conductivity $K(h)$ is highly dependent on the amount of water in the soil and also on the thickness of the films or channels that are filled. For many soils, $K(h)$ drops to a small fraction of its value at saturation after the loss of only a small percentage of the total water content. This property of $K(h)$ completely dominates important flow processes under some conditions. Although real flow systems in soil are too complex to analyze quantitatively with the Poiseuille equation (3.10), a simple model of soil water flow may be developed using bundles of capillary tubes of different sizes. This capillary bundle model of soil contains many of the same properties as those of a real soil water system and may be analyzed to produce a relationship between the unsaturated hydraulic conductivity and the geometry of the solid phase.

The number and sizes of the capillary tubes in the bundle are chosen so as to reproduce the matric potential–water content function of the soil they represent. However, the flow system of this bundle of tubes differs from the water flow network of a real soil in several ways:

- Each capillary tube is continuous and contains no dead end or stagnant regions, although it may have a twisted or tortuous shape.
- There are no narrow necks constricting flow at points along the way from input to output.
- All capillary tubes in the model have the same length.
- The water flow boundaries of the model are composed entirely of solid–water interfaces, with empty capillaries constituting the air space, whereas in soil, the flow boundaries are a combination of solid–water and solid–air interfaces.
- The radius of the capillary tube entirely governs the thickness of the water films, whereas the water film thickness on surfaces in soil is governed by the matric potential.
- Flow is steady state, which is contrary to the usual situation in soil.

The hydraulic conductivity of the capillary bundle model is derived as follows. The network of tubes, each of which is assumed to have a length L_c, is twisted internally and ultimately bundled to form a soil column of length $L < L_c$ and cross-sectional area A. A hydraulic head gradient ΔH is placed across the ends of the "column," causing water to flow through each of the capillary tubes in accordance with Poiseuille's law (3.10). The total flux through the column is equal to the sum of the volume flow rates out of each tube divided by the cross-sectional area of the column. Thus, a single capillary of radius R_J has a volume flow rate Q_J given by (3.10) as

$$Q_J = \frac{\rho_w g \pi R_J^4}{8\nu} \frac{\Delta H}{L_c} \tag{3.30}$$

and the total volume flow rate Q_T through the column when all tubes are filled is

$$Q_T = \sum_{J=1}^{M} N_J Q_J = \frac{\pi \rho_w g}{8\nu} \frac{\Delta H}{L_c} \sum_{J=1}^{M} N_J R_J^4 \tag{3.31}$$

where N_J represents the number of capillaries of radius R_J in the bundle and M is the number of different capillary size classes in the bundle of tubes making up the column. The flux J_w through the column is given by

$$J_w = \frac{Q_T}{A} = \frac{\pi \rho_w g}{8\nu} \frac{\Delta H}{L_c} \sum_{J=1}^{M} n_J R_J^4 \tag{3.32}$$

where $n_J = N_J / A$ is the number of tubes per unit area of radius R_J in the bundle.

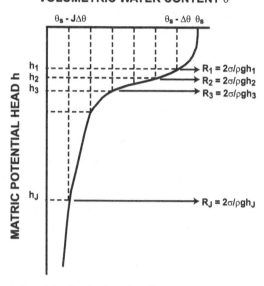

Figure 3.10   Calculation of the distribution of capillary tube radii required to produce a matric potential–water content curve that is equivalent to the real one.

A correspondence between a real soil and the capillary bundle model is established by using the matric potential–water content curve of the soil (Fig. 3.10) to calculate the distribution of capillary tubes that would have equivalent water-retention characteristics. This is achieved using the following procedure:

1. The matric potential–water content curve $h(\theta)$ of the real soil is divided into a number of equally spaced water content intervals of width $\Delta\theta$.
2. The matric potential associated with each decrease in θ is determined from the $h(\theta)$ relation. Thus, $h_1 = h(\theta_s - \Delta\theta)$, $h_2 = h(\theta_s - 2\Delta\theta)$, and so on.
3. All tubes of radius $r > R_J$ are assumed to be drained when $h = h_J$, where R_J is given by the capillarity equation

$$R_J = -\frac{2\sigma}{\rho_w g h_J} \tag{3.33}$$

Thus, when the water content changes from θ_s (when all pores are filled) to $\theta_s - \Delta\theta$, we assume that all of the tubes of radius $R_1 = -2\sigma/\rho_w h_1$ in the capillary bundle model will drain. Therefore, the number n_1 of these tubes per unit area[5] must correspond to the decrease $\Delta\theta$ in the water content of the real soil.

---

[5] Because the tubes are cylinders, the number of tubes per area and tubes per volume is the same.

If we assume that in a unit volume the capillary bundle contains n_1 cylinders per unit area, each one of which has cross-sectional area πR_1^2 and a unit length, then the change $\Delta\theta$ in water volume per total volume when these cylinders drain is

$$\Delta\theta = n_1 \pi R_1^2 \tag{3.34}$$

Therefore, the number of tubes of radius R_1 per unit area is

$$n_1 = \frac{\Delta\theta}{\pi R_1^2} \tag{3.35}$$

Similarly, $n_J = \Delta\theta/\pi R_J^2$ in each water content interval, where R_J is related to the soil by (3.33) and $h_J = h(\theta_s - J\Delta\theta)$.

We may now insert (3.33) and (3.35) into (3.32), so that

$$J_w = \frac{\rho_w g}{8\nu} \frac{\Delta H}{L_c} \Delta\theta \sum_{J=1}^{M} R_J^2 \tag{3.36}$$

$$= -\left(\frac{\sigma^2 \Delta\theta}{2\nu\rho_w g} \sum_{J=1}^{M} \frac{1}{h_J^2} \frac{L}{L_c} \right) \frac{\Delta H}{\Delta z} \tag{3.37}$$

where $\Delta z = 0 - L$, in accordance with the sign convention used for Darcy's law (3.14). Equation (3.37) now is in the form of Darcy's law (3.14), so that we may associate the quantity in parentheses with the saturated hydraulic conductivity K_s of the capillary bundle:

$$K_s = \frac{\tau \sigma^2 \Delta\theta}{2\nu\rho_w g} \sum_{J=1}^{M} \frac{1}{h_J^2} \tag{3.38}$$

where

$$\tau = \frac{L}{L_c} \tag{3.39}$$

The factor τ, the *tortuosity*, is the ratio between the column length and the capillary or water path length.

As the water content decreases from θ_s to $\theta_s - \Delta\theta$, the largest tubes of radius R_1 drain and no longer contribute to the flow. Thus,

$$K(\theta_s - \Delta\theta) = \frac{\tau \sigma^2 \Delta\theta}{2\nu\rho_w g} \sum_{J=2}^{M} \frac{1}{h_J^2} \tag{3.40}$$

and in general,

$$K(\theta_s - i\,\Delta\theta) = \frac{\tau\sigma^2\,\Delta\theta}{2\nu\rho_w g} \sum_{J=i+1}^{M} \frac{1}{h_J^2} \tag{3.41}$$

Equation (3.41) represents the predicted unsaturated hydraulic conductivity of the capillary bundle, calculated from the measured $h(\theta)$ curve. It was first derived by Childs and Collis-George (1950), with later improvements by Marshall (1958) and Millington and Quirk (1959). Equation (3.41) still contains the unknown tortuosity τ, which must be evaluated by fitting the model to measured $K(\theta)$ data. In the preceding derivation, it was assumed that the length L_c of each of the capillaries was the same, so that the tortuosity factor τ should be a constant. Proper selection of the point of calibration has produced reasonable agreement with the model in certain media (Green and Corey, 1971; Jackson, 1972). However, the model has been able to represent real soil hydraulic conductivities of many soils only when τ is allowed to vary as a function of θ (Fatt and Dykstra, 1951). Since τ can be evaluated only by calibration, this restriction makes the model useful primarily as a conceptual aid in visualizing the relationship between conductivity, pore sizes, and water content.

There have been a number of alternative approaches used to model the unsaturated hydraulic conductivity function. The most versatile of these are statistical models that express the conductivity as a function of the pore-size distribution, which is assumed to be related to the matric potential through the capillary equation (2.5). Mualem (1992) reviews these models and the assumptions used to derive them in detail. Brooks and Corey (1966) developed generalized functional forms for $h(\theta)$ and $K(h)$ in which the model parameters were linked together. This approach has proven to be useful in field measurements of these functions (Russo and Bresler, 1981). Mualem (1976a) has extended the Brooks and Corey (1966) model into a general model for predicting the hydraulic conductivities of different soils. This method was used by Van Genuchten (1980) to construct a widely used parametric form for $K(\theta)$ (see Table 3.6).

Recent work on characterizing the transport and retention functions in terms of idealized pore-scale geometry has replaced the capillary bundle with a representation of soil as a network of connected pores with narrow necks (Fig. 3.11). Reeves and Celia (1996) constructed a model porous medium consisting of a lattice of spherical pores

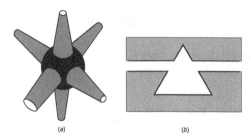

(a)   (b)

Figure 3.11   Pore models used by (a) Reeves and Celia (1996) and (b) Tuller and Or (2001). In each model, a large central pore is connected to adjacent pore through a narrow channel.

connected to six adjacent pores through conical throats (Fig. 3.11*a*). Water is placed in the "medium" and redistributes through the pore network by satisfying pressure equilibrium (2.5) at all the interfaces. The final configuration is then averaged to produce a pressure–water content relationship. Tuller and Or (2001) use an angular pore shape connected by slits (Fig. 3.11*b*) and separate the forces on the liquid into adsorptive and capillary contributions. Adsorptive forces assemble water onto thin films covering the surfaces at low saturation, and these eventually join as water content increases, producing curved interfaces that lower the liquid pressure according to (2.5).

3.3.4 Steady-State Water Flow Problems

When a matric potential difference $\Delta h = h_2 - h_1$ is maintained across a vertical soil column of length $L = z_2 - z_1$, the flow will eventually reach a steady state, at which time the Buckingham–Darcy flux law may be written as

$$J_w = -K(h)\left(\frac{dh}{dz} + 1\right) \tag{3.42}$$

where now an ordinary derivative is used because h depends only on z.

Integral Form of Darcy's Law   We may calculate the value of the water flux J_w in steady state between any two points z_1 and z_2 in the soil by turning (3.42) into an integral. This is accomplished by moving all factors that depend explicitly on h to the same side of the equation and those that depend on z to the other.

$$-\frac{J_w}{K(h)} = \frac{dh}{dz} + 1$$

$$-\left[1 + \frac{J_w}{K(h)}\right] = \frac{dh}{dz}$$

$$\frac{dh}{1 + J_w/K(h)} = -dz$$

Now we may integrate this last expression from $h_1 = h(z_1)$ to $h_2 = h(z_2)$:

$$\int_{h_1}^{h_2} \frac{dh}{1 + J_w/K(h)} = -\int_{z_1}^{z_2} = z_1 - z_2 \tag{3.43}$$

Equation (3.43) is called the integral form of the Buckingham–Darcy equation. If the unsaturated hydraulic conductivity $K(h)$ has a known functional form, the integral on the left side may sometimes be integrated exactly to produce an analytic expression for the water flux. This procedure is illustrated in the next section.

Evaporation from a Water Table   Although water evaporation in the field is not a steady-state process, a nearly steady upward flow from a water table to a bare soil surface may be established if the daily evaporative demand is reasonably uniform for a long period of time. A classic solution to this problem adapted from Gardner (1958) and Gardner and Fireman (1958) is given in the next example.

Example 3.7   Calculate the maximum possible water evaporation rate above a water table located at a distance $z = L$ below the soil surface ($z = 0$) for a soil that has a hydraulic conductivity–matric potential that may be represented well by the following functional form:

$$K(h) = \frac{K_s}{1 + (h/a)^N} \tag{3.44}$$

where K_s is the saturated hydraulic conductivity and $N > 0$, $a < 0$ are constant parameters representing the shape of the function.

SOLUTION: In this problem, $h_1 = 0$ at $z = -L$ (the water table) and $h_2 = -\infty$ at $z = 0$, representing the maximum possible attraction to the soil surface and therefore the maximum evaporation rate. Letting $J_w = +E$ represent the evaporation rate, we may rewrite (3.43) as

$$-L = \int_0^{-\infty} \frac{dh}{1 + (E/K_s)[1 + (h/a)^N]} \tag{3.45}$$

Letting $y = h/a > 0$ in (3.45) produces the following change in the integral:

$$-L = a \int_0^{\infty} \frac{dy}{1 + E/K_s + (E/K_s)y^N} \approx a \int_0^{\infty} \frac{dy}{1 + (E/K_s)y^N} \tag{3.46}$$

where we have simplified (3.46) by assuming that $E/K_s \ll 1$. Letting $z = y(E/K_s)^{1/N}$ in (3.46) produces the following change in the integral:

$$-L = a \left(\frac{K_s}{E}\right)^{1/N} \int_0^{\infty} \frac{dz}{1 + z^N} \tag{3.47}$$

The integral in (3.47) may be looked up in any standard table of definite integrals, or it may be evaluated by the calculus of residues. It has the value (Abramowitz and Stegun, 1970)

$$\int_0^{\infty} \frac{dz}{1 + z^N} = \frac{\pi}{N \sin(\pi/N)} \tag{3.48}$$

After plugging (3.48) into (3.47), we can solve for the maximum evaporation rate

$$E = K_s \left(\frac{-a\pi}{LN \sin(\pi/N)}\right)^N \tag{3.49}$$

Equation (3.49) gives the maximum evaporation rate E as a function of the distance L between the soil surface and the water table, and the soil parameters K_s, a, and N are used to model the hydraulic conductivity function (3.44).

Gardner and Fireman (1958) conducted a laboratory experimental test of (3.49) on two soils, a Pachappa fine sandy loam and a Chino clay, by maintaining a hanging water column connected to the bottom of soil columns packed with samples of the two soils. The water column was placed at a specified position below the column to correspond to the water table depth. They found very good agreement between predicted and measured evaporation rate.

This simple model calculation demonstrates some fundamental principles that are useful in interpreting water flow behavior in the field environment. Coarse-textured soils, containing mostly large pores that drain at modest suctions, are characterized in the model expression by larger values of N than the finer-textured soils, which have a broader distribution of pore sizes. As a consequence, when the distance between the water table and the surface is great enough, the coarse-textured soils offer more resistance to upward water flow than do finer-textured soils.

A major problem associated with agriculture in areas overlying shallow groundwater that is high in salt content is that saline water will move upward to the surface and evaporate, leaving behind salt deposits that can damage crops or deteriorate soil structure. As shown by this model, the upward flow of water in a fine-textured soil can be more significant than in a coarse-textured soil. Thus, to avoid salinization, water tables must be kept lower for fine-textured soils than for coarse-textured soils.

Example 3.7 illustrates a fundamental difference between saturated and unsaturated flow. In saturated flow, the flux through a finite layer of soil is proportional to the potential difference ΔH across the layer. In unsaturated flow, the conductivity $K(h)$ depends on the matric potential of the soil, so that the flux is not a linear function of ΔH. For example, in the maximum evaporation calculation, an infinite potential difference led to a finite evaporation rate. The reason for this result is that the conductivity in the dry (upper) end of the soil is decreasing rapidly and ultimately approaches zero as $h \to -\infty$. In fact, as shown by Gardner (1958), the maximum evaporation rate is approached for relatively modest degrees of drying at the surface (i.e., $h \sim -1000$ cm). Further decreases in h at the surface do not increase E.

Steady-State Downward Water Flow   Steady-state downward water flow, although never actually achieved in the field, is nevertheless an approximation of certain flows, such as subsurface drainage in environments under high-frequency irrigation or frequent rainfall. Under these conditions, the Buckingham–Darcy flux equation (3.28) may be replaced by a far simpler expression. This is illustrated in the next example.

Example 3.8   Calculate the steady-state downward flux $J_w = -i$ from the surface to a water table at a depth L below the surface. Assume that the soil hydraulic conductivity is described by (3.44) with $N = 2$.

SOLUTION: With this substitution, the integral form of the Buckingham–Darcy flux law (3.43) may be written as $h_1 = 0$ at $z = -L$ and $J_w = -i < 0$:

$$- (z + L) = \int_0^h \frac{dh}{1 - (i/K_s)[1 + (h/a)^2]} \tag{3.50}$$

where the upper limit of the integral has been evaluated at an arbitrary depth z where the matric potential is $h = h(z)$. Letting $y = h/a$ as before, the integral simplifies to

$$- (z + L) = \int_0^{h/a} \frac{dy}{1 - i/K_s - (i/K_s)y^2} \tag{3.51}$$

This integral is a standard form that may be looked up in integral tables (Abramowitz and Stegun, 1970):

$$\int \frac{dy}{\alpha - \beta y^2} = \frac{1}{\sqrt{\alpha\beta}} \tanh^{-1} \frac{y\sqrt{\alpha\beta}}{\alpha} \tag{3.52}$$

where $\alpha = 1 - i/K_s$, $\beta = i/K_s$, and $\tanh^{-1}$ is the hyperbolic arctangent. Thus, applying (3.52) to (3.51) yields

$$- (z + L) = \frac{a}{\sqrt{(1 - i/K_s)i/K_s}} \tanh^{-1} \frac{h\sqrt{(1 - i/K_s)i/K_s}}{a(1 - i/K_s)} \tag{3.53}$$

or

$$h = a\sqrt{\frac{K_s}{i} - 1} \tanh \left[\frac{-\sqrt{(1 - i/K_s)i/K_s}}{a}(z + L) \right] \tag{3.54}$$

Figure 3.12 shows a graph of relative matric potential head h/a versus relative depth z/L for various values of the parameter i/K_s. A significant feature of this curve is that the matric potential approaches a constant value for any flux rate i as the surface is approached, provided that the water table is not too shallow. This important result shows that when water is flowing downward at a constant rate, matric potential gradients dh/dz approach zero and water flows under the influence of gravity alone. Thus, provided that the water table is far below the surface, one may approximate the Buckingham–Darcy law (3.28) for downward flow as

$$J_w \approx -K(h) \tag{3.55}$$

This approximation is known as *gravity flow*. A general discussion of steady upward and downward flows is given in Raats and Gardner (1975).

Measurement of Unsaturated Hydraulic Conductivity   The unsaturated hydraulic conductivity–matric potential function may be measured in the laboratory in steady-state flow experiments using the principles developed in the preceding. The next

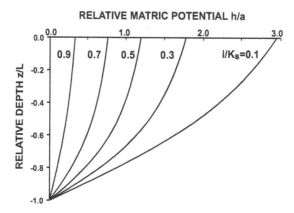

RELATIVE MATRIC POTENTIAL h/a

Figure 3.12 Matric potential–depth profiles for steady downward flow calculated with (3.53), with $L/a = 5.0$.

example demonstrates how a steady-state evaporation experiment may be used to calculate average values of $K(h)$ in a soil column instrumented with tensiometers.

Example 3.9   A 50-cm-high soil column is placed above a water source maintaining the bottom of the column ($z = 0$) at saturation ($h = 0$). The top of the column ($z = 50$ cm) is exposed to constant evaporative conditions. Tensiometers are located at z values of 10, 20, 30, and 40 cm within the column. The evaporation rate E in steady state (when the tensiometers and water flux do not change with time) is equal to the flux through the column and is measured by the rate of water entry per unit area into the column from the water source. In steady state the tensiometer readings are as shown in Table 3.3 and the evaporation rate is 0.5 cm day^{-1}. Calculate the unsaturated hydraulic conductivity from the data given in Table 3.3.

SOLUTION: To calculate $K(h)$, the steady-state form of the Buckingham–Darcy flux law (3.28) is written between two depths z_1 and z_2 where the measured matric potentials are h_1 and h_2, respectively. As an approximation, $K(h)$ is replaced by its average value $K(\bar{h})$ between z_1 and z_2, where $\bar{h} = (h_1 + h_2)/2$. Thus, (3.28) becomes

TABLE 3.3   Tensiometer Readings at Steady State in an Evaporation Experiment

z (cm)	h (cm)
40	-125
30	-75
20	-40
10	-15
0	0

TABLE 3.4   Sequence of Steps Used to Calculate $K(h)$ from a Steady-State Evaporation Experiment

Step 1	Calculate H at each depth.
Step 2	Calculate ΔH between each depth interval.
Step 3	Calculate $\Delta H / \Delta z$ between each depth interval.
Step 4	Calculate $K(\bar{h})$ between each depth interval using (3.57).
Step 5	Calculate $\bar{h}$ between each depth interval.

$$J_w = E = -K(h)\frac{dH}{dz} \approx K(\bar{h})\frac{\Delta H}{\Delta z} \tag{3.56}$$

where $\Delta z = z_2 - z_1$ and $\Delta H = H_2 - H_1 = (h_2 + z_2) - (h_1 + z_1)$. Thus,

$$K(\bar{h}) = -E\frac{\Delta z}{\Delta H} \tag{3.57}$$

The data in Table 3.3 may be evaluated using (3.57) by following the sequence of steps given in Table 3.4. Table 3.5 summarizes the calculation of $K(h)$ using this procedure and the data in Table 3.3. Thus, this experiment yielded four values of the $K(h)$ function, averaged over the range of matric potentials between the successive depth increments.

The resolution of the $K(h)$ measurement could be improved by increasing the density of tensiometers. The $K(h)$ function may also be measured in steady-state downward flow experiments. However, if this method is used, only a small range of matric potentials is covered in a single flow experiment.

More recently, with the advent of high-speed computational tools, inverse methods have become popular for estimating water flow and retention properties. With these methods, a standard water flow experiment with well-defined boundary conditions is conducted and the numerical solution of the differential equation describing flow

TABLE 3.5   Evaluation of $K(h)$ by the Procedure from Table 3.4 Using Data from Table 3.3

z	h	H	ΔH	$\Delta H / \Delta z$	$K(\bar{h})$	$\bar{h}$
40	−125	−85				
			−40	−4.0	0.125	−100.0
30	−75	−45				
			−25	−2.5	0.200	−57.5
20	−40	−20				
			−15	−1.5	0.333	−27.5
10	−15	−5				
			−5	−0.5	1.000	−7.5
0	0	−0				

(see Section 3.3.6) is solved repeatedly over a range of values of the parameters describing the hydraulic functions to determine the best agreement between theory and observation. A discussion of this approach and its application to determination of hydraulic properties is given in Hopmans et al. (2002).

Steady-State Downward Water Flow through a Crop Root Zone   Steady state rarely occurs in unsaturated soil, particularly in the field, because temperature, evaporation rate, water uptake, irrigation, and other processes that affect water transport tend to be influenced by the diurnal cycle of radiant energy to Earth's surface. Nevertheless, if the external influences are repeated at regular intervals at approximately the same intensity, the soil water will develop a characteristic response that varies mostly at the surface and becomes relatively steady below it. For this reason, useful information can be obtained by approximating the flow regime by a steady-state model. This is illustrated in the next example.

Example 3.10   A fully developed crop root zone extending from $z = 0$ to $z = -L$ is irrigated daily at an irrigation rate i_0. Water uptake within the root zone is spatially and temporally uniform at a rate r_w, and the total removed each day is equal to the evapotranspiration ET $< i_0$. Calculate the steady water flux as a function of depth.

SOLUTION: In steady state, the rate of decrease of water flux J_w over distance is equal to minus the water uptake per unit volume per unit time r_w. Thus,

$$\frac{dJ_w}{dz} + r_w = 0 \tag{3.58}$$

The water uptake $r_w = \text{ET}/L$ when the rate is constant. Thus we may integrate (3.58) to obtain

$$J_w(z) = -i_0 - \frac{\text{ET} \cdot z}{L} \qquad \text{for } -L < z < 0 \tag{3.59}$$

Thus the flux in the root zone decreases linearly when the water uptake is constant.

3.3.5   Water Conservation Equation

The steady-state water flow condition discussed in previous sections describes only a special subset of the possible water transport processes in soil. In general, wetting or drying of the soil will occur as water flows, and the matric potential and water content will be functions of time as well as of space. Such flows, called *transient* or *time dependent*, require a more complete mathematical description than steady-state flows. The first step in a complete transient water flow description is to specify the *water conservation equation*, also called the *mass balance* or *continuity equation*.

As in the formulation of the flux laws, we do not attempt to describe water flow at the pore scale but will produce a volume-averaged description of water conservation. We focus our analysis on a cube of soil occupying a volume $V = \Delta x \, \Delta y \, \Delta z$ at a point

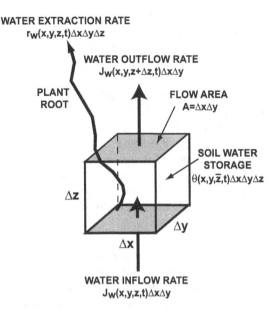

WATER EXTRACTION RATE
$r_w(x,y,z,t)\Delta x\Delta y\Delta z$

WATER OUTFLOW RATE
$J_w(x,y,z+\Delta z,t)\Delta x\Delta y$

PLANT
ROOT

FLOW AREA
$A=\Delta x\Delta y$

SOIL WATER
STORAGE
$\theta(x,y,\bar{z},t)\Delta x\Delta y\Delta z$

Δz

Δy

Δx

WATER INFLOW RATE
$J_w(x,y,z,t)\Delta x\Delta y$

Figure 3.13   Unit volume of soil used to derive the water mass balance equation.

(x,y,z) in the flow field of a porous medium (Fig. 3.13). For simplicity, we assume that water is flowing in the z direction through homogeneous, unsaturated soil.

In the most general case, water may be flowing in and out of the soil volume at different rates, causing the water content of the volume to change. In addition, if plant roots are present inside the volume, some of the water that enters will "disappear" from the volume by entering the plant root and moving toward the leaf within the root xylem. Our statement of water conservation must account for all of these processes.

The water conservation equation is derived by calculating the mass balance for the system during an arbitrarily small time period Δt between t and $t + \Delta t$. Over this time period, the water conservation equation may be stated in words as follows:

volume of water entering soil volume during Δt

$\quad$ = volume of water leaving soil volume during Δt

$\quad\quad$ + increase of water volume stored in soil volume during Δt

$\quad\quad$ + volume of water extracted from soil volume by plant roots during Δt

$\hfill$ (3.60)

For the one-dimensional vertical flow process shown in Fig. 3.14, the volume of water entering the soil volume during Δt is given by

volume of water entering soil volume $= J_w(x, y, z, t + \Delta t/2)\, \Delta x\, \Delta y\, \Delta t$ $\quad$ (3.61)

where J_w (flow per unit area per unit time) is the water flux in the z direction and $\Delta x\, \Delta y$ is the cross-sectional area of the inflow surface. The flux is expressed as a function of x, y, z, and t, although it will not depend on x or y if the flow is in the z direction only. It is assigned an average value[6] at the midpoint $t + \Delta t$ of the time interval.

Similarly, at the exit boundary,

volume of water leaving soil volume

$$= J_w(x, y, z + \Delta z, t + \Delta t/2) \, \Delta x \, \Delta y \, \Delta t \qquad (3.62)$$

where now the water flux is evaluated at the point $z + \Delta z$, the location of the outflow surface.

The increase in water volume stored in the soil volume during Δt may be expressed in terms of the volumetric water content[7] θ as

increase in water volume

$$= [\theta(x, y, z + \Delta z/2, t + \Delta t) - \theta(x, y, z + \Delta z/2, t)] \, \Delta x \, \Delta y \, \Delta z \qquad (3.63)$$

The water content is assigned an average value at the midpoint $z + \Delta z$ of the volume.

The volume of water extracted by plant roots is given by

volume of water extracted from volume

$$= r_w(x, y, z + \Delta z/2, t + \Delta t/2) \, \Delta x \, \Delta y \, \Delta z \qquad (3.64)$$

where r_w is the rate of loss of water per unit volume. The function r_w is called a *sink term* in the conservation equation.

Inserting (3.61)–(3.64) into (3.61) produces

$$[J_w(x, y, z, t + \Delta t/2) - J_w(x, y, z + \Delta z, t + \Delta t/2)] \, \Delta x \, \Delta y \, \Delta t$$
$$= [\theta(x, y, z + \Delta z/2, t + \Delta t) - \theta(x, y, z + \Delta z/2, t)] \, \Delta x \, \Delta y \, \Delta z$$
$$+ r_w \, \Delta x \, \Delta y \, \Delta z \, \Delta t \qquad (3.65)$$

After dividing by $\Delta x \, \Delta y \, \Delta z \, \Delta t$ and rearranging terms, (3.65) may be reexpressed as

$$\frac{J_w(x, y, z + \Delta z, t + \Delta t/2) - J_w(x, y, z, t + \Delta t/2)}{\Delta z}$$

$$+ \frac{\theta(x, y, z + \Delta z/2, t + \Delta t) - \theta(x, y, z + \Delta z/2, t)}{\Delta t} + r_w = 0 \qquad (3.66)$$

Finally, in the limit as $\Delta z \to 0$ and $\Delta t \to 0$ and using the definition (3.29) of the partial derivative, (3.66) becomes

---

[6] Since Δt is arbitrarily small, this will not affect the accuracy of the result.
[7] The subscript v on the volumetric water content will be omitted from now on.

$$\frac{\partial J_w}{\partial z} + \frac{\partial \theta}{\partial t} + r_w = 0 \qquad (3.67)$$

Equation (3.67) is called the soil water conservation or continuity equation. If we had allowed water to flow in an arbitrary direction in Fig. 3.14, the water conservation equation would be written as

$$\frac{\partial J_{wx}}{\partial x} + \frac{\partial J_{wy}}{\partial y} + \frac{\partial J_{wz}}{\partial z} + \frac{\partial \theta}{\partial t} + r_w = 0 \qquad (3.68)$$

where J_{wx}, J_{wy}, and J_{wz} are the components of the water flux vector

$$\vec{J}_w = J_{wx}\hat{i} + J_{wy}\hat{j} + J_{wz}\hat{k} \qquad (3.69)$$

and $\hat{i}$, $\hat{j}$, and $\hat{k}$ are unit vectors in the x, y, and z directions, respectively.

The water uptake term r_w is included for completeness in (3.67) and (3.68). It is equal to zero when there are no plant roots or other sinks of water present and must be specified with a model when water uptake is occurring. Processes involving water uptake will be discussed later in the book.

3.3.6   The Richards Equation for Transient Water Flow

The water conservation equation relates water fluxes, storage changes, and sources or sinks of water. When it is combined with the Buckingham–Darcy flux equation (3.28), an equation may be derived to predict the water content or matric potential in soil during transient flow. We assume for simplicity that the flow is vertical and that no plant roots are present ($r_w = 0$).

Inserting (3.28) into (3.67) produces

$$\frac{\partial \theta}{\partial t} = \frac{\partial}{\partial z}\left[K(h)\left(\frac{\partial h}{\partial z} + 1 \right) \right] \qquad (3.70)$$

Equation (3.70) may not be solved in the form it is in, because it contains two unknowns (θ and h) and only one equation. This difficulty may be overcome by using the water characteristic or matric potential–water content function $h(\theta)$ to eliminate either θ or h from (3.70). Since either variable may be eliminated, there are two forms of the equation.

Water Content Form of the Richards Equation   The flux equation (3.28) may be reexpressed as a function of θ alone by the following transformations:

1. Since $K(h)$ is a function of h and $h(\theta)$ is a function of θ, K may be written directly as a function of θ:

$$K(h(\theta)) \equiv K(\theta) \qquad (3.71)$$

2. The partial derivative $\partial h/\partial z$ may be rewritten by the chain rule of differentiation (Kaplan, 1984) as

$$\frac{\partial h(\theta)}{\partial z} = \frac{dh}{d\theta}\frac{\partial\theta}{\partial z} \qquad (3.72)$$

where $dh/d\theta$ is the slope of the matric potential–water content function.

Inserting (3.71) and (3.72) into (3.28) produces

$$J_w = -K(\theta)\frac{dh}{d\theta}\frac{\partial\theta}{\partial z} - K(\theta) \equiv -D_w(\theta)\frac{\partial\theta}{\partial z} - K(\theta) \qquad (3.73)$$

where

$$D_w(\theta) = K(\theta)\frac{dh}{d\theta} \qquad (3.74)$$

is called the *soil water diffusivity*.

After this transformation, (3.70) may be written as

$$\frac{\partial\theta}{\partial t} = \frac{\partial}{\partial z}\left[D_w(\theta)\frac{\partial h}{\partial z}\right] + \frac{\partial K(\theta)}{\partial z} \qquad (3.75)$$

Equation (3.75) is called the *water content form* of the Richards equation. It is a second-order nonlinear partial differential equation called a *Fokker–Planck equation* and can generally be solved only by numerical methods. Assuming that $D_w(\theta)$ and $K(\theta)$ are known, (3.75) requires two boundary conditions (i.e., the soil surface and deep in the soil), where the behavior of θ or J_w is known as a function of time. It also requires specification of an initial condition describing the water content of the entire profile at $t = 0$.

Modern high-speed computers have made the task of solving (3.75) almost routine when the soil water functions and boundary conditions are known. Use of this equation as a predictive tool is limited primarily by difficulties in measuring D_w and K accurately over the region where the simulation is to be run, particularly when the soil is heterogeneous and each location in the soil has a different $D_w(\theta)$ and $K(\theta)$ function.

Equation (3.75) has also been derived by ignoring hysteresis. The slope $dh/d\theta$ is defined only for a uniform wetting or drying process in which θ is described uniquely by a single curve (see Fig. 2.16). When repeated wetting and drying cycles are present, (3.75) is invalid and the water diffusivity D_w cannot even be defined. Only a limited amount of study of the influence of hysteresis on water flow has been performed, either theoretically or experimentally, although there are some indications that it may have a substantial influence in certain cases (Curtis and Watson, 1984; Jones and Watson, 1987; Russo et al., 1989).

Matric Potential Form of the Richards Equation   The time derivative of the water content in (3.70) may be rewritten using the chain rule as

$$\frac{\partial \theta}{\partial t} = \frac{d\theta}{dh}\frac{\partial h}{\partial t} \equiv C_w(h)\frac{\partial h}{\partial t} \tag{3.76}$$

where

$$C_w(h) = \frac{d\theta}{dh} \tag{3.77}$$

is called the *water capacity function*. It is equal to the inverse slope of $h(\theta)$. As with the diffusivity, it is defined only for uniform wetting or drying processes.

Inserting (3.76) into (3.70) produces

$$C_w(h)\frac{\partial h}{\partial t} = \frac{\partial}{\partial z}\left[K(h)\left(\frac{\partial h}{\partial z} + 1 \right) \right] \tag{3.78}$$

Equation (3.78) is called the *matric potential form* of the Richards equation. It may be solved if two boundary conditions and an initial condition are specified, provided that $C_w(h)$ and $K(h)$ are known.

Water Diffusivity Function $D_w(\theta)$ The water diffusivity function may be calculated from $K(\theta)$ and $h(\theta)$ using the definition in (3.74). There are also direct methods for evaluating it in transient flow experiments using theoretical analyses. Useful references for these methods include Bruce and Klute (1956), Clothier et al. (1983), Dirksen (1975, 1979), Klute and Dirksen (1986), and Whisler et al. (1968). A review of current methodologies is given by Klute and Dirksen (1986).

Approximate methods may also be used to determine $D(\theta)$. Shao and Horton (1996, 2000) presented an analytical solution of the diffusivity form of the Richards equation for horizontal water flow. They showed that the solution could describe soil water content as a function of position and time in a horizontal soil column and described a simple experiment for determining soil water diffusivity. They took two horizontal soil columns, one with wet soil and one with dry soil, and connected the columns together. Measurements were made of the wetting front position, x_f, as a function time while water moved from the wet soil into the dry soil. Shao and Horton (1996) showed that the following equation described the wetting front data:

$$x_f(t) = at^b \tag{3.79}$$

where a and b are empirical parameters determined from regression by fitting the equation to the observed wetting front data. The a and b parameters are used to evaluate the γ and D_0 coefficients in the following power function form of soil water diffusivity for the soil:

$$D(\theta) = D_0\theta^\gamma \tag{3.80}$$

Parlange and Hogarth (1997) and Shao and Horton (1997) offer additional insights into this method for determining soil water diffusivity.

Water Capacity Function $C_w(h)$ The water capacity function $C_w(h)$ defined in (3.77) expresses the increase in matric potential per unit increase in water content. Figure 3.14 shows typical $C_w(h)$ curves for two different soil types. The largest values of C_w are associated with the wet end of the capillary region of the retention curve, where the larger pores are emptying with modest changes in suction. The function is equal to zero at saturation, prior to the onset of desaturation at the air-entry suction. It approaches zero at extreme dryness when enormous changes in suction occur with small changes in water content.

3.3.7   Model Functional Forms

Considerable effort has been spent developing versatile functional forms that have a sufficient number of parameters to represent the range of shapes of the soil water hydraulic functions that might be found in different soils (Brooks and Corey, 1964; Mualem, 1976b; van Genuchten, 1980). Table 3.6 gives the analytic functional forms for $K(\theta)$ and $h(\theta)$ derived by Brooks and Corey (1964) and van Genuchten (1980), which have been used extensively in numerical modeling. The expression (3.84) for $K(\theta)$ in the van Genuchten model was derived from (3.82) using the theoretical model of Mualem (1976b). A review of general methods for calculating hydraulic conductivity model forms from water characteristic functions is given by van Genuchten and Nielsen (1985).

Parameters in the hydraulic property functions are often determined by curve fitting appropriate equations to measured water retention data. Alternatively, inverse methods have been used to determine parameters by fitting numerical solutions of the Richards equation to one-step or multistep outflow data. A third way to determine hydraulic function parameters is to apply analytical solutions of the Richards equation to horizontal infiltration data. Advantages of the horizontal infiltration methods are the simplicity of the experiments and the short time required to obtain data.

Shao and Horton (1998) describe a procedure for determining van Genuchten hydraulic property parameters in Table 3.6 from a horizontal infiltration experiment. The method requires measurement of infiltration rate as a function of time and measurement of the wetting front position. From this the authors were able to derive explicit expressions for the α and N parameters in Table 3.6. Wang et al. (2002a) extend this method to evaluation of the Brooks and Corey parameters. Shao and Horton (1998) showed that the water retention functions derived from horizontal infiltration

TABLE 3.6   Matric Potential–Water Content and Hydraulic Conductivity–Water Content Functional Forms

Function	Brooks and Corey (1964)		Van Genuchten (1980)	
$h(\theta)$	$\Theta(h) = (h_d/h)^\lambda$	(3.81)	$\Theta(h) = [1 + \alpha(-h)^N]^{-M}$	(3.82)
$K(\theta)$	$K(h[\theta]) = (h_d/h)^\omega = \Theta^{\omega/\lambda}$	(3.83)	$K(\theta) = K_s\Theta^{1/2}[1 - (1 - \Theta^{1/M})^M]^2$	(3.84)
	where $\omega = 2 + 3\lambda$		where $\Theta = (\theta - \theta_r)/(\theta_s - \theta_r)$	
			and $M = 1 - 1/N$	

experiments reasonably matched independently measured water retention curves. Additional research is needed to evaluate how well the unsaturated hydraulic conductivity can be determined.

Equations (3.82) and (3.84) use fitted parameters N, α, θ_s, and θ_r where the subscript s denotes saturation and θ_r is called the *residual water content*, referring to the region of θ where adsorptive forces are dominant and h is decreasing rapidly with little change in θ. The water capacity function (3.77) must be compatible with the definitions for $h(\theta)$. This is illustrated in the next example.

Example 3.11   Derive a model functional form for $C_w(h)$ using (3.82).

SOLUTION: By definition, $C_w(h) = d\theta/dh$. Thus, (3.82) is differentiated as follows:

$$\frac{d\theta}{dh} = (\theta_s - \theta_r)\frac{d}{dh}[1 + \alpha(-h)^N]^{-M}$$

$$= \frac{(\theta_s - \theta_r)(-M)}{[1 + \alpha(-h)^N]^{1+M}}\frac{d}{dh}[1 + \alpha(-h)^N]$$

$$= \frac{(\theta_s - \theta_r)(-M)\alpha N(-1)(-h)^{N-1}}{[1 + \alpha(-h)^N]^{1+M}}$$

or since $M = 1 - 1/N$,

$$C_w(h) = \frac{\alpha(\theta_s - \theta_r)(N - 1)(-h)^{N-1}}{[1 + \alpha(-h)^N]^{2-1/N}} \tag{3.85}$$

The water capacity function (3.85) is plotted in Fig. 3.14 as a function of h for two different model soil types, a coarse-textured sandy loam and a silt soil. Their parameters are given in Table 3.7.

3.3.8   Transient Flow Calculations

As we have seen, the mechanisms governing transient unsaturated flow are more complex than those of saturated flow. Consequently, soils with different hydraulic and retention properties will behave in a different manner when subjected to the same initial and boundary conditions. In this section we illustrate the principles of transient flow using several different flow problems on two distinct soil types, a coarse-textured sandy loam and a finer-textured silt soil, which are described by the van Genuchten parametric models (3.82) and (3.84). Their properties are given in Table 3.7 and plotted in Figs. 3.15 and 3.16. All calculations are performed with the model code HYDRUS-1 (Simunek et al., 1997).

Infiltration at Constant Potential   Infiltration at constant potential is a classic problem first solved in 1957 by Philip (1957a–e; 1958a,b) in the days before computers

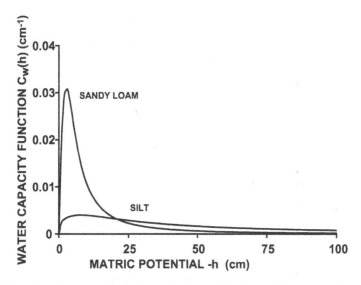

Figure 3.14   Matric potential–water capacity relations for the prototype sandy loam and silt soils provided in the HYDRUS-1 code of Simunek et al. (1997).

made numerical solution of (3.75) or (3.78) possible. The boundary and initial conditions to the problem are

$$h(0, t) = h_0 \tag{3.86}$$

$$h(-\infty, t) = h_i \tag{3.87}$$

$$h(z, 0) = h_i \tag{3.88}$$

Figure 3.17 shows matric potential head profiles during a 5-day period following the initiation of infiltration in the two soils, each of which is initially at $h_i = -40$ cm and has the surface held at $h_0 = -15$ cm. Several features in Fig. 3.17 are worth noting. First, the transition between the wet and dry zones is abrupt and significantly narrower in the coarse-textured sandy loam soil than the finer silt soil. Second, in each soil the wetting front moves faster at first and slows down to a more constant

TABLE 3.7   Parameter Values for the van Genuchten Model Functions (3.76)–(3.77) of Two Prototype Soils Used in the HYDRUS-1 Code

Soil Type	θ_s	θ_r	α $(\text{cm}^{1/N})$	K_s (cm h^{-1})	N
Sandy loam	0.41	0.065	0.075	106.1	1.89
Silt	0.46	0.034	0.016	6.0	1.37

Source: Simunek et al. (1997).

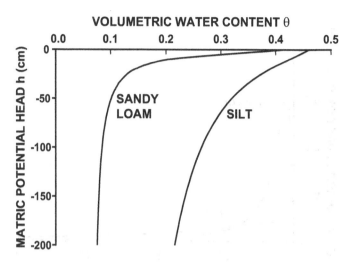

Figure 3.15 Matric potential–water content relations for the prototype sandy loam and silt soils provided in the HYDRUS-1 code of Simunek et al. (1997).

speed at longer times. Finally, except at the beginning, the wetting-front shape does not change as it moves downward. All of these characteristics were demonstrated by Philip in his classic series of papers in 1957 (see Philip, 1969).

Figure 3.18 shows the infiltration rate versus time for the two soils. The rate is high at early times, because water is drawn into the soil by capillary attraction as well as

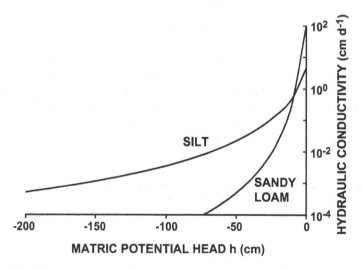

Figure 3.16 Matric potential–hydraulic conductivity relations for the prototype sandy loam and silt soils provided in the HYDRUS-1 code of Simunek et al. (1997).

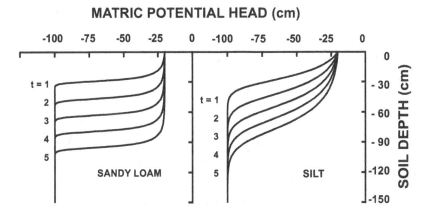

Figure 3.17   Matric potential profiles during infiltration at constant potential in a sandy loam and a silt soil, each with $h_i = -40$ cm and $h_0 = -15$ cm.

being pulled down by gravity. At longer times, only gravity is significant. The sandy loam has a higher infiltration rate than the silt because its hydraulic conductivity is higher at the entry matric potential $h_0 = -15$ cm.

Upward Flow from a Water Table   Figure 3.19 shows matric potential–depth profiles during upward flow from a water table $h = 0$ at $z = -100$ cm over a 20-day period in the two soils. Each soil is initially at a matric potential of $h_i = -200$ cm, and the surface is held at that value throughout the simulations. In contrast to the infiltration example, water moves faster in the upward direction in the silt than in

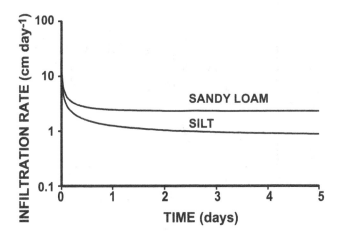

Figure 3.18   Infiltration rate at constant potential in a sandy loam and a silt soil, each with $h_i = -40$ cm and $h_0 = -15$ cm.

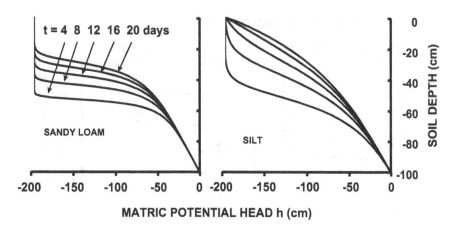

Figure 3.19 Matric potential profiles during upward flow from a water table in a sandy loam and a silt soil, each with $h_i = -200$ cm, $h(-100) = 0$, and $h(0) = -200$ cm.

the sandy loam, because the conductivity of the latter becomes very low in much of the soil profile. In fact, the final steady evaporation rates calculated for these soils by running the model to long times are 0.076 cm day^{-1} for the silt and 0.0066 cm day^{-1} for the sandy loam. This confirms the conclusion drawn in Section 3.3.4 with the simple parametric $K(h)$ model of Gardner (1958) that finer-textured soils move water upward more easily than do coarse-textured soils.

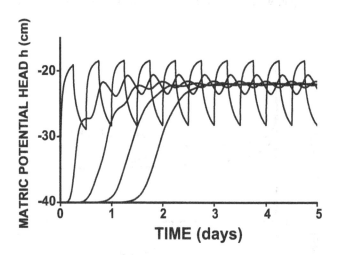

Figure 3.20 Matric potential versus time at $z = -3, -25, -50, -75, -100$ cm for the sandy loam soil when the surface flux is cycled every 12 h between -3 cm day^{-1} (infiltration) and 0.5 cm day^{-1} (evaporation).

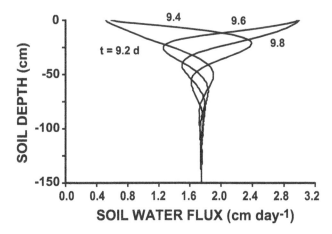

Figure 3.21   Water flux versus depth over a 1-day interval for the sandy loam soil under the conditions described in Fig. 3.20.

Infiltration–Evaporation Cycles   Numerical calculations are useful for checking the accuracy of assumptions made in simplifying the flow process. One such assumption is the validity of the gravity flow approximation (3.55). Water never enters the soil at a constant rate for a prolonged period of time, except in laboratory experiments. A far more common circumstance is occasional infiltration, with cycles of evaporation in between, such as with agricultural irrigation. Figure 3.20 shows the matric potential values as a function of time generated at five depths ($z = -3, -25, -50, -75, -100$ cm) for the sandy loam soil when the surface flux is cycled every 12 h between -3 cm day^{-1} (infiltration) and 0.5 cm day^{-1} (evaporation). The matric potential fluctuations are large near the surface, but attenuate greatly at increasing depth until the potential is essentially constant. The water flux is shown over one day of change in Fig. 3.21. Below about 75 cm, the flux is essentially constant, showing the validity of the approximation (3.55).

The amount of attenuation of a wave amplitude moving into the soil depends on the frequency of the oscillations (see Section 5.3.4). The shorter is the period of the cycle, the less the depth of penetration. Thus, daily changes damp out more rapidly than weekly changes, and so on. Regardless of the structure of the water application, however, as long as there is a net downward flow over a long period of time, (3.55) will eventually be valid at sufficiently great depths below the surface.

PROBLEMS

3.1   A long soil column containing tensiometers is in steady-state evaporation, with a water source at the bottom. The area of the column is 100 cm^2. If 150 cm^3 of water must be added to the water source at the bottom each day, calculate the evaporation flux and $K(h)$ from the data given in Table 3.8. Plot $K(h)$ versus h on semilogarithmic paper.

TABLE 3.8   Data for Problem 3.1

z	h
120	−750
100	−300
70	−175
40	−70
20	−30
5	−6
0	0

3.2 A saturated soil column (Fig. 3.22) contains two soil layers, each 10 cm thick, with sand of $K_s = 10 \text{ cm } h^{-1}$ underneath loam of $K_s = 5 \text{ cm } h^{-1}$. The bottom of the soil column is open to the atmosphere ($p = 0$). At $t = 0$, water of height $d = 10$ cm above the top of the column is ponded on the surface.

   (a) Assuming steady state, calculate the flux through the column.

   (b) Assuming steady state, calculate the water pressure at the sand–loam interface (point A).

   (c) If this were a falling-head permeameter, what would probably happen as d decreased from 10 cm to zero?

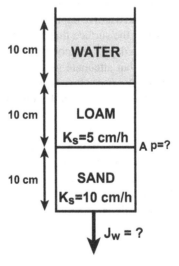

Figure 3.22   Layered soil column in Problem 3.2.

3.3 A soil column contains 50 cm of sand over 50 cm of clay (Fig. 3.23). A piezometer at $z = 50$ cm measures the hydrostatic pressure head $p = 50$ cm at the interface. The column is saturated and 20 cm of water is ponded on the top while the bottom is open to the atmosphere ($p = 0$). The steady measured flux rate is $J_w = -20$ cm h^{-1}. Calculate the saturated hydraulic conductivity of the sand and the clay and the effective conductivity of the column.

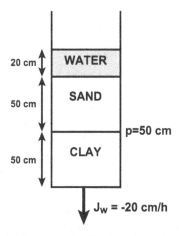

Figure 3.23   Layered soil column in Problem 3.3.

3.4 A cylindrical soil column of 100-cm^2 cross-sectional area and 50-cm height is filled with homogeneous soil that is saturated, and 10 cm of water is continuously ponded on it. The steady-state volume flow rate Q through the soil is 1000 cm^3 h^{-1} (downward).

  (a) Calculate the steady-state flux through the column.

  (b) Calculate the saturated hydraulic conductivity of the soil.

3.5 A 1-mm-diameter tube is pushed through the column described in Problem 3.4 and hollowed out. In steady state, water flows through the soil in accordance with Darcy's law (3.14) and through the tube in accordance with Poiseuille's law (3.10).

  (a) Calculate the volume flow rate through the tube.

  (b) Calculate the flux (Q_{tot}/A_{tot}) through the column–tube system.

  (c) Calculate the effective saturated hydraulic conductivity of the column–tube system.

3.6 A clay lens of saturated hydraulic conductivity $K_c = 0.1$ cm day^{-1} is sandwiched somewhere inside a soil column of height 100 cm that is otherwise filled with sand of $K_s = 200$ cm day^{-1}. A height of water $d = 10$ cm is ponded above the surface.

  (a) If the clay lens were not present, what would the steady flux rate be?

  (b) If the actual flux rate is $J_w = -15$ cm day^{-1}, how thick is the lens?

  (c) Calculate the pressure at the clay–sand interface for two special cases:

(i) Clay on top of the sand

(ii) Clay on the bottom of the sand

(d) Discuss the physical implications of your results.

3.7 A matric potential–water content function is fit to the following theoretical curve:

$$h(\theta) = -30 \left(\frac{0.25}{\theta^2} \right)^{1/2} \qquad \text{for } 0.1 \le \theta \le 0.5$$

(a) Calculate $h(\theta)$ at $\theta = 0.1, 0.2, 0.3, 0.4,$ and 0.5.

(b) Using these five regions of water content, calculate the tube radii and the number of tubes per area of a model porous medium of capillary tubes that would have the same $h(\theta)$ curve ($\sigma = 72$ dyn cm^{-1}).

(c) Under a unit hydraulic head gradient, calculate the fraction of the total flow carried by each group of tubes if the model porous medium is saturated.

3.8 Using the approach followed in Section 3.3.4 for steady flow problems:

(a) Calculate the integral form of the Buckingham–Darcy flux law for steady-state horizontal flow of water from the differential form

$$J_w = -K(h) \frac{dh}{dx}$$

in a manner analogous to the derivation of (3.43).

(b) Calculate an equation relating the maximum horizontal evaporation rate E to the column length L if the left side of the column is kept saturated and

$$K(h) = \frac{K_s}{1 + (h/a)^2}$$

(c) Compare this answer with the corresponding result for the vertical evaporation case [(3.49) with $N = 2$]. Which E is greater for a given L? Why?

3.9 A root zone of thickness L is receiving high-frequency water irrigation at a rate $J_w = -i_0$ at the surface $z = 0$. Assuming that the water uptake distribution is

$$r_w(z) = a(z + L) \qquad -L \le z \le 0$$

and that steady state has developed, calculate the water flux as a function of depth z. Relate the parameter a to the total water loss rate ET (evapotranspiration) in the root zone.

3.10 A vertical soil column is initially saturated to $\theta_s = 0.4$ and drains uniformly [$\theta(z, t)$ is same at all locations z] over its entire height $L = 50$ cm. The soil has a hydraulic conductivity–water content functional form given by

$$K(\theta) = K_s \exp[\beta(\theta - \theta_s)]$$

where $K_s = 100$ cm day^{-1} and $\beta = 20$. Use the gravity flow model,

$$J_w \approx -K(\theta)$$

in the following exercises.

(a) Calculate the average water content of the soil as a function of time.

(b) What is the predicted water content at $t = \infty$?

(c) How long does it take to drain the column to $\theta = 0$?

(d) Discuss the physical significance of your findings. Why did the final water content have the value it did?

3.11 A falling-head permeameter is constructed by ponding 20 cm of water at $t = 0$ above a saturated soil column of height $L = 100$ cm. The ponded water, which is held above the column in a chamber of the same area as the column, falls to 0 cm in 1 h.

(a) Calculate K_s for this column.

(b) Next repeat the analysis approximately by pretending that the water flux through the column was constant and equal to $J_w = -20$ cm h^{-1} and that the height of ponded water was held at 10 cm for the entire experiment.

(c) Compare the result calculated by Darcy's law with these approximations to the exact solution and calculate the percentage of error.

3.12 Repeat Problem 3.11 for the case where the soil column is only 5 cm high. Explain the reason for the difference in these two results.

3.13 Calculate the effective hydraulic conductivity of a saturated soil column of length L that has a saturated hydraulic conductivity $K_s(z)$ that varies arbitrarily over its length. Show that (3.26) is a special case of the formula you derive.

4 Water Flow under Natural Conditions

Natural field soils offer a formidable challenge for modelers of water, chemical, and energy transport processes. The geometrical arrangement of the soil matrix varies from one location to the next and often contains complex structural voids. The surface zone is riddled with biological channels such as worm holes, insect burrows, and plant roots. Organic material is deposited throughout the near subsurface in varying degrees of decomposition. The upper boundary between the soil and air is subjected to continuous variations in water and energy input.

In the face of this complexity, modifications will have to be made to some of the approaches we have developed for describing transport processes. However, the general principles and foundations of our theory still hold and will guide us in the development of descriptions appropriate for the field regime.

4.1 SPATIAL VARIABILITY AND TRANSPORT

In 1973, Nielsen and co-workers published the results of a comprehensive field experiment that measured infiltration rate and soil water properties at 20 sites over a 150-ha area (Nielsen et al., 1973). They ponded water at 20 random locations on 6.5-m^2 plots to measure steady infiltration rate and then allowed the plots to drain while they monitored matric potential at six depths. From this information and associated laboratory measurements, they calculated a number of soil properties. They discovered extensive spatial variation in transport and retention properties among the 20 sites, and at different depths between 30 and 180 cm in the same sites. Table 4.1 summarizes the mean and coefficient of variation (CV) of various properties at the 120 locations. While static soil properties such as bulk density and porosity varied little among the replicate measurements, the transport properties spanned a large range of values, with CVs the order of 100% or more.

The landmark study by Nielsen at al. (1973) clearly illustrates the challenge facing soil physicists at the field scale. Natural soil is highly variable, and any in situ measurement of a transport property will only provide information about the immediate vicinity of the measurement. Yet many of the important applications of soil physics such as water and chemical movement in agricultural fields involve characterizing transport over large areas. Since only a limited number of measurements can be taken, information about soil properties at places where no measurements are taken has to

TABLE 4.1   Mean and CV of Various Properties Measured by Nielsen et al. (1973) at Their 150-ha Field Site

Property	Units	Mean Value	CV (%)
Infiltration rate i	cm day^{-1}	14.6	94
Saturated hydraulic conductivity K_s	cm day^{-1}	20.6	120
Saturated water content θ_s		0.43	13
Bulk density ρ_b	g cm^{-3}	1.36	7
Water content $\theta(h)$ at $h = -90$ cm	—	0.35	13
Diffusivity $D(\theta)$ at $\theta/\theta_s = 0.8$	cm^2 day^{-1}	1115	258

be inferred from the known values at the measurement sites. Soil physics thus has become a statistical discipline (see the Appendix).

4.2   FIELD WATER BALANCE

The zone comprising the top several meters of soil is the most important region for agriculture. It contains the majority of plant roots, most of the microbial population, and all of the topsoil necessary for growth. Since water cannot flow upward over great distances, water and chemicals reaching the bottom of this zone will generally not return to it, either remaining in storage or migrating to groundwater. Thus it is of interest to treat this surface zone as a unit by conducting a field water balance over it.

The general equation describing the water balance in the surface soil zone may be expressed as

$$P + I - R - \mathrm{ET} = D + \Delta W \qquad (4.1)$$

where the terms on the left-hand side of (4.1) are the precipitation P (including dew and frost), applied irrigation water I, evapotranspiration ET (the combined loss of water to the atmosphere from evaporation and plant water uptake or transpiration), and the surface runoff R. These four terms represent the net addition of water to the soil profile over a time period of interest. On the right-hand side of (4.1) are drainage or deep percolation D below the soil zone and the increase in water storage ΔW in the soil profile. Each of the terms in (4.1) represent water flows or storage changes over some arbitrary time interval (e.g., 1 day). The terms on the left side of (4.1) are positive, while D and ΔW may be either positive or negative. A negative value for the drainage term[1] implies that water is flowing upward into the profile where the water balance is conducted, which will occur whenever the hydraulic head gradient is negative. This upward flux could be quite significant in certain situations, such as if a water table is located at a shallow depth below the surface profile and there is no input of water to the soil for a prolonged time period (see Example 3.8).

---

[1] This sign convention differs from the one used in this book for fluxes because the word *drainage* connotes downward flow. Thus, a positive drainage flux is a negative water flux.

The various terms in the water balance equation (4.1) are of considerable importance in disciplines other than soil science. For example, hydrologists must know what fraction of the incoming precipitation will result in direct runoff and in deep percolation to groundwater. Meteorologists are very interested in the estimate of evapotranspiration, since this component of the water balance has a significant influence on the energy budget of the earth. Plant scientists and agronomists are concerned both with evapotranspiration losses by crops and with root zone profile storage changes to determine optimum moisture conditions for plant roots.

In this chapter we consider each of the terms in the water balance equation (4.1). Wherever possible, models will be used for these processes that are derived from the soil water transport equations introduced in Chapter 3. Much of the analysis will involve making simplifications to construct approximate models of the flow regime. However, these simplified approaches are based on sound physical principles. Therefore, results of the approximate analysis will provide valuable insight into the behavior of the real system. This is illustrated in the next example.

Example 4.1   A 9×9 m^2 level field plot is divided into nine 3×3 m^2 subplots, each of which is blocked off and separated from the others. All plots are ponded with water until the infiltration rate is constant. The nine values of infiltration rate measured are 0.5, 2.0, 7.0, 6.5, 5.0, 13.0, 31.0, 12.0, and 4.0 cm h^{-1}. If the entire field plot had been ponded, what would the steady infiltration rate have been? What fraction of the infiltrating water would have gone into the subplot area having a rate of 31.0 cm h^{-1}? What would happen if the entire field plot were irrigated by a sprinkler system at 1.0 cm h^{-1}?

SOLUTION: The mean infiltration rate $< i >$ of the field plot is simply the average of the rates through the subplots, since each has the same cross-sectional area. Thus, $< i > = 9.0$ cm h^{-1}. The total volume of water entering the field plot is $0.09 \times 81.0 = 7.29$ m^3 h^{-1}. The volume flow rate of the water entering the subplot with the high infiltration rate of 0.31 m h^{-1} is $0.31 \times 9.0 = 2.79$ m^3 h^{-1}. Thus, $2.79/7.29 \times 100 = 38.27\%$ of the water enters the one-ninth of the field having the highest rate. Similarly, only $0.005 \times 9.0 = 0.045$ m^3 h$^{-1} = 0.62\%$ of the water enters the lowest one-ninth of the field. Under a steady application of 1.0 cm h^{-1}, this low part of the field would pond, and runoff would send the water to other parts of the plot.

4.2.1   Analysis of Field Water Content and Matric Potential Profiles

The last term in (4.1), which represents the water storage change over some time interval, is limited in magnitude by the maximum amount of water (porosity times profile thickness) that can be held by the soil profile. Moreover, if an average water balance is calculated over a long period of time, this term will be less significant than the others. In fact, as long as the input of water from rainfall or irrigation is reasonably regular (i.e., no prolonged drying cycles), the actual soil profile water content in the field will tend to fluctuate about an equilibrium or steady-state value (see Figs. 3.20 and 3.21); consequently, over long time periods the storage change may be neglected. Although fluctuations in water flux and water storage occur continuously near the surface, changes occurring at greater depths are frequently very small, and the steady-state model provides a reasonable approximation to the real system.

It is usually possible to measure P and I accurately in the field. Runoff R from basin or furrow irrigation in agricultural fields can be monitored at the downstream end of the field. Stormwater runoff, however, poses more of a challenge to measure. Evapotranspiration ET is straightforward to estimate when it occurs at the maximum allowable rate given the external energy and mixing conditions available, less so when the limiting resistance to loss resides in the soil. Drainage D is by far the most difficult to measure or calculate of all the components of the water balance.

The difficulties that may be encountered in the interpretation of field data are shown in Fig. 4.1. In this figure, the soil water content depth profiles under two forested sites are shown at several different times of the year. Fluctuations in soil water content in response to rainfall and evaporation are very pronounced near the soil surface but diminish significantly at greater depths. However, both profiles show some evidence of a storage change deep in the soil during the monitoring period.

The water content measurements do not allow quantitative conclusions to be drawn about the flow processes occurring deep in the soil. However, when they are combined with matric potential measurements, the water content data present a clearer picture of the wetting and drying cycles in the soil profile. Despite the complexity of the data in Figs. 4.1 and 4.2, the flow principles we have learned in earlier chapters will assist us in identifying significant flow features.

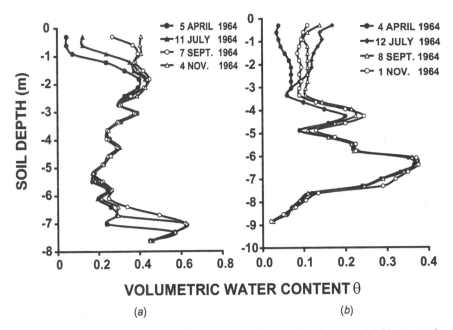

Figure 4.1 Soil water content profiles at various times during the year: (*a*) Young sand, Mount Gambier forest; (*b*) Kalangadoo sand, Penola forest. (After Holmes and Colville, 1970.)

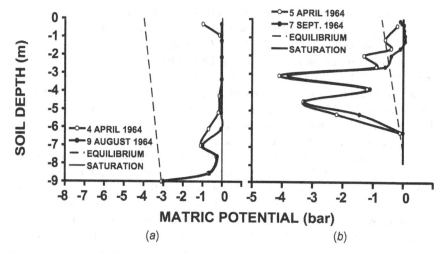

Figure 4.2   Matric potential profiles for the wettest and driest times of sampling for the sites in Fig. 4.1: (a) Young sand, Mount Gambier forest; (b) Kalangadoo sand, Penola forest. (After Holmes and Colville, 1970.)

In Chapter 2 we analyzed soil water systems at equilibrium (i.e., those in which no water flow occurs). In a vertical unsaturated profile at equilibrium, with no solute membranes or air pressure changes, the total potential head h_T is given by

$$h_T = h + z = \text{const} \tag{4.2}$$

In particular, with a water table located at $z = -L$, the matric potential distribution for an equilibrium soil profile is given by

$$h = -(z + L) \qquad -L \leq z \leq 0 \tag{4.3}$$

Thus, the matric potential at any known height above a water table may be used to indicate whether water is flowing upward or downward or the profile is at equilibrium. For example, if a water table is located at a depth $z = -200$ cm and the matric potential at $z = -100$ cm is greater (more positive) than -100 cm, the flow will be downward. As was shown in Example 3.9, matric potential gradients are close to zero during steady downward water flow, particularly far away from the water table (see Fig. 3.22). In this situation, the Buckingham–Darcy flux equation (3.28) reduces to the gravity flow equation

$$h = \text{const} \rightarrow J_w \approx -K(\theta) \tag{4.4}$$

where $K(\theta)$ is the unsaturated hydraulic conductivity. Thus, during prolonged downward flow, the soil water content converges to the value satisfying (4.4) required to drain the soil under gravity at the imposed rate.

Figure 4.2 shows the matric potential profiles at the two sites for the wettest and driest sampling dates. Also plotted on the figure is the equilibrium line (4.3), showing the values of $h(z)$ that would produce zero flow. Slopes of $h(z)$ greater than this line imply upward flow, and the converse. At the Mount Gambier Forest site, it is obvious that there is very little change in matric potential between 1 and 6 m. Hence, the hydraulic head gradient is almost equal to unity [i.e., gravity flow (4.4)] and the direction of water movement must be downward. The hydraulic head gradient is upward near the soil surface in April, implying that water is moving toward a dry surface and evaporating. The increase in matric potential between 7 and 9 m probably represents a pulse of water that entered the profile at an earlier time.

The situation in the Penola Forest is significantly different. The hydraulic head gradient is downward between the surface and the 3-m depth and upward between 5 and 6 m. Furthermore, the matric potential reaches a maximum at about 4 m and separate minima at 5 and 6 m. This profile is probably the consequence of two prolonged drying periods with a period of excessive rainfall between them. The site is obviously experiencing transient water flow, because the water content and matric potential at a given point vary significantly between the sampling periods. Moreover, water is clearly moving upward through some regions and downward through others. Consequently, some of the profile dried out between May and July, and part of it became wetter. At the Mount Gambier site, on the other hand, the profile appears to be much closer to a steady state, having a net downward flow of water that causes only minor storage changes over time. Although in the absence of any other evidence one might be tempted to conclude that no water was flowing through the 3- to 5-m region where the water content was not changing with time, it is far more likely that a steady downward flux of water is continually moving through this region, causing recharge of the dry profile below.

Equilibrium profile analysis has proven to be a valuable tool for assessing water flow in arid environments, where flow velocities may be very small and difficult to measure. Figure 4.3 shows data from four arid-zone sites: in Nevada (NTS), Washington (HANF.), Texas (EF), and Australia (MB). The two sites on the left side of the figure are experiencing subsurface downward flow, while the two on the right show evidence of upward flow according to the equilibrium model.

4.3 INFILTRATION

Infiltration refers to the entry of water into a soil profile from the boundary. Generally, it refers to vertical infiltration, where water moves downward from the soil surface. Since infiltration causes the soil to become wetter with time, water at the leading edge of the wetting pattern advances into the drier soil region ahead of the front under the influence of matric potential gradients as well as gravity (if infiltration is vertical). During the early stages of infiltration when the wetting front is near the surface, the matric potential gradients predominate over the gravitational force.

The mathematical theory of vertical infiltration, based on the solution of the Richards equation (3.73), was described in detail by Philip (1969). Since this mechanistic infiltration model is derived from the physically based water flow equation, it

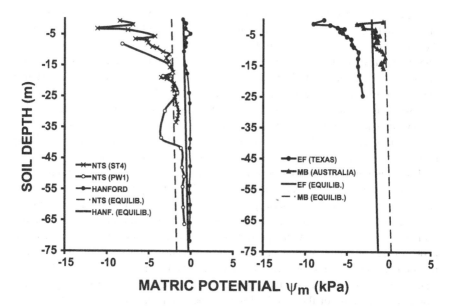

Figure 4.3 Evaluation of the direction of water movement according to the relationship between water potential profiles and the equilibrium line. Data are from Hanford, Washington, the Nevada Test Site, Nevada, Eagle Flat, Texas, and Murray Basin, South Australia. The equilibrium line refers to the equilibrium matric potential that balances gravitational potential (Nevada Test Site data shown as an example). (After Scanlon et al., 1997.)

gives considerable insight into the processes governing infiltration and is presented in some detail in this chapter. However, there are many other vertical infiltration models in use today, the majority of which have been derived empirically from field data. These models all share the common feature that the infiltration rate is highest when water first enters the soil and decreases with time as the wetting front moves away from the surface. The empirical approach to the development of field infiltration equations consists of first finding a mathematical function whose shape as a function of time matches the observed features of the infiltration rate and then attempting a physical explanation of the process. In contrast, the mechanistic approach consists of solving the water flow equation to derive an expression for the infiltration rate. The time dependence of this expression can be interpreted physically as a consequence of the decreasing influence of the matric potential gradient as the wetting front moves farther from the surface during infiltration (Philip, 1969).

4.3.1 Infiltration Models

Horton Equation   Horton (1933, 1939) was one of the pioneers in the study of infiltration in the field and developed an equation that he felt both described the general features of infiltration in different soils and was consistent with his physical concept of the process. The infiltration rate is given in Horton's model by the equation

$$i = i_f + (i_0 - i_f) \exp(-\beta t) \tag{4.5}$$

where i_0 is the initial infiltration rate at $t = 0$, i_f is the final constant infiltration rate that is achieved at large times, and β is a soil parameter that describes the rate of decrease of infiltration. Equation (4.5) may be integrated to produce the formula for cumulative infiltration:

$$I = i_f t + \frac{i_0 - i_f}{\beta} [1 - \exp(-\beta t)] \tag{4.6}$$

Horton (1940) felt that the reduction in infiltration rate with time after the initiation of infiltration was largely controlled by factors operating at the soil surface. They included swelling of soil colloids and the closing of small cracks that progressively sealed the soil surface. Compaction of the soil surface by raindrop action was also considered important when it was not prevented by crop cover. Horton's field data, similar to those of many other workers, indicated a decreasing infiltration rate for 2 or 3 h after the initiation of storm runoff. The infiltration rate eventually approached a constant value that was often somewhat smaller than the saturated hydraulic conductivity of the soil. Air entrapment and incomplete saturation of the soil were assumed to be responsible for the latter finding. Horton used an exponential function to describe the decreasing infiltration rate since it fit the data reasonably well.

Green–Ampt Infiltration Model   Green and Ampt (1911) derived an approximate mechanistic model of the infiltration process by making several simplifying assumptions about the wetting process during water infiltration. These assumptions were based on approximations of real soil behavior obtained from experience, so that the model they created still has instructive value today.

During an actual infiltration event in which the soil surface is held at a constant matric potential head h_0 with associated water content θ_0 (e.g., by ponding water over it), water enters the soil behind a sharply defined wetting front that moves downward with time (Fig. 4.4*a*). Green and Ampt replaced this process with one that has a discontinuous change in water content at the wetting front (Fig. 4.4*b*). In addition, they made the following assumptions:

- The soil in the wetted region has constant properties (K_0, D_0, θ_0, h_0).
- The matric potential head at the moving front is constant and equal to h_F.

These two assumptions allowed Green and Ampt to solve for the infiltration rate exactly, as shown in the next two examples.

Example 4.2   Use the Green–Ampt model to calculate the infiltration rate into a horizontal soil column initially at a uniform water content θ_i that has a water content $\theta_0 > \theta_i$ and an associated matric potential h_0 maintained at the entry surface for all $t > 0$.

SOLUTION: From the assumptions of the model, we may replace the wetted soil profile at time t with a uniformly wet region of thickness L. Because the hydraulic

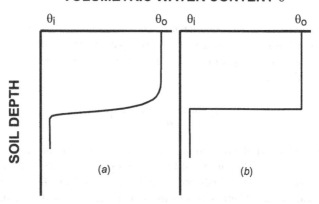

VOLUMETRIC WATER CONTENT θ

Figure 4.4 (*a*) Soil water content profile during infiltration; (*b*) corresponding Green–Ampt profile.

conductivity of the entire wetted zone is constant, the infiltration rate i may be calculated using Darcy's law (3.15) as

$$i = J_w = -K_0\frac{h_F - h_0}{L} = K_0\frac{\Delta h}{L} \tag{4.7}$$

where K_0 is the constant hydraulic conductivity of the "wet" region $0 < x < L$, $\Delta h = h_0 - h_F > 0$, and h_F is the matric potential of the moving front. Furthermore, the infiltration rate is equal to the time rate of change of water storage in the soil, or

$$i = \frac{d}{dt}[(\theta_0 - \theta_i)L] = \Delta\theta\frac{dL}{dt} \tag{4.8}$$

where $\Delta\theta = \theta_0 - \theta_i > 0$. When (4.8) is inserted into (4.7), we obtain a differential equation for the position L of the wetting front:

$$\Delta\theta\frac{dL}{dt} = K_0\frac{\Delta h}{L} \tag{4.9}$$

This equation may be integrated after all factors that depend explicitly on L are placed on one side. Thus,

$$\int_0^L L\,dL = K_0\frac{\Delta h}{\Delta\theta}\int_0^t dt \tag{4.10}$$

These integrals are easily evaluated, with the result that

$$\frac{L^2}{2} = K_0\frac{\Delta h}{\Delta\theta}t \equiv D_0 t \tag{4.11}$$

where $D_0 = K_0 \, \Delta h / \Delta \theta$ [see (3.74)] is the soil water diffusivity of the wet soil region $0 < x < L$. The cumulative infiltration $I = L \, \Delta \theta$ is thus equal to

$$I = \Delta \theta \sqrt{2 D_0 t} \tag{4.12}$$

and the infiltration rate is

$$i = \frac{dI}{dt} = \Delta \theta \sqrt{\frac{D_0}{2t}} \tag{4.13}$$

In this model the infiltration rate into the soil is proportional to $t^{-1/2}$.

Example 4.3 Repeat Example 4.2 for vertical infiltration. Let $z = 0$ at the soil surface.

SOLUTION: The hydraulic head at the surface $z = 0$ is equal to $H_0 = h_0$, whereas at the front, $z = -L$ and $H_L = h_F - L$. Thus, Darcy's law across the wetted region is given by

$$J_w = -i = -K_0 \frac{h_F - L - h_0}{-L - 0} = -\frac{K_0}{L}(\Delta h + L) \tag{4.14}$$

where, as before, $\Delta h = h_0 - h_F > 0$. Note that since the infiltration rate i and the downward distance L refer to positive quantities, we must use a minus sign to indicate that J_w is in the downward direction. Thus,

$$i = \Delta \theta \frac{dL}{dt} = \frac{K_0}{L}(\Delta h + L) \tag{4.15}$$

which may be rearranged and integrated as follows:

$$\int_0^L \frac{L \, dL}{\Delta h + L} = \frac{K_0}{\Delta \theta} \int_0^t dt = \frac{K_0 t}{\Delta \theta} \tag{4.16}$$

The integral on the left side may be looked up in standard tables. It is equal to (Lide, 2002)

$$\int_0^L \frac{L \, dL}{\Delta h + L} = L - \Delta h \ln \left(1 + \frac{L}{\Delta h}\right) \tag{4.17}$$

Thus, since $I = L \, \Delta \theta$, (4.16) may be rewritten as

$$I - \xi \ln \left(1 + \frac{I}{\xi}\right) = K_0 t \tag{4.18}$$

where $\xi = \Delta h \, \Delta \theta$. For short times, soon after infiltration begins, I is small and the logarithm in (4.18) may be approximated by

$$\ln\left(1 + \frac{I}{\xi}\right) \approx \frac{I}{\xi} - \frac{1}{2}\left(\frac{I}{\xi}\right)^2 \tag{4.19}$$

(Lide, 2002). Therefore, at short times (4.18) reduces to

$$I \approx \sqrt{2\xi K_0 t} = \Delta\theta\sqrt{2D_0 t} \tag{4.20}$$

and

$$i \approx \Delta\theta\sqrt{\frac{D_0}{2t}} \tag{4.21}$$

which is the same as the horizontal infiltration result (4.12)–(4.14). At very large times, the rate of change of the second term in (4.18) is small compared with the first and may be neglected. Therefore,

$$i \approx K_0 \tag{4.22}$$

These results will be of interest in what follows when the exact model for infiltration is discussed.

Because it uses an approximate description of the actual flow regime, the Green–Ampt model has parameters like h_F which cannot be measured directly. Therefore, it has been used primarily as a conceptual aid in visualizing a complex process, although indirect evaluation of h_F has permitted the Green–Ampt model to be used in certain practical applications (Chong et al., 1982, Gomez et al., 2001).

Philip Infiltration Model   Although the mathematical theory of water flow was fully developed by the early part of the twentieth century, the Richards equation (3.73) could not be solved without approximations because the equation was nonlinear. Thus, many of the subtleties of unsaturated water flow could not be investigated theoretically. In 1957, J. R. Philip devised a numerical technique for solving the flow equation exactly for a mathematical study of an important practical problem: infiltration into an infinitely deep homogeneous porous medium at a uniform initial water content θ_i that has its boundary (i.e., the soil surface) held at a higher water content $\theta_0 > \theta_i$ (Philip, 1957a–f).

The first problem that Philip addressed was that of horizontal infiltration (no gravity). Using the Richards equation (3.73) [with $K(\theta)$ removed] he showed that the infiltration rate i is given exactly by

$$i = \tfrac{1}{2}St^{-1/2} \tag{4.23}$$

where $S = S(\theta_0, \theta_i)$, is called the *sorptivity*, is a function of the boundary and initial water contents θ_0, θ_i. It is constant during a given experiment in which the inflow

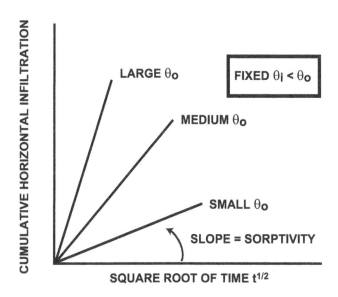

Figure 4.5   Cumulative infiltration into a horizontal soil column at fixed initial water content θ_i and boundary water content θ_0. The slope is equal to the sorptivity as defined in (4.24).

end of the uniform horizontal soil column is held at a constant water content. For a given initial water content, S increases as θ_0 increases (Fig. 4.5). Note that the exact solution for this problem is identical to the Green–Ampt solution (4.12).

Since S is constant over time, cumulative horizontal infiltration I is given by

$$I = St^{1/2} \tag{4.24}$$

Thus, the sorptivity S may be measured simply by determining the slope of I versus $\sqrt{t}$ in a horizontal infiltration experiment (Fig. 4.5).

Philip's vertical infiltration solution appears as two separate expressions appropriate for short and long times after infiltration commences. The short-time solution is expressed as an infinite series in powers of $t^{1/2}$ as

$$I = \tfrac{1}{2}St^{1/2} + A_1t + A_2t^{3/2} + \cdots \tag{4.25}$$

where S is the sorptivity, which appears in the horizontal infiltration solution (4.24), and $A_1, A_2, \ldots$ are constants that depend on the soil properties and θ_0 and θ_i. Philip (1957a) indicated how these constants (and S) could be calculated from $D(\theta)$ and $K(\theta)$. The Philip model is generally approximated by just the first two terms of (4.25),

$$I = St^{1/2} + At \tag{4.26}$$

with a corresponding infiltration rate

$$i = \frac{1}{2}\frac{S}{\sqrt{t}} + A \qquad (4.27)$$

where S and A are constant for a given soil during constant potential infiltration with a uniform initial matric potential.

The long-time infiltration solution calculated by Philip (1957b) has the following properties:

- The infiltration rate approaches a constant equal to the value $K(\theta_0)$ of the unsaturated hydraulic conductivity at the surface water content:

$$i \rightarrow K(\theta_0) \neq A \qquad (4.28)$$

Note that $K(\theta_0)$ is not the same as A in (4.27). This is because the solutions (4.27)–(4.28) are valid over different time ranges.

- The wetting front advances without changing its shape.
- The velocity V_F of the moving wetting front approaches a constant value given by

$$V_F = \frac{K(\theta_0) - K(\theta_i)}{\theta_0 - \theta_i} \qquad (4.29)$$

The features of the infiltration solution derived by Philip are illustrated clearly in the numerical solution to the Richards equation shown in Figs. 3.17 and 3.18.

The Richards equation solution to the constant potential problem has been tested experimentally under laboratory conditions in homogeneous soil, with good agreement. Figure 4.6 shows predicted and measured water content profiles from an infiltration experiment reported by Davidson et al. (1963). A zone of almost constant water content extends immediately down from the soil surface. This is sometimes referred to as the *transmission zone*. Both the observed and predicted wetting fronts are extremely steep below the transmission zone and move downward without changing shape. Such a steep wetting front is characteristic of infiltration into relatively dry soil and is easily discernible by eye.

It is obvious from this analysis why the assumption of a uniform water content and a constant hydraulic conductivity above a sharp wetting front was used in the approximate infiltration model of Green and Ampt. Because these assumptions maintained a realistic approximate picture of the process, the infiltration equations (4.12) and (4.18) predicted by the Green–Ampt model have properties that are very similar to those calculated by the Richards equation by an exact analysis.

Other Infiltration Models   Several other formulas based on a physical analysis of the flow equation have been proposed. Haverkamp et al. (1994) use the same two-term functional form

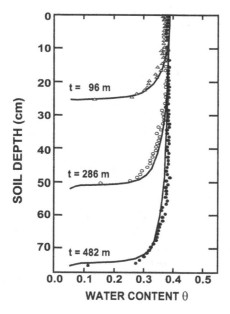

Figure 4.6 Measured and predicted soil water profiles for air-dry Hesperia soil allowed to wet at $\theta_0 = 0.385$. (After Davidson et al., 1963.)

$$I \approx C_1 t^{1/2} + C_2 t \tag{4.30}$$

as the Philip equation but have different theoretical expressions for the coefficients C_1 and C_2. Swartzendruber (1987) developed a single infiltration solution to apply to the short- and long-term infiltration stages of Philip's model as

$$I = [1 - \exp(-A_s \sqrt{t})]\frac{S}{A_s} + K(\theta_0)t \tag{4.31}$$

where A_s is an empirical constant. Barry et al. (1995) presented a complicated physically based model that can be represented with four parameters.

Clausnitzer et al. (1998) conducted a test of the parameter uncertainty of a number of infiltration models using numerically simulated data with the Richards equation, finding poor fits by the estimates of the empirical Horton equation (4.6) and good agreement with the higher-parameter models such as (4.31).

Field versus Laboratory Infiltration   The rigorous analysis that leads to (4.26) neglects several important factors that cannot always be overlooked in the field. These include possible entrapment of air in the soil profile during infiltration, the development of surface crusts that decrease the hydraulic conductivity, microbial growth, and vertical movement of earthworms that may lead to increasing hydraulic conductivity and infiltration rate with time. In fact, Horton (1940) attributed the falling infiltration rate to surface sealing. The Philip analysis predicts that the infiltration rate must decrease with time due to the decreasing matric potential gradient, and his model (4.26) tends to predict a much slower rate of decrease that continues for a much longer period of time than is usually found with field data. For example, the Horton equation (4.5) generally decreases to a constant value in 2 to 3 h when used with field-calibrated parameters, whereas the Philip equation (4.26) may continue to decrease for days (Philip, 1969). Green et al. (1970) conducted a theoretical analysis of infiltration that took air entrapment into account and found that this could explain discrepancies between theory and experiment. Another reason that field infiltration data frequently show different characteristics than the models based on theoretical calculations is that field soil profiles are seldom uniform with depth, nor is the water content distribution uniform at the initiation of infiltration. This can have a significant influence if the water table is near the surface (Freeze, 1969). These two effects usually tend to reduce the infiltration rate more rapidly than would be predicted from a model that assumes that the soil is homogeneous. Wang et al. (1999) presented a physically based model that included surface sealing caused by sediment deposit from muddy water irrigation. As yet, however, no one has proposed a mechanistic infiltration model that includes the effects of earthworms or other burrowing organisms.

4.3.2 Infiltration When Rainfall Is Limiting

The infiltration models described thus far have all assumed that the surface is maintained at a fixed potential rather than the more usual case where water flux is held at a fixed rate. Under rainfall irrigation, for example, the only time that the surface would be exposed to a fixed potential is if the rainfall rate is so intense that ponding (i.e., positive-pressure head) begins immediately. However, the initial capacity of soil to absorb incoming water is generally very high, and ponding will not begin immediately. In fact, if the rainfall rate never exceeds the final gravity-dominated infiltration rate of a soil, there will never be any runoff.

When the applied rainfall rate is less than the initial infiltration capacity of the soil but greater than the final steady-state gravity-dominated rate, a transition must occur. The soil will absorb the incoming rainfall until the matric potential gradients near the surface diminish to a point where the water cannot be taken up by the soil profile as fast as it is entering through the surface. At this time, the soil near the surface saturates and ponding develops.

There is no easy way to estimate the time at which runoff begins when a dry soil profile is exposed to prolonged rainfall, except by direct numerical calculation of the time required to saturate the surface. Figure 4.7 shows a plot of the predicted infiltration rate as a function of time using a numerical solution of the water flow equation for both ponded surface conditions (dashed curve) and constant-rainfall inputs (solid curves). As shown in this figure, ponding will not occur under constant-rainfall input until some time after the applied rate exceeds the value the infiltration rate would have if the soil were continually ponded. Furthermore, more water will enter the soil prior to ponding when the rainfall rate is lower than when it is higher. Various approximate models have been proposed to determine the ponding time (Boulier et al., 1987; Kutilek, 1980; Parlange and Smith, 1976; Smith et al., 1993).

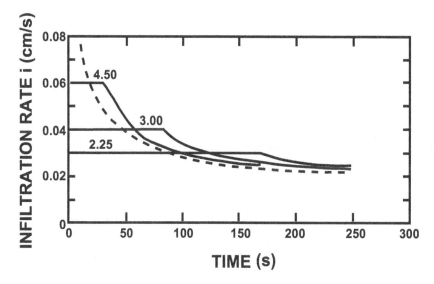

Figure 4.7 Comparison of constant flux infiltration (solid lines) at $i > K_s$ with maximum potential infiltration by continuous ponding (dashed line). Numbers on the figure refer to values of i/K_s. (After Rubin, 1968.)

However, it is quite likely that factors not taken into account by models based on the Richards equation in homogeneous soil (e.g., surface sealing) will have a significant influence on the time to ponding.

Example 4.4   Assuming that the ponded (maximum possible) infiltration rate into a soil is given by the Philip model (4.27), calculate the time t_1 at which the ponded infiltration rate equals the rainfall rate P and the amount of time t_2 required for a process with a constant rainfall rate P to add the same total amount of infiltration water to the soil as in the ponded process.

SOLUTION: The infiltration rate in the Philip model is given by (4.27). Therefore, the time at which this equals the rainfall rate P is given by the solution to

$$\frac{S}{2\sqrt{t}} + A = P \tag{4.32}$$

Solving for t in (4.32), we obtain the time t_1 at which the two rates are equal:

$$t_1 = \frac{S^2}{4(P - A)^2} = t_{min} \tag{4.33}$$

Since this time is less than the ponding time (see Fig. 4.7), it will be called t_{min}. The cumulative infiltration under ponding is given by (4.26). This equals the cumulative rainfall $P \times t$ at a time t_2 satisfying

$$S\sqrt{t} + At = Pt \tag{4.34}$$

Solving for t in (4.34), we obtain the time at which the two processes have infiltrated the same amount of water:

$$t_2 = \frac{S^2}{(P - A)^2} = t_{max} = 4 \times t_{min} \tag{4.35}$$

Because continuous ponding of the soil surface results in the maximum possible cumulative infiltration, ponding will occur under constant-rainfall infiltration prior to this time. Thus, the time given in (4.34) is an overestimate and will be called t_{max}.

The maximum infiltration rate curve shown in Fig. 4.7 may be represented adequately by the Philip model (4.27) with $S = 0.4512$ cm s$^{-1/2}$ and $A = 0.0082$ cm s^{-1}. Table 4.2 summarizes the maximum and minimum ponding time estimates using (4.33) and (4.38) together with the time at which ponding was predicted to occur using the numerical model. This time is intermediate between the two extremes and is somewhat closer to the minimum time than the maximum.

4.3.3   Infiltration into a Layered Soil Profile

When a soil layer of different texture and permeability from the surface layer is present in the soil profile, it will reduce the infiltration rate, regardless of whether it is coarser or finer than the surface layer. If the texture is finer, the reduction in infiltration is due directly to its lower permeability. In contrast, a subsurface coarse-textured layer generally has a saturated hydraulic conductivity that is greater than the finer-textured layer above it. However, if the matric potential at the wetting front is low enough, it will prevent the large, highly conducting pores of the coarse-textured region from filling [see (2.5)]. The unsaturated conductivity of the resulting partially saturated, coarse-textured region is actually lower than the wetter finer-textured region above, and the infiltration rate will decrease as the front reaches the interface.

If the coarse soil in the lower region has a negligible conductivity at the matric potential of the wetting front, unstable flow may result (Raats, 1973). During infiltration into dry soil layered in this manner, water cannot enter the coarse-textured zone until the pressure has built up sufficiently to wet the larger pores. If this occurs at discrete locations along the wetting front, the new wetted channels in the coarse-textured zone may become conduits for all of the water entering from above. These

TABLE 4.2   Comparison of Minimum (4.33) and Maximum (4.34) Ponding Times with the Actual Calculated Times t_p Shown in Fig. 4.4

P (cm s^{-1})	t_{max} (s)	t_{min} (s)	t_p (s)
0.06	72	18	30
0.04	188	47	84
0.03	392	98	170

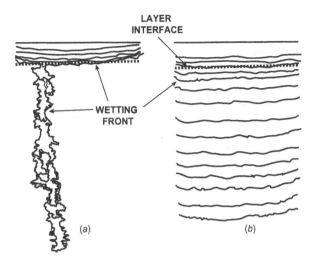

Figure 4.8   Infiltration and movement of water in a layered soil with the finer-textured soil on top: (*a*) Yolo clay over quartz sand; (*b*) Yolo clay over Oakley sand. (Redrawn from Chang et al., 1994.)

narrow flow channels, called *fingers*, can persist through the entire coarse-textured zone. Since the local flux in these channels can be much higher than the average flux of the continuous front in the fine-textured zone above, the water velocity can be appreciable. Figure 4.8*a* shows a trace of the wetting front before and after it reaches the interface between a fine-textured Yolo clay and a coarse-textured quartz sand. The front stops temporarily at the interface until the pressure rises enough to fill the large pores of the sand at one location, after which all of the flow from above was carried in a single channel of the highly conducting wetted sand. Chang et al. (1994) also showed experimentally that the front could remain stable at the interface between an overlying fine layer and underlying coarse layer, providing that the underlying layer maintained a sufficiently high conductivity at the matric potential of the wetting front. Thus, the Oakley sand shown in Fig. 4.8*b* temporarily slowed down the infiltrating front from the overlying Yolo clay, but allowed water to enter over the entire interface boundary, and no fingers formed.

4.3.4   Measurement of Infiltration

Ring Infiltrometer   The infiltration rate is usually measured in the field with an infiltrometer, which is a device for recording the entry of water into the soil over a known surface. The simplest infiltrometer is merely a thin ring that is pushed partly into the soil and then ponded with water. The device may use a single ring or may have a larger buffer ring surrounding the inner ring. A Marriotte reservoir may be used to keep the height of ponded water constant and to measure the water volume

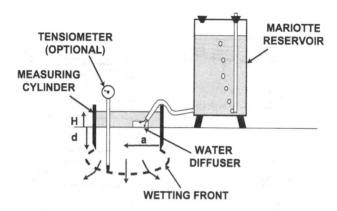

Figure 4.9   Schematic cross section of the single-ring infiltrometer. (After Reynolds et al., 2002.)

entering the soil (Reynolds et al., 2002). Figure 4.9 shows a schematic cross section of the single-ring infiltrometer device.

The infiltrometer can be used to measure the saturated hydraulic conductivity K_s, which must be inferred from the steady infiltration rate using a model of the ring geometry. The simplest model is the approximate steady-state infiltration rate solution of Wooding (1968) for a shallow circular pond on the surface:

$$i = K_s \left(1 + \frac{4\lambda_c}{\pi r}\right) \tag{4.36}$$

where r is the radius of the pond and $\lambda_c = 1/a$ in the hydraulic conductivity function

$$K(h) = K_s \exp(ah) \tag{4.37}$$

Equation (4.36) thus applies approximately to the case of a single-ring infiltrometer that is not pushed far into the ground and does not have much water ponded on the surface. Reynolds and Elrick (1990) introduced the following formula for the single-ring infiltrometer shown in Fig. 4.9:

$$i = \frac{Q}{\pi a^2} = K_s \left(1 + \frac{H + 1/\lambda^*}{C_1 d + C_2 a}\right) \tag{4.38}$$

where Q is the volume flow rate entering the soil, $C_1 = 0.316\pi$, $C_2 = 0.184\pi$, and λ^* is the *macroscopic capillary length* (Philip, 1983), which is equal to λ_c if (4.37) is valid and can be estimated from the general $K(h)$ relation if it is not. The formula (4.38) takes into account the three-dimensional flow underneath the infiltrometer, the ponding height, and the depth of penetration of the ring.

Tension Disk Infiltrometer   When water is ponded on the surface of a heterogeneous soil containing cracks or other structural voids, the water intake rate may be dominated by the flow through the voids. However, when the soil is even moderately unsaturated, these voids will empty and flow will occur through the matrix. In some natural field soils, the saturated hydraulic conductivity may be orders of magnitude higher than the conductivity of the same soil under a few centimeters of suction sufficient to drain the large macropores (Auzet et al., 1999).

True saturation is rare in the field because of air entrapment. Even in the laboratory, special care is required to wet a medium to complete water saturation. Thus the maximum hydraulic conductivity attained naturally is often less than the true saturated value. This circumstance has given rise to terms such as *near-saturated* or *quasi-saturated hydraulic conductivity* (Faybishenko, 1999) to denote the field value. The tension disk infiltrometer (Perroux and White, 1988) is designed to sort out the contribution of macropores to the infiltration rate and conductivity at saturation. The device, shown in Fig. 4.10, consists of chamber with a permeable porous plate or cloth at the bottom that will support a modest suction, which is maintained by a Marriotte bottle feeding the supply reservoir. The porous plate is placed on the soil surface (leveled to produce good contact), allowing water to infiltrate under a specified suction that is sufficient to prevent entry into macropores that reach the surface. The infiltration value thus obtained can be compared to the value from the infiltrometer under positive

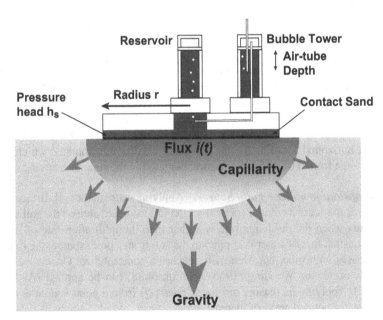

Figure 4.10   Schematic cross section of the tension disk infiltrometer. (After Clothier and Scotter, 2002.)

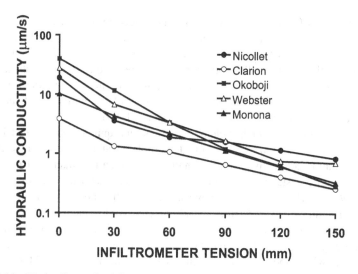

Figure 4.11 Hydraulic conductivity values of five Iowa field soils measured with the tension disk infiltrometer at various values of infiltrometer tension in millimeters. (Plotted from table in Logsdon, 2002.)

pressure. In addition, suction can be varied in the tension infiltrometer to delineate macropores in various tension ranges (Mohanty et al., 1996).

Ankeny et al. (1988) presented a design for an automated tension disk infiltrometer, in which two pressure transducers were used to determine and record infiltration rates at different tensions. This was modified by Casey and Derby (2002), who showed how to make the infiltration measurements with a single pressure transducer. Ankeny et al. (1991) determined K_s and λ_c from a series of steady infiltration measurements at different tensions.

Figure 4.11 shows the extreme reduction in hydraulic conductivity measured from saturation to slightly below saturation in five field soils from central and southwestern Iowa. The horizontal coordinate is the tension of the disk infiltrometer, which covers only the range from 0 to −15 cm of matric potential in the soils measured.

Ponded Infiltration under a Point Source Applying water to the soil surface from a point source produces a circular (or nearly circular) saturated area on the soil surface. This occurs when the discharge rate is higher than the infiltration rate of the soil. When a constant flux of water is applied to the soil from a point source, the saturated area increases with time, but eventually reaches a constant size. Once steady-state conditions occur, the Wooding (1968) solution (4.38) can be applied (Al-Jabri et al., 2002a). Applying increasing discharge rates (Q) from a point source at the soil surface creates ponded areas of increasing r_0.

As seen from (4.36), plotting the values of i versus the corresponding $1/r$ results in a straight line with an intercept equivalent to K_s, and λ_c can be determined from the resulting slope $s = 4K_s\lambda_c/\pi$. Figure 4.12 shows the setup of a dripper system for

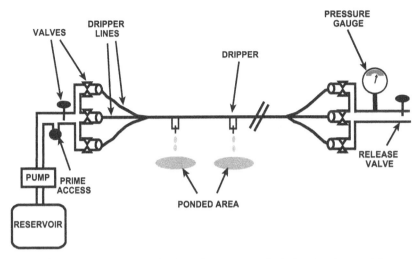

Figure 4.12   Experimental setup for the point source method, showing the ponded areas on the soil surface. (After Al-Jabri et al., 2002a.)

ponded infiltration. The apparatus is useful for making simultaneous measurements at a number of surface locations. Al-Jabri et al. (2002b) made 50 concurrent measurements of surface infiltration, and thus determined 50 values of K_s and λ_c. Figure 4.13 shows a set of infiltration data along with (4.36) fitted to the data.

4.3.5   Preferential Flow of Water and Solutes

A field soil is full of local pathways such as structural voids or biological channels that can carry water at velocities much greater than those of the surrounding matrix, even when the entire surface is watered uniformly. Moreover, local obstacles in a heterogeneous soil can cause water to funnel into narrow plumes under certain circumstances, even when moving within the soil matrix (Kung, 1990). Finally, the fluid phase can become unstable if conditions support the growth of flow perturbations, creating fingers of flow that move ahead of a wetting front. All of these phenomena have been linked under the descriptive name of preferential flow, referring to the significantly greater than average downward movement of water through part of the soil during an infiltration or drainage event. However, they arise from distinctly different physical processes and therefore must be analyzed individually. For purposes of discussion, preferential flow may be subdivided into three groups: macropore flow, funnel flow, and unstable flow.

Macropore Flow   *Macropore flow* is flow through a part of the soil that contains a structural void or other conduit with a large aperture (White, 1985). Macropores are formed in various ways, including soil shrinkage by drying (Brewer, 1964), chemical weathering (Reeves, 1980), cycles of freezing and thawing, or biological activity.

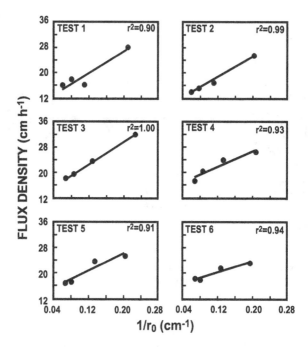

Figure 4.13 Flux density versus $1/r_0$ for all observation sites from the point source method. (After Al-Jabri et al., 2002a.)

Tillage generally destroys macropores to some degree in the immediate subsurface (Thomas and Phillips (1979), although they may still act as flow channels below the plow layer (Quisenberry and Phillips, 1978). Macropore flow is the dominant transport mechanism in fractured rock because of the extremely low permeability of the medium surrounding the fractures. In soil, macropore flow may occur in tandem with matrix flow, or matrix flow may occur in the absence of flow through nearby voids. Beven and Germann (1982) reviewed a large body of experimental evidence on infiltration and redistribution of water in soils containing macropores, and concluded that the theories of flow based on homogeneous media did not adequately describe the behavior observed. They also felt that most soils have at least some macropores.

Macropore flow is easily explained with Poiseuille's law (3.10), which shows the nonlinear dependence of water flow on aperture size for flow through a cylinder, and by the capillarity equation (2.5). If water is held in a surrounding matrix at a matric potential h next to a macropore channel of effective radius R, water cannot flow from the matrix into the channel unless

$$h > \frac{-2\sigma}{R}$$

The potential importance and limitations of macropore flow are illustrated in the next example.

Example 4.5   A 2.0-mm-diameter cylindrical macropore is drilled through a permeable soil block $1 \times 1 \times 1 \text{ m}^3$ whose matrix has a hydraulic conductivity $K_s = 100$ cm day^{-1}. The block and macropore are saturated and have 10 cm of water continuously ponded on top, while the bottom is open to the atmosphere. Assuming that the flow in the matrix and macropore occur in parallel, calculate the total volume flow and the fraction through the matrix and the macropore. Assuming a porosity ϕ of 0.4 in the matrix, calculate the water velocity in each region.

SOLUTION: The water flux and flow volume Q_s through the soil matrix are calculated by Darcy's law (3.14):

$$J_w = -100 \left(\frac{10 + 100 - 0}{100 - 0} \right) = -110 \text{ cm/day}^{-1} \rightarrow Q_s = |J_w|A$$

$$\approx 1.1 \text{ m}^3/\text{day}^{-1}$$

The flow volume Q_m through the macropore is calculated by Poiseuille's law (3.10):

$$Q_m = \frac{\rho_w g \pi R^4}{8 \nu} \frac{\Delta H}{\Delta z} = 0.366 \text{ m}^3/\text{day}^{-1}$$

Thus, the small hole occupying an infinitesimal part of the 1-m^2 surface area carries over one-third of the volume flow. The water velocity through the soil matrix is $V_s = J_w/\phi = 2.75$ m day^{-1}, whereas the velocity through the macropore is $V_m = Q_m/\pi R^2 = 134$ cm s^{-1}! The small hole is essentially a short circuit under positive pressure–driven flow. However, this conclusion needs to be modified if the flow conditions are slightly different. Suppose, for example, that the wormhole extends from the surface to 90 cm and that the last 10 cm of the distance to the bottom, where $P = P_{\text{atm}}$, contains soil. The resistance of the wormhole is so small that the last 10 cm of soil essentially feels the full weight of the 100 cm of water on top of it. Thus, this small part of the medium has a flux

$$J_w = -100 \left(\frac{10 + 100 - 0}{10 - 0} \right) = -1100 \text{ cm/day}^{-1} \rightarrow Q_m = |J_w|\pi R^2$$

$$= 3.5 \times 10^{-6} \text{ m}^3/\text{day}^{-1}$$

In this case the effect of the macropore on flow volume is insignificant. In fact, the water in the cylinder would disperse in all directions. Finally, we might ask what would happen if the macropore cylinder had soil lying over it and the surrounding soil was unsaturated. Under what conditions would the cylinder fill up with water? Here we use the capillarity equation (2.5) with $R = 0.1$ cm, and calculate $h = P_{\text{liq}} - P_{\text{atm}} \approx -1.5$ cm. Thus, unless the soil matrix is essentially saturated, no water will enter the macropore. This also demonstrates that if the tension disk infiltrometer is set on a tension of 3 cm, it would measure an infiltration rate in this soil that would reflect the matrix properties even if it were centered over the macropore.

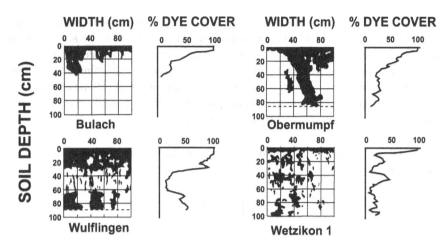

Figure 4.14   Vertical flow patterns of Brilliant Blue FCF and one-dimensional profiles of dye coverage after a 40-mm sprinkler application in four soils that were not covered prior to irrigation. The dashed lines represent the maximum depth of sampling. (After Flury et al., 1994.)

In a field study, Flury et al. (1994) gave a dramatic demonstration of the importance of macropore flow. They surveyed the 14 major agricultural soil types in Switzerland for their potential to experience macropore flow by adding 40 mm of water containing the dye Brilliant Blue FCF (C.I. Food Blue 2) onto 2×2 m^2 plots with a sprinkling apparatus. Each soil type had two plots, one in a natural state and a second that had been covered to prevent water entry for two months prior to the experiment. One day after irrigation the plots were excavated, and the stained pattern was examined on a vertical 1×1 m^2 soil profile. The flow patterns showed that water bypassed the soil matrix in most soils, even penetrating beyond the 1-m depth in some cases. Structured soils were more prone than nonstructured soils to produce bypass flow, deep dye penetration, and pulse splitting. Figure 4.14 shows the dye trace pattern revealed along the trench face in four of the 14 plots, revealing flow channeling that varied from moderate to extreme. The graphs on the right side of the figure show the fraction of the trench face at a given depth that is covered by dye. Figure 4.15 summarizes the maximum observed depth of penetration of the dye in all plots for both wet and dry initial conditions. Since 4 cm of water would penetrate only 16 cm in a soil with $\theta = 0.25$ if the entire wetted soil volume were active in transport, it is clear that most of the soils studied by Flury et al. (1994) displayed significant preferential flow.

Funnel Flow   A comprehensive preferential flow study was conducted by Kung (1990), who added dye to the surface of a potato field on Plainfield sand over a 4-m^2 surface area and later exposed the profile to the 6.6-m depth. Dye patterns showed very little lateral displacement in the uniform top layer, but significant lateral movement was observed in the deeper unsaturated zone along textural discontinuities.

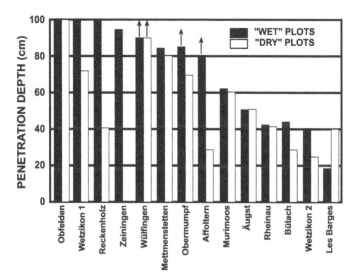

Figure 4.15 Maximum dye penetration depth in each soil for plots in natural state prior to irrigation ("wet") and plots that were covered for two months prior to the study ("dry") Arrows indicate plots not sampled to 1-m depth. (After Flury et al., 1994.)

Water channeled into preferential pathways, and at 5.6 to 6.6 m, water flowed through less than 1% of the soil matrix. The top 1.1 m of the soil was very uniform, but from 1.1 to 7.4 m depth the soil had distinct layers made of fluvioglacial deposits. These layers were neither horizontal nor continuous, and consisted of coarse sand lenses interbedded in finer sand layers. The coarse lenses acted as barriers when dry, channeling the water around them into smaller and smaller pathways (Fig. 4.16). Since the conductivity of the surrounding matrix was high, these pathways were still able to transport water without lateral spreading, a phenomenon Kung described as funnel flow.

Unstable Flow   Unstable flow during infiltration in unsaturated porous media has been studied for many years and is known to be associated with a number of existing conditions, including vertical flow from a fine-textured layer into a coarse one (Baker and Hillel, 1990; Hill and Parlange, 1972), vertical flow into a compressed air phase (Peck, 1965), infiltration into water-repellent soil (Hendrickx et al., 1993), and two-phase flow involving two fluids of contrasting density and viscosity (Chuoke et al., 1959). More recently, it has been demonstrated to occur during infiltration into homogeneous soil at flux rates substantially less than the saturated hydraulic conductivity (Geiger and Durnford, 2000; Selker et al., 1992), and during redistribution following infiltration in homogeneous soil (Diment and Watson, 1985; Wang et al., 2003a). Unstable flow is distinct from other forms of preferential flow in that it is a fluid phenomenon whose extreme flow location is not a consequence of permeability variations in the porous medium.

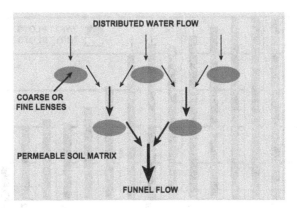

Figure 4.16   Schematic illustration of funnel flow, showing discrete lenses of coarser or finer material than the surrounding permeable matrix. Water will move around both kinds of lens, causing the water to channel into a narrow plume.

Raats (1973) and Philip (1975) applied perturbation flow theory to porous media and demonstrated that the condition for instability at a wetting front was that the total hydraulic gradient behind the front is less than unity (i.e., the water pressure decreases toward the surface). The physical explanation for this condition is that fingers that are initiated at the wetting front for whatever reason will continue to be supplied by water as they grow.

Figure 4.17 illustrates the propagation of the finger when the matric potential decreases behind the front. The water pressure at the interface between the wet and dry zones is at the water entry potential h_{we}, which allows water to enter the dry region. As a perturbation forms, the depth of penetration becomes slightly greater, which shifts the water pressure distribution downward above that location. As a result, regions of the surrounding matrix begin to supply the zone above the finger, because the horizontal pressure distribution induces lateral flow. Subsequently, the water pressure in the surrounding matrix decreases and the pressure at the wetting front drops below the water entry pressure h_{we}. The profile drainage then proceeds exclusively through the propagating finger and continues until flow stops. Capillary flow will eventually cause the finger to dissipate and surmount the barrier imposed by the wetting front suction, but this occurs on a time scale long compared to the propagation event.

Figure 4.17 demonstrates a second necessary condition for finger growth. There must be a pronounced threshold water-entry pressure between the wet and dry regions so that downward flow in the matrix adjacent to the finger will become insignificant as the pressure is lowered above the front by lateral flow. Distinct water-entry pressures tend to be most prominent in coarse-textured sands which have a narrow pore-size distribution (Wang et al., 1998). The water entry matric potential has been measured by a number of authors, including in layered media by Baker and Hillel (1990) and in homogeneous media by Geiger and Durnford (2000) and Wang et al. (2000, 2003a).

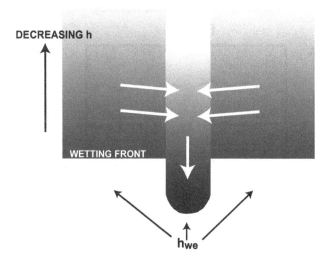

Figure 4.17 Development of a fluid instability during redistribution, when the pressure distribution decreases toward the surface. When the front advances ahead at one location, the pressure distribution above it shifts downward, creating a lateral flow gradient from adjacent regions toward the finger (arrows). Darker shade indicates wetter soil at higher matric potential. (After Jury et al., 2003.)

The final condition necessary for substantial finger propagation is that the fingers do not dissipate rapidly by lateral flow into the surrounding dry matrix. Glass et al. (1989) offered a physical explanation involving hysteresis in the matric-potential water content function (see Fig. 2.16) for why the fingers don't disperse by lateral diffusion. The interior of the finger first wets up to near saturation and then drains as the tip of the finger moves downward. At the same time, a narrow zone of wetting is created at the outside fringe of the finger adjacent to the inner core. The interior of the finger thus follows the drainage loop of the water characteristic curve, whereas the fringe is on the wetting curve. Since at the same water content the drying curve is at a lower matric potential than the wetting curve, equilibrium is reached at different water contents inside the core and at the fringe and the fingers remain narrow.

Figure 4.18 shows a cross section of the draining front from a redistribution experiment conducted in the 1.2-m-wide 1.0-m-deep 1-cm-thick Hele–Shaw cell described in detail in Wang et al. (2003a). In this study, 5 cm of water was ponded on the surface of the cell containing a dry coarse sand, and redistribution began at about $t = 1$ min. Several characteristics are evident in the figure. First, although perturbations in the shape of the front are present from the outset, instabilities do not form and cause fingers to grow until after redistribution begins. The reason for this is that the matric potential increases behind the front during ponding but reverses during redistribution. Second, once the fingering process begins, downward flow stops in the matrix region between the fingers, because lateral flow has lowered the matric potential below h_{we}. Third, the initial finger area fraction is substantial (about 50% in

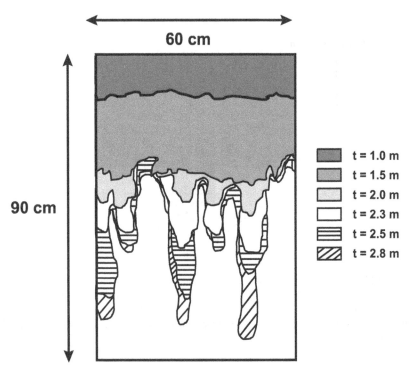

Figure 4.18   Advancing unstable redistribution profile in a Hele–Shaw cell as a function of time after infiltration ceases. (After Jury et al., 2003.)

this experiment), but the fingers still are able to move a substantial distance beyond the front. The reason for this is that much of the finger volume is at low saturation, because the finger drains behind the tip. Selker et al. (1992) have observed this finger profile during unstable flow. Thus, a small amount of water can create a long finger.

4.3.6   Two- and Three-Dimensional Infiltration

The previous discussion dealt with one-dimensional infiltration, in which water is assumed to flow vertically into the soil. Several two- and three-dimensional infiltration models are of interest since they are more closely related to certain field infiltration measurements and irrigation methods than are the one-dimensional models.

Infiltration from a water-filled semicircular furrow and from a hemispherical cavity were studied in some detail by Philip (1968). Using solutions to the water transport equation, he showed that in both cases the infiltration rate is very high initially and decreases until it approaches a constant rate. This constant rate can be estimated if the dimensions of the cavity or furrow are specified and the hydraulic conductivity of the soil is known. Talsma (1969) found good agreement between field measurements and the predicted rate using Philip's treatment of the problem. The shape of the wetting front below a local source of water such as a furrow depends very much upon

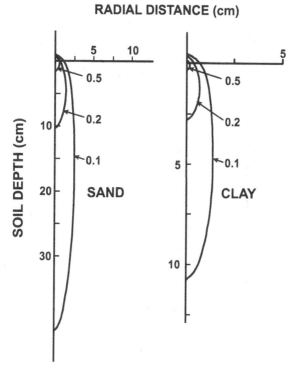

Figure 4.19 Limiting moisture profiles as $t \rightarrow \infty$ during infiltration from a cylindrical cavity. Numbers on the curves represent values of $(\theta - \theta_0)/(\theta_1 - \theta_0)$. (After Philip, 1968.)

the relative importance of matric and gravitational forces during infiltration. If the matric forces predominate, the wetting front tends to be symmetrical and moves as much laterally as vertically. If gravity is more important, as in the case of very coarse textured soils, the wetting front is elongated and is more nearly ellipsoid in shape. This behavior is illustrated in Fig. 4.19, which shows lines of equal water content for large times during infiltration from a cylindrical cavity for sandy and clay soils.

4.4   REDISTRIBUTION

4.4.1   Redistribution of Water in Soil Profiles

The term *redistribution* refers to the continued downward movement of water through a soil profile after irrigation has ceased at the soil surface. This is a complex process, because the lower part of the profile ahead of the front will increase its water content and the upper part of the profile near the surface will decrease its water content after infiltration ceases. Thus, hysteresis can have an effect on the overall shape and dynamic behavior of the water content profile.

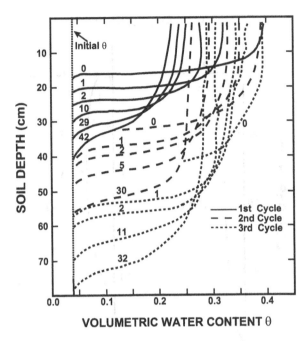

Figure 4.20 Successive water content profiles during redistribution cycles following one, two, and three irrigations of 5 cm each. (After Gardner et al., 1970.)

Figure 4.20 shows experimental water content profiles during redistribution stages following three different irrigation cycles. Several general features of the water content profile are apparent from an examination of this figure. First, the water content of the wetted soil profile decreases relatively uniformly over space; there is no abrupt drying of the surface layer. Second, the front of the profile continues to move downward with a sharp boundary between the wet and dry soil regions. The overall change of profile shape is similar to that of a rectangle (whose height is wetted profile thickness and whose width is the difference between wet and dry region water contents) that becomes progressively taller and thinner over time while maintaining equal area. Also, because the water content is very uniform within the wetted region, the drainage water across a fixed plane in the profile may be described reasonably well by gravity flow. The *gravity-drained rectangle model* was used by Jury et al. (1976) to describe water redistribution during intermittent irrigation. Because the water content of the wetted region is decreasing continually, the drainage rate also decreases and the profile slowly approaches equilibrium. Contrary to the equilibrium profile reached after a uniform wetting or drying process, in which water content increases with depth, the water content profile of a redistribution process is more uniform with depth because of hysteresis. As shown in Fig. 4.20, the equilibrium water content for a given matric potential is higher for a drying process than for a wetting process. Thus, the equilibrium profile, where $dH/dz = 0$ or $h = -z$, will have a smaller (more negative) value of matric potential near the surface than deeper in the soil. However, the water content of the surface layer results from a drying process and therefore is larger for a given h than is the water content in the lower region, which results from a wetting process.

Although hysteresis has made exact analysis of redistribution difficult, some approximate models (e.g., Jury et al., 1976; Sander et al., 1991) have been developed to produce simple expressions for water content–time relations that have similar features to those of the experimental profiles. For example, it has been observed by a number of workers that the decrease in average water content θ over time in the zone above the wetting front is described accurately by an empirical equation of the form

$$\overline{\theta} = a(t + c)^{-b} \tag{4.39}$$

where a, b, and c are constants. Gardner et al. (1970) have shown that these constants can be related to the soil water diffusivity and permeability if certain simplifying assumptions are made. When the time t in (4.39) is longer than a day or so, the constant c can usually be neglected, allowing (4.39) to be rewritten in the form

$$\log \overline{\theta} = \log a - b \log t \tag{4.40}$$

Hence, values of the constants a and b can be obtained directly from the slope and intercept of a logarithmic graph if the relationship in (4.39) is reasonably accurate.

Values of average water content in various layers of a sandy loam soil profile during drainage after irrigation (Ogata and Richards, 1957) are shown plotted in a log-log graph in Fig. 4.21. The solid lines represent the best fit of the data to (4.40).

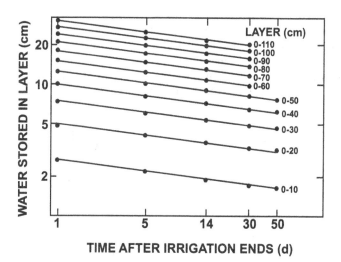

Figure 4.21   Depth of water stored in various layers as a function of time during redistribution (logarithmic scale). Solid lines are the best fit to (4.40). (After Ogata and Richards, 1957.)

The rate of water loss from the wetted profile of thickness d, which is equal to the drainage rate, is obtained by differentiating (4.39) and multiplying by d, or

$$d\frac{d\overline{\theta}}{dt} = -dba(t+c)^{-(1+b)} \tag{4.41}$$

Example 4.6 The gravity-drained rectangle model treats the redistributing profile as a zone of spatially uniform water content $\overline{\theta}(t)$ and assumes that the drainage rate at the wet–dry interface is a unique function of the water stored in the profile. Calculate this drainage rate assuming that the drainage rate is equal to the unsaturated hydraulic conductivity, which is given by

$$K(\overline{\theta}) = K_s\left(\frac{\overline{\theta}}{\theta_s}\right)^N \tag{4.42}$$

where the subscript s refers to saturation and $N > 1$ is a constant. Assume initially that the water content profile is saturated to a depth $z = -L$ and that the water content θ_i of the dry layer beneath the wetting front is negligible. Thus, the water content at the onset of redistribution is equal to θ_s if $z < -L$ and zero below that depth.

SOLUTION: The gravity-drained rectangle model of redistribution uses two simplifying assumptions. First, the area within the profile "rectangle" is constant. Thus, for $t > 0$ when the water content has reached depth z and has fallen to $\overline{\theta} < \theta_s$, the water stored in the rectangle is given by

$$L\theta_s = z\overline{\theta} \tag{4.43}$$

This is shown in Fig. 4.22. Second, the instantaneous drainage rate at the interface between the wet and dry zones is equal to the gravity flux (4.42). Thus, when the profile has reached depth z, we may write the mass balance equation for the rectangle by stating that the rate of change of water storage in a fixed soil volume of thickness z is equal to the flux out of the volume at z. This is given by

$$z\frac{d\overline{\theta}}{dt} = \frac{L\theta_s}{\overline{\theta}}\frac{d\overline{\theta}}{dt} = -K(\overline{\theta}) = -K_s\left(\frac{\overline{\theta}}{\theta_s}\right)^N \tag{4.44}$$

Equation (4.44) may be integrated by placing all factors that depend on $\overline{\theta}$ on the same side. Thus,

$$\int_{\theta_s}^{\overline{\theta}} \overline{\theta}^{-(1+N)}\, d\overline{\theta} = -\int_0^t \frac{K_s}{L\theta_s^{1+N}}\, dt \tag{4.45}$$

These integrals may be looked up in any standard table. The result is

AVERAGE WATER CONTENT

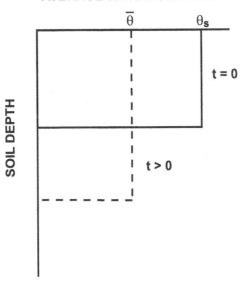

Figure 4.22   Rectangular approximation to the redistribution process.

$$-\frac{1}{N}\left(\frac{1}{\overline{\theta}^N} - \frac{1}{\theta_s^N}\right) = -\frac{K_s t}{L\theta_s^{1+N}} \qquad (4.46)$$

We may solve for $\overline{\theta}$ in (4.46) with the result

$$\overline{\theta}(t) = a(t + c)^{-b} \qquad (4.47)$$

where

$$a = \left(\frac{L\theta_s^{1+N}}{NK_s}\right)^{1/N}$$

$$b = \frac{1}{N}$$

$$c = \frac{L\theta_s}{NK_s}$$

The position z of the front is given by (4.43) and (4.47). Note that (4.47) is identical to (4.39).

This example illustrates how an empirical model (4.39) can be reproduced with a physically based model (gravity flow). It can be shown that for a fixed water table

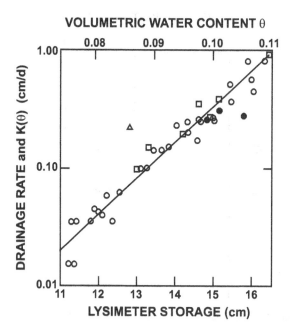

Figure 4.23   Lysimeter drainage rate (circles) as a function of water storage and measured unsaturated hydraulic conductivity (squares and triangle) as a function of water content. (After Black et al., 1969.)

depth even very nonhomogeneous soils tend to drain in such a way that the rate of drainage out of the profile can be related in a simple fashion to the water content of the soil. This provides a useful simplification for the calculation of the field water balance where the assumption of a unique relation between the soil profile water content and the drainage rate is justified. Figure 4.23 shows the drainage rate measured with a weighing lysimeter plotted as a function of the average soil water content of the lysimeter for Plainfield sand (Black et al., 1969). The unsaturated hydraulic conductivity is also shown in the same graph and can be seen to be very nearly equal to the drainage rate, implying that the unit gradient drainage approximation works very well in this soil. Figure 4.24 compares the drainage rate for the summer season for the soil profile predicted from the line in Fig. 4.23 with the actual drainage rate measured in the lysimeter.

4.4.2   Unstable Flow during Redistribution

As discussed in Section 4.4.3, redistribution tends to produce a reverse matric potential gradient (more negative toward the surface) that is conducive to the development of unstable flow. Wang et al. (2003b) conducted a field study of redistribution in two

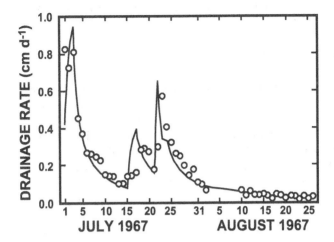

Figure 4.24   Predicted drainage (solid line) from bare Plainfield sand compared with that measured by a lysimeter (circles). (After Black et al., 1969.)

soils (a sandy loam and a sand), using dye tracing and trench face exposure to monitor for preferential flow during the redistribution process. Figure 4.25 shows the shape of the draining front at the beginning of the redistribution phase, and approximately 60 h later. These two profiles were observed from trench faces at different locations within the plots.

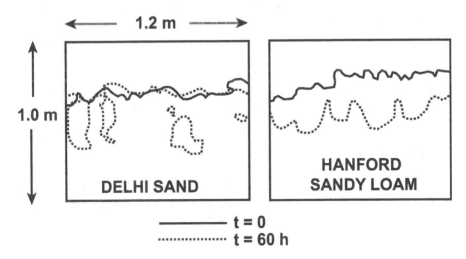

Figure 4.25   Wetting front shapes at the beginning and 60 h after redistribution starts in two field soils. Profiles were measured at different locations within the plots. (After Wang et al., 2003b.)

4.5   FIELD MEASUREMENT OF UNSATURATED HYDRAULIC CONDUCTIVITY

In Section 4.3.4 we described steady-state methods for determining saturated and unsaturated hydraulic conductivity of surface soil. Since it is difficult to create steady-state unsaturated flow in subsurface soil, the standard methods for measuring unsaturated hydraulic conductivity in situ often involve transient flow experiments. Early field methods (Rose et al., 1965) involved simultaneous measurement of matric potential profiles with tensiometers and water content profiles with neutron probes on field plots undergoing redistribution following initial saturation. The Richards equation (3.70) was written in a finite-difference form and solved for the average hydraulic conductivity of discrete layers that provided the best agreement between the observed changes in h and θ. A problem with the preceding method is that the field soil profile may be heterogeneous, so that using the homogeneous Richards equation (3.68) to interpret the data may lead to inaccurate or even absurd (i.e., negative) values for K in certain locations within the profile. For this reason, $K(\theta)$ is often evaluated with simpler models that are not as sensitive to soil heterogeneity.

As shown in the preceding, experimental observations of redistribution profiles in homogeneous soils show a very uniform water content through the wetted zone. This suggests that the gravity flow model may be used to describe profile drainage during redistribution without evaporation as

$$L\frac{d\bar{\theta}}{dt} = -J_w(L) = -K(\bar{\theta}) \tag{4.48}$$

where $\bar{\theta}$ is the average water content and L is the thickness of the profile. Nielsen et al. (1973) and Libardi et al. (1980) used (4.48) to measure $K(\bar{\theta})$ under field conditions. The latter authors also compared this method to the procedure used by Rose et al. (1965).

Example 4.7   A soil profile of $L = 1.0$ m thickness is wetted up to saturation and the surface is covered. Neutron probe readings of water content are taken over time and are used to calculate the quantity of water contained in the profile. Using the data given in Table 4.3, calculate the hydraulic conductivity of the soil as a function of average water content.

SOLUTION: According to (4.48), the hydraulic conductivity is equal to the rate of change of water stored, provided that the gravity flow approximation (4.4) is reasonably valid. The average water content $\bar{\theta}$ is given by

$$\bar{\theta} = \frac{1}{L}\int_0^L \theta(z)\,dz \tag{4.49}$$

Thus, from Table 4.3 we can calculate each of the terms in (4.48) and (4.49), and finally, $K(\bar{\theta})$, with the result shown in Table 4.4.

TABLE 4.3   Water Storage Values as a Function
of Time in a 1-m Profile

Water Stored $\int_0^L \theta \, dz$ (cm)	Time (h)
50	0
45	1
39	5
35	10
32	20
31	30

Other indirect field methods for measuring unsaturated hydraulic conductivity involve using model functions of the conductivity and an analytic solution of an experimental procedure. Equation (4.36) may be used to estimate the parameter a in (4.37) if K_s is known (Clothier and Scotter, 2002). Russo and Bresler (1980) used a modified version of the Green–Ampt (1911) infiltration model (4.18) to interpret the advance of the wetting front beneath an air-entry hydrometer (Bouwer, 1966). By solving the infiltration problem with model functions for $K(h)$ and $h(\theta)$, they were able to estimate the model parameters by forcing agreement between observed and predicted wetting-front position.

Inverse Methods   Inverse methods refer to numerical fitting of the appropriate solution to the Richards equation (3.70) by varying parameters of model functional forms until minimum sum of squares deviation between model and data is achieved (Hopmans et al., 2002). Although used principally on laboratory data, this method has been employed to interpret the readings of tension disk infiltrometers (Simunek et al., 1998). To achieve a more significant evaluation of model parameters, it is recommended that values of matric potential or water content in the soil be obtained during the infiltration experiment.

4.6   EVAPORATION

The rate of evaporation from a wet, bare soil surface is limited by external meteorological conditions such as wind speed, relative humidity, and the flux of radiant energy

TABLE 4.4   Evaluation of $K(\bar{\theta})$ Using (4.48) and (4.49) and Table 4.3

$L\bar{\theta}$ (cm)	t (h)	$L\,\Delta\theta$ (cm)	Δt (h)	$\bar{\theta}$	$K(\bar{\theta})$ (cm/h^{-1})
50	0	−5	1	0.475	5.0
45	1	−6	4	0.420	1.5
39	5	−4	5	0.370	0.8
35	10	−3	10	0.335	0.3
32	20	−1	10	0.315	0.1
31	30				

to the surface (Penman, 1948). In contrast, water loss from a soil with a dry surface layer is regulated primarily by soil water resistances that limit the rate at which water moves upward to the evaporating surface (Philip, 1957f). In the latter case, the water evaporation rate will be less than the maximum potential loss rate dictated by the external conditions.

During transient drying of a soil, control of the evaporation rate can pass from the meteorological factors to the soil resistance. Initially, when the soil surface is wet, evaporation occurs at the potential rate, which is limited by the amount of energy available at the soil surface. If the evaporation rate is intense, the rate of supply of water to the surface from the soil below may not be able to match the rate of loss, causing the surface water content to become progressively drier. This period, when the evaporation is proceeding at the maximum rate, is called the *first stage* or *constant-rate stage of drying*. Eventually, the surface approaches a low water content whose vapor pressure is very nearly equal to the atmospheric vapor pressure at the soil boundary. At this time, the surface layer cannot continue to provide water from storage, and the evaporation rate becomes limited by the rate of water movement to the soil surface. This is the *second stage* or *falling-rate stage of evaporation*, when the evaporation rate decreases continuously with time. It is very difficult to model the evaporation process during the first and second stages of drying, as it requires simultaneous evaluation of water and heat flow (Jury, 1973). However, if one is concerned primarily with average daily evaporation, the solution of the flow equation under isothermal conditions provides a reasonable first approximation to the evaporation rate. Evaporation losses from initially wetted soil as a function of time were modeled by Ritchie (1972) and in a field study by Shouse et al. (1982) using the following two-stage evaporation model.

- *Stage 1: Potential evaporative loss.* During the first stage, the soil surface is wet and the upward flow of water is assumed to be high enough to match the external rate, which is regulated by external conditions at a rate $E = E_p$, the potential loss rate. This lasts for a period of time t_c after irrigation or rainfall ceases. Gradually, gravity-driven drainage and water loss by evaporation deplete the surface layer, and it dries out to a point where regulation of subsequent loss of water shifts to the soil. This triggers the onset of the second stage of drying.
- *Stage 2: Soil-regulated evaporative loss.* In the second stage of evaporation, soil water flow theory predicts that if gravity is neglected, cumulative loss of water will be proportional to the square root of time. [See (4.12) with $\theta_0 < \theta_i$.] This is the model used by Black et al. (1969) and Ritchie (1972).

Thus, the two-stage evaporation model may be described mathematically as

$$E = E_p \qquad \text{if } 0 < t < t_c \tag{4.50}$$

$$E_{\text{cum}} = a(t - t_c)^{1/2} \qquad \text{if } t > t_c \tag{4.51}$$

where a is a constant. This model was field calibrated in the study of Shouse et al. (1982) on prewetted soil plots that were instrumented with neutron access tubes.

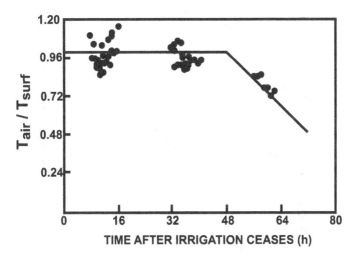

Figure 4.26   Ratio of measured air temperature to measured soil surface temperature as a function of time after irrigation ceased for two bare soil plots. (After Shouse et al., 1982.)

The authors monitored the soil surface and air temperatures simultaneously to determine the time t_c when the soil surface dried and the second stage of drying began. As shown in Fig. 4.26, the soil temperature of the bare plots increased significantly compared to the air temperature after 2 days of drying during summer in the sandy loam field where the experiment was conducted. The authors took the value $t_c = 2$ days to be a constant for the field and season. Idso et al. (1974) found that this time to drying increased significantly in other seasons, to as long as 2 weeks during winter.

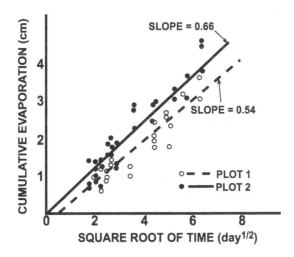

Figure 4.27   Cumulative evaporation loss on two bare soil plots during soil-limited phase of water loss. (After Shouse et al., 1982.)

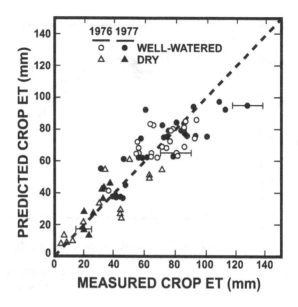

Figure 4.28 Predicted versus measured ET over a 2-week interval of potential and stress-limited water loss. (After Shouse et al., 1982.)

After the second stage of drying began, Shouse et al. (1982) determined the constant a in (4.51) by equating the cumulative evaporative loss to the measured storage change of two soil plots over time. A mean value of a was selected by averaging the replicates (Fig. 4.27).

Shouse et al. (1982) used the evaporation model thus calibrated as part of a model to predict evapotranspiration losses from a growing crop, with good agreement found between predicted and measured evapotranspiration losses over a season (Fig. 4.28).

PROBLEMS

4.1 Use the Green–Ampt model to calculate upward infiltration above a water table. What happens to the cumulative infiltration as $t \rightarrow \infty$? Discuss the possible differences between the behavior of infiltration in a Green–Ampt soil and a real soil.

4.2 A soil column of height L is saturated everywhere ($\theta = \theta_s$). At $t = 0$, the bottom is exposed to the air and drainage begins. If the soil has a hydraulic conductivity–water content relation that fits the model

$$K(\theta) = K_s \left(\frac{\theta}{\theta_s} \right)^N$$

and the column drains according to gravity flow, calculate the average water content and flux $J_w = -K(\theta)$ from the column as a function of time. What is the water content at $t = \infty$? Discuss the difference in long-time behavior between this model solution and the one derived in Problem 3.10.

4.3 The data in Table 4.5 were recorded by tensiometers and gamma-ray scanning in a 190-cm vertical soil column with steady upward flow from a water table and a steady evaporation rate of 0.5 cm day^{-1}. From the data given, construct and plot $K(h)$, $K(\theta)$, $D(\theta)$, $C(h)$, and $h(\theta)$.

TABLE 4.5   Data for Problem 4.3

Height (cm)	Water Content θ	Matric Potential Head h (cm)
180	0.11	−780
160	0.13	−480
140	0.15	−340
120	0.17	−240
100	0.20	−180
80	0.25	−125
60	0.30	−75
40	0.40	−45
20	0.48	−22
0	0.50	−0

4.4 A weighing lysimeter of surface area 4 m^2 and depth 1.5 m is cropped to wheat and under rainfall irrigation. The data in Table 4.6 are recorded by the lysimeter instruments, which record the weight as 0 kg at $t = 0$. Calculate the evapotranspiration rate as a function of time, and plot irrigation rate, drainage rate, and ET rate on the same graph in the same units of cm/day^{-1}. What is the leaching fraction of this system averaged over the 20-day period?

TABLE 4.6   Data for Problem 4.4

Time (days)	Cumulative Rainfall (cm)	Lysimeter Weight (kg)	Cumulative Drainage (m^3)
0	0	0	0.00
2	2	0	0.04
4	3	0	0.04
6	5	+40	0.04
8	7	+40	0.08
10	7	+10	0.08
12	8	+30	0.08
14	9	+50	0.08
16	14	+100	0.20
18	17	+80	0.32
20	18	+60	0.36

4.5   Assuming that the unsaturated hydraulic conductivity of a soil is given by the model function

$$K(\theta) = K_s \exp[\beta(\theta - \theta_s)]$$

where $K_s = 100$ cm day^{-1}, $\beta = 40$, and $\theta_s = 0.5$, calculate the long-time velocity of the infiltrating water front V_F under continuous ponding with zero head using (4.29) for initial water contents of $\theta_i = 0.0, 0.1, 0.2, 0.3, 0.4$, and 0.49. Assuming that this stage began at $t = 0$ (i.e., no capillary attraction), calculate the position of the front after 12 h, and calculate the amount of water that has entered the soil at the different initial water contents.

4.6   When water is infiltrated into the soil while holding the water potential and water content of the surface at a constant value, the infiltration rate changes as a function of time but the water content profile is relatively uniform above the front (see Fig. 3.17). Assuming that the soil is initially very dry, sketch plausible water content profiles for various times during constant-flux infiltration, and discuss the major differences between this case and the case of constant-water-content infiltration.

4.7   Calculate the minimum and maximum times to ponding under a rainfall rate of 2, 4, and 6 cm h^{-1} as in Example 4.3 for a soil whose maximum infiltration rate obeys the Horton model (4.5), with $i_0 = 10$ cm h^{-1}, $i_f = 1$ cm h^{-1}, and $\beta = 1.0$ h^{-1}. You will have to calculate the value of t_{max} numerically or graphically.

5 Soil Thermal Regime

Knowledge of soil temperature and heat flow is essential to a number of disciplines. Three mechanisms—radiation, convection, and conduction—are responsible for the transfer of heat in soil. Radiative energy transfer includes direct and diffuse short-wave solar radiation and long-wave sky radiation to the soil surface, and long-wave radiation emitted outward from the soil surface. Radiation transfer is generally important only at or near at the soil surface.

Convective heat transfer in soil is by definition the movement of heat associated with a net fluid flux. Constantz et al. (2003), Ren et al. (2000), Shao et al. (1998), and Wang et al. (2002b) have characterized the importance of convective heat transfer associated with liquid water flow in soil. As a general rule, convection due to water flow is significant only if the flow velocity is large (e.g., during rainfall, irrigation, seepage below streams, etc.) or if the fluid temperature differs significantly from the nearby soil. Westcott and Wierenga (1974) studied convective heat transfer associated with vapor flow in soil, which is insignificant unless the vapor temperature is quite high (e.g., steam injection). The most significant influence that water vapor has on heat flow is through absorption or release of heat energy during evaporation or condensation processes in the soil. Although convection may be important in specific circumstances, conduction of heat by molecular exchanges of kinetic energy is the mechanism most responsible for subsurface soil heat transfer and associated soil temperatures.

Soil temperature is one of the most critical factors that influence important physical, chemical, and biological processes in soil and plant science. Soil hydraulic properties are affected by soil temperature (Bachmann et al., 2002). Bacterial growth and plant production are both strongly temperature dependent, as are organic matter decomposition and mineralization. Other important microbiological rate processes, such as biodegradation of pesticides and other organic chemicals, also vary in their intensity with temperature. Many of these rate processes reach maximum levels at some particular range of temperatures and decrease both above and below that point.

Soil temperature affects plant growth first during seed germination. Although seeds of different plants vary in their ability to germinate at low temperatures, all species show a marked decrease in germination rate in soils with low surface temperatures (Russell, 1973). The germination rate will increase significantly with temperature up to a certain point, above which the rate falls off again. Since rapid seed germination ensures an early crop, the temperature of the soil at spring planting thus has a major influence on when the growth stages will occur.

Plant growth after germination is also influenced by soil temperature. Metabolically regulated plant processes, such as water and nutrient uptake, can be diminished below optimum rates at both low and high temperatures, resulting in temperature-dependent growth and yield patterns. For example, corn yields in Iowa were observed to increase almost linearly as a function of soil temperature at 100 cm between the range of 60 and 81.3°F (Allmaras et al., 1964). Above 81.3°F the yields decreased.

In this chapter we discuss the major factors influencing soil temperature and soil heat flux, beginning with a description of the radiant energy flux to the soil. Subsequently, heat transport equations will be derived to predict the distribution and changes of temperature within the soil as a function both of the external radiation striking the surface and of the principal soil thermal properties.

5.1 ATMOSPHERIC ENERGY BALANCE

5.1.1 Extraterrestrial Radiation

Stefan–Boltzmann Law   The source of all radiant energy for the Earth is the sun, which continually emits short-wave electromagnetic radiation as a consequence of its high temperature (≈ 5700 K at the surface). The general formula expressing the energy flux Σ (in W m^{-2}) from a body at a temperature T (in kelvin) is given by the *Stefan–Boltzmann law* (Halliday et al., 2001):

$$\Sigma = \epsilon \sigma T^4 \tag{5.1}$$

where σ is the Stefan–Boltzmann constant, which has the value 5.67×10^{-8} W m^{-2} K^{-4}, and ϵ is the emissivity, which equals 1 for a blackbody and has values in the range $0 \leq \epsilon \leq 1$ for other radiant surfaces.

Example 5.1   Calculate the radiant energy flux from the sun striking the outer edge of Earth's atmosphere assuming that the sun radiates as a blackbody ($\epsilon = 1$). Use the following data:

Distance from Earth to the sun: $X = 1.50 \times 10^{11}$ m
Radius of the sun: $R = 6.97 \times 10^8$ m
Surface temperature of the sun: $T = 5760$ K

SOLUTION: If we assume that the radiant energy from the sun expands radially outward isotropically, the same quantity of energy per unit time will pass through the surface of any sphere enclosing the sun at its center. The total quantity of energy Q emitted (in watts) is, using (5.1),

$$Q = \Sigma A = (\sigma T^4)(4\pi R^2) \tag{5.2}$$

since $A = 4\pi R^2$ for a sphere. At Earth's atmosphere, let the radius of the expanding sphere enclosing the sun at its center and having Earth at the surface of the sphere be X. The energy flux Σ_E (W m^{-2}) at Earth is therefore equal to, by (5.1) and (5.2),

$$\Sigma_E = \frac{Q}{4\pi X^2} = \frac{R^2 \Sigma}{X^2} = \frac{R^2 \sigma T^4}{X^2} \tag{5.3}$$

Plugging in the preceding values, we obtain

$$\Sigma_E = 1350 \text{ W m}^{-2} = 1.94 \text{ cal cm}^{-2} \text{ min}^{-1} \tag{5.4}$$

This value is known as the *solar constant*.

Energy–Wavelength Laws   The radiation that is emitted from the sun is not monochromatic; it has a range of wavelengths. The radiant flux density per unit wavelength B_λ for a blackbody is given by *Planck's radiation law*,

$$B_\lambda = \frac{2\pi c^2 h}{\lambda^5 [\exp(hc/\lambda kT) - 1]} \tag{5.5}$$

where λ is the wavelength of the radiation in meters, $h = 6.63 \times 10^{-34}$ J $\cdot$ s is Planck's constant, $c = 2.99 \times 10^8$ m s^{-1} is the speed of light in vacuum, and $k = 1.380 \times 10^{-23}$ J K^{-1} is Boltzmann's constant (Halliday et al., 2001). This distribution has a maximum at a wavelength λ_m that obeys the equation known as *Wien's law*:

$$\lambda_m T = 2.898 \times 10^{-3} \text{ m K} \tag{5.6}$$

Figure 5.1 shows a plot of the radiant flux density (5.5) integrated over a small wavelength interval along with actual measurements of radiation flux density taken with a spectrophotometer at the outer edge of Earth's atmosphere. Two values (5760 and 6090 K) were used in (5.5) for the surface temperature of the sun. The good agreement between the model and the measurement indicates that the sun may be treated to a first approximation as a blackbody. Note that less than half of the total energy flux is in the visible range from 0.4 to 0.76 μm. However, over 99% of the total radiation is contained in wavelengths of 0.3 to 4.0 μm (Chang, 1968). For this reason, solar radiation is known as *short-wave radiation*.

5.1.2 Solar Radiation

Numerous studies have been conducted in ecology, agriculture, soil science, hydrology, architecture, design, urban planning, solar engineering, and other disciplines to assess components of solar radiation. One of the challenges that early researchers faced in their radiation investigations was to determine the radiation actually reaching the soil surface.

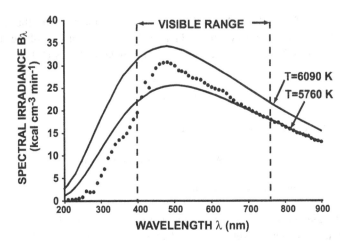

Figure 5.1 Measured (circles) and calculated (lines) results using values of energy flux from (5.5) at the outer edge of Earth's atmosphere. (After van Wijk, 1963.)

Interactions with Atmosphere Much of the short-wave solar radiation that reaches the outer edge of Earth's atmosphere is dissipated before it strikes the soil surface. Some of it is reflected back into space by clouds. A portion of it is absorbed by water vapor, oxygen, ozone, and carbon dioxide molecules in the atmosphere. Part of the radiation is scattered diffusely by molecules and particles in the air; some of this radiation will strike Earth's surface afterward as scattered, diffuse radiation. The remainder of the radiation reaches Earth's surface by passing directly through the atmosphere (Fig. 5.2). Thus, the fraction of the radiation from the sun that reaches Earth, the *global solar radiation* R_s, is the sum of direct and diffuse radiation.

Net Radiation The solar radiation transmitted through the atmosphere is partitioned further when it strikes the soil or crop surface (Fig. 5.3). A fraction of the solar radiation, the *albedo*, a, is reflected at the surface and returns to space. In addition, thermal radiation from the sky, R_{sky} (in the long-wavelength range), strikes the surface, and thermal radiation from the soil or canopy surface, R_{Earth}, radiates into space [see (5.1)]. The net downward radiation component is thus

$$R_n = (1 - a)R_s + R_{nl} \tag{5.7}$$

where R_{nl} is the net long-wave thermal radiation equal to

$$R_{nl} = R_{sky} - R_{Earth} \tag{5.8}$$

Each of the terms in (5.8) may be calculated with the Stefan–Boltzmann equation (5.1) using appropriate values for the emissivity ϵ. The atmospheric emissivity increases as vapor density increases. The soil surface emissivity usually ranges from 0.9 to 0.99 and increases as water content increases.

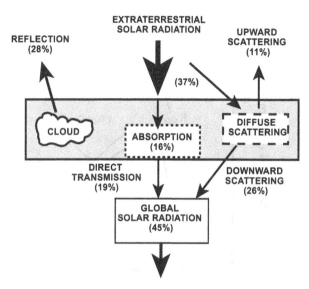

Figure 5.2 Typical partitioning of the extraterrestrial radiation as it passes through the atmosphere.

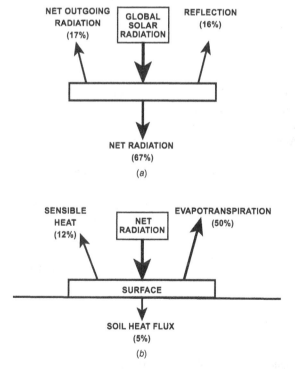

Figure 5.3 Typical partitioning of the global solar radiation as it reaches the land surface: (a) contributions to the net radiation; (b) net radiation partitioned into its components.

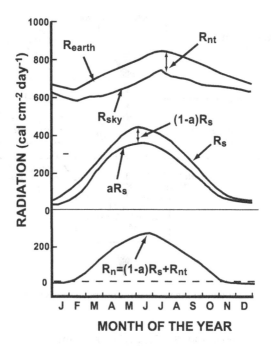

Figure 5.4 Radiation balances throughout the year at Hamburg, Germany. (After Fleischer according to Geiger, 1965.)

The size of the various terms in the radiation balance (5.7) throughout the year at Hamburg, Germany, is illustrated in Fig. 5.4. The net thermal radiation loss ranged from a low of about 6% in January to a high of about 14% of the extraterrestrial radiation in August. However, it is a dominant component of the net radiation in winter when solar radiation decreases and actually causes R_n to become negative. The albedo varies from a high of about 0.2 in June to a low near zero in December.

Example 5.2 Calculate the long-wave thermal radiation energy flux R_{Earth} from Earth assuming that $T = 300$ K and $\epsilon = 0.95$. Estimate the wavelength at which the maximum energy flux occurs.

SOLUTION: The energy flux density is calculated from the Stefan–Boltzmann equation (5.1):

$$R_{Earth} = \epsilon \sigma T^4 = 436 \text{ W m}^{-2} = 0.63 \text{ cal cm}^{-2} \text{ min}^{-l}$$

This is about 32% of the value of the solar constant. The maximum radiation wavelength is calculated from Wien's law (5.6):

$$\lambda_m = 9.66 \times 10^{-6} \text{ m} = 9.66 \times 10^4 \text{ Å}$$

This wavelength is in the infrared region of the electromagnetic spectrum. Thus, we do not see Earth glowing as it radiates.

The fraction of the solar radiation that remains as net radiation at the surface varies significantly with climate, latitude, and surface cover. Table 5.1 summarizes data for different climates describing the fraction of global radiation that is retained as net radiation (Chang, 1968). Since the global radiation decreases significantly during the winter months away from the equator, there is a greater seasonal variation in this ratio in northern and southern latitudes than in equatorial regions. In fact, as shown in Fig. 5.4, net radiation may even be negative in winter months.

5.1.3 Physical Factors Affecting Solar Radiation

Albedo   The fraction or percentage of incoming solar radiation that is reflected at the crop or soil surface is called the *albedo*. It is dependent on the nature of the surface, the angle of the sun, and latitude. The albedo increases significantly with distance from the equator, as shown in Fig. 5.5, where the percentage of incoming solar radiation that is reflected ranges from a low of 7% near the equator to a high of 56% near the north pole (Houghton, 1954). The albedo coefficients of various surfaces are given in Table 5.2. In general, water surfaces have a lower albedo than cropped or soil surfaces, reflecting less than 10% of the radiation striking them. Canopy surfaces have albedos that vary between 5 and 25%. The lower values are characteristic of forest canopies, and the higher values are more representative of nonequatorial crops at full ground cover. Crops at full cover in the tropics, however, have much lower albedos than at higher latitudes. The high position of the sun in the sky in tropical climates causes the radiation to strike the surface of the Earth normally, which causes less reflection than when the sunlight arrives at an angle (Chang, 1968). For this reason, the albedo is higher in the early morning and later afternoon because of the lower angle of the sun (Chang, 1961).

TABLE 5.1   Ratio of Net Radiation to Global Radiation during 24-h Periods

Location	Cover	R_s (cal cm^{-2} day^{-1})	R_n/R_s
Denmark (55° 44' N)	Grass	463 (July)	0.51
England (51° 48' N)	Grass	550	0.41
California (38° 30' N)	Grass	750 (July)	0.56
	Grass	175 (December)	0.33
Australia (38° 2' S)	Grass	181 (July)	0.18
	Grass	689 (January)	0.69
Hawaii (21° 18' N)	Sugarcane	725 (summer)	0.69
	Sugarcane	400 (winter)	0.65
	Pineapple	710 (May–June)	0.66
	Pineapple	500 (December)	0.53

Source: Chang (1968).

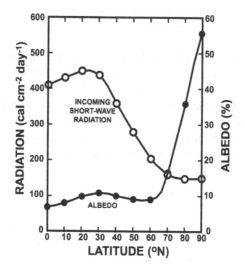

Figure 5.5   Radiation and albedo distributions in the northern hemisphere in relation to latitude. (After Houghton, 1954.)

The principal factors influencing the albedo of a bare ground surface are the soil type, surface roughness, and moisture level of the surface soil. For example, dry, light-colored desert soils have twice the albedo of dry, dark-colored soils (Table 5.2). Therefore, adding water to light soils, which darkens the surface, can decrease the solar reflection by up to 50%. Other alterations of soil surface color will also affect the albedo. Stanhill (1965) showed that application of white $MgCO_3$ to the surface of a bare soil doubled its albedo, causing a subsequent 10°C decrease in soil temperature.

TABLE 5.2   Albedo of Various Surfaces

Type of Surface	Albedo (%)
Fresh snow	75–95
Light sand dunes	30–60
Meadows and fields	12–30
Forests	5–20
Water surface	3–10
Dry soil	14–30
Wet soil	8–16
Corn crop	23.5
Sugarcane	5–18
Pineapple	5–8
Potatoes	15–25

Source: Data from Chang (1968), Chudnovski (1966), and Geiger (1965).

Net radiation is larger for a rough soil surface than for a smooth soil surface (Cary and Evans, 1975), due in part to a lower albedo (Idso et al., 1975). Multiple reflections can occur from a rough soil surface, providing several opportunities to absorb radiation.

Latitude   The angle at which the sun's rays meet Earth greatly influences the amount of radiation received per unit area for two reasons. First, the radiation approaching Earth at an angle moves through more of the atmosphere and is subjected to greater scattering, reflection, and adsorption than radiation moving directly through the atmosphere. Second, radiation striking Earth at an angle to the vertical has a higher albedo than radiation coming in normally. The amount of radiation per unit area reaching the Earth is proportional to the cosine of the angle made between the perpendicular to the surface and the direction of the radiation as it strikes the surface.

The radiation distribution in the northern hemisphere in relation to latitude is illustrated in Fig. 5.5 (Houghton, 1954). The global radiation decreases rapidly at latitudes above 30° and reaches a constant amount above 60°. The albedo is lowest (<10%) in tropical regions, increases slightly in the middle latitudes, and rises sharply in the polar areas. The relatively constant daily global radiation and the tremendous increase in the percentage of reflection at the higher latitudes are the result of the lower solar elevation. Thus, in the tropical latitudes solar radiation is high and albedo low, producing extreme warming conditions for all 12 months of the year.

Exposure   The effects of latitude may be simulated on a small scale within certain latitudes by changes in the direction of exposure and the degree of slope of the land. For example, the angle at which the rays of the sun strike a steep south slope is entirely different from that on a steep north slope; in the northern hemisphere the southern slope receives more solar radiation per unit area. These differences in exposure have great ecological and agricultural significance, inasmuch as in the northern hemisphere the temperature of the soil is always higher on south than on north-facing slopes. Wollney (1878) pioneered the study of the effect of exposure on the temperature of surface soils. He found that southern exposures were always several degrees warmer than north-facing slopes. Because this increased warmth was due entirely to the direct action of the sun's rays, greater temperature variations from night to day were observed on the southern exposure. He also found that the temperature differences between exposures increased as the slopes became steeper. The direction of exposure, however, was of greater significance than the degree of slope.

Swift (1976) published a generalized algorithm providing the daily total of solar radiation for a point on any sloping surface at any latitude. However, this algorithm did not take into account shadows cast by topographic features surrounding the point of interest. Horton et al. (1984b), in a study on soil temperature in a row crop with incomplete surface cover, developed an algorithm to track the shadow cast by plants of simplified (hexagonal) shape. Brinsfield et al. (1984) included cloud cover effects in their model to predict solar radiation on a horizontal surface. Benjamin et al. (1990) reported the effects of exposure due to ridge tillage on soil temperature. The German Association of Engineers (VDI, 1994) provided a compilation of equations to determine short-wave radiation from the orientation of Earth relative to the sun, the orientation of the point of interest on any slope, shadows caused by topographic features, and values of cloud cover and visibility.

Exposure has very little influence on the surface heat balance in the tropics because of the high elevation of the sun. It also has little influence in polar climates, because much of the radiation striking the surface is diffuse sky radiation, which reaches north and south slopes at the same intensity. Consequently, exposure is of greatest consequence in climates at intermediate latitudes, where the solar elevation is significant, and diffuse sky radiation is not the dominant form of solar energy.

Distribution of Land and Water   In general, island climates are less variable than continental climates. The presence of large bodies of water tends to stabilize the temperature because of the high specific heat of water, which is responsible for the absorption of large amounts of heat. In addition, the atmosphere over land masses surrounded by bodies of water is highly saturated with water vapor, which reduces the amount of radiant energy reaching Earth.

In contrast, continental climates can undergo greater seasonal extremes in temperature. For example, the central plains region and corn belt of the United States are characterized by hot summers and cold winters. Even the nights are warm in summer, because of the large amount of thermal radiation from Earth. The climate of the coastal regions bordering large land masses is influenced by both land and water. Characteristic ocean currents in the area can have a major influence on any land mass that comes in contact with them. The Gulf current along the coast of Great Britain and the Japanese current along the shores of the northwestern United States are examples of warm currents that significantly affect the climate of the land they touch.

Vegetation   Vegetation is an important factor governing soil temperature because of the insulating properties of plant cover. Bare soil is unprotected from the direct rays of the sun and becomes very warm during the hottest part of the day. When cold seasons arrive, exposed soil rapidly loses its heat to the atmosphere. On the other hand, a good vegetative cover intercepts a considerable portion of the sun's radiant energy, which prevents the soil beneath from becoming as warm as bare soil during the summer. By preventing heat loss at night and shielding the surface during the day, vegetative cover reduces the daily variations in soil temperature relative to those in bare soil. In winter, the vegetation acts as an insulator to reduce the rate of heat loss from the soil. For this reason, frost penetration is more rapid and the depth of freezing is greater in bare soils than in those under a vegetative cover.

The major ways in which vegetation affects the soil energy balance are by (1) altering the albedo of the surface (Table 5.2), (2) decreasing the depth of penetration of global radiation through the canopy (Chang et al., 1965), (3) increasing the removal of latent heat by evapotranspiration, and (4) decreasing the rate of heat loss from the soil by insulating the surface.

Special attention has been given to row crops. From the time of planting to the time of full canopy the surface condition can change from that of bare soil to full canopy cover. Before full canopy cover is attained, soil in the interrow can be exposed to direct sunlight while soil in the plant rows is shaded. Perpendicular to the crop rows the soil surface heat fluxes and temperatures can vary sharply (Horton et al., 1984a). Horton (1989) published a numerical model for predicting net radiation, energy partitioning, and soil temperatures in row crops that provide incomplete cover.

Mulches can affect the thermal regime of the soil in several ways. Light-colored plastic mulches transmit short-wave solar energy to the soil but prevent the loss of long-wave thermal radiation, thereby producing a greenhouse effect that warms the soil underneath. Mulches that have low thermal conductivities decrease conduction of heat into and out of the soil, causing mulch-covered soil to be cooler during the day and warmer during the night than bare soil. For this reason, mulches are often used to protect the soil from excessive cooling during the winter and to prevent the soil from warming up early in the spring.

Ham and Kluitenberg (1994) discussed how optical properties of various plastic mulches can affect their radiation balance and the temperature of underlying soil. A review of studies of the influence of crop residue mulches on surface radiation balance, surface energy balance, and soil temperature was published by Horton et al. (1996). Various numerical simulation models have been developed that calculate the effects of mulches on surface energy partitioning and soil temperature. These are based either on energy balance principles (Bristow et al.,1986; Chung and Horton, 1987; Hares and Novak, 1992) or empirical correlations (Gupta et al., 1981).

5.2   SOIL SURFACE ENERGY BALANCE

5.2.1   Energy Balance Equation

The net radiation (5.7) arriving at the soil or crop canopy surface will partition in a number of different ways, depending on the surface and atmospheric conditions. If we disregard lateral inputs of heat to the surface and neglect transient energy changes caused by heating or cooling the surface or canopy, we may write a one-dimensional steady-state heat energy balance at the surface as follows:

$$\text{net heat energy arriving at surface} = \text{net heat energy leaving surface} \qquad (5.9)$$

Components of Energy Balance   There are three major transport processes carrying heat away from the soil surface. The first, the *sensible* or *convective heat flux S*, represents the vertical transport of warm air from the surface zone to the atmosphere above. This occurs predominantly by turbulent convection (bulk flow) of air. The second, the *soil heat flux* J_H, represents the vertical transport of heat into the soil. The third process is the use of heat for evaporation with the subsequent transport of water vapor from the surface zone to the atmosphere. This is the *latent heat flux* $H_v \cdot ET$, where ET is the evapotranspiration (the water vapor flux to the atmosphere by evaporation and plant transpiration) and H_v is the latent heat of vaporization. The vapor transport away from the surface also occurs predominantly by convection. Thus, the steady-state heat balance equation is given by

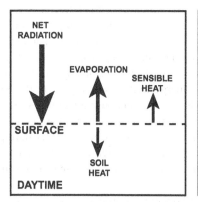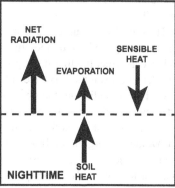

Figure 5.6   Components of the surface energy balance.

$$R_n = S + H_v \cdot \mathrm{ET} + J_H \qquad (5.10)$$

The terms on the right side of (5.10) are positive when they move away from the surface (Fig. 5.6).

At first glance it may seem strange to have a water vapor flux in the heat balance equation. However, it requires energy (585 cal g^{-1} at 20°C) to evaporate liquid water, and therefore the water vapor carries away a portion of the energy that arrives at the surface. Thus, the presence or absence of evapotranspiration at a surface can significantly influence the size of the other heat flow terms on the right side of (5.10). Table 5.3 represents two extremes in the partitioning of the components of (5.10) and clearly illustrates the importance of evapotranspiration on the partitioning of radiation.

Figure 5.6 and Table 5.3 apply to daily averages or to daylight periods when radiation is incident on the surface. The surface energy balance is quite different at night, when the net transfer of energy is usually away from the soil to the air (Fig. 5.6). Figure 5.7 shows diurnal patterns of energy balance terms for a no-till cornfield in central Iowa (Sauer et al., 1998). The data in Figure 5.7 were obtained in November when the soil and fresh residue layer were dry and the mean surface temperature was 6.8°C. Daytime J_H averaged 14.4% and 16.8% of R_n for days 310 and 311, respectively.

TABLE 5.3   Representative Relative Daytime Values of Components of Surface Energy Balance Equation (5.10) for Bare Dry Soil and Well-Watered Crop under Full Cover

Energy Component	Dry Bare Soil	Watered Full-Cover Crop
S/R_n	0.45	0.30
$H_v \cdot \mathrm{ET}/R_n$	~ 0	0.70
J_H/R_n	0.55	~ 0

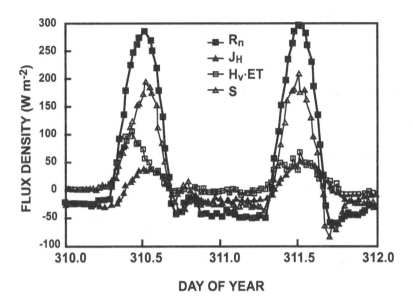

Figure 5.7 Diurnal patterns of energy balance terms for a no-till corn field in central Iowa. (After Sauer et al., 1998.)

5.2.2 Measurement of Evapotranspiration

There are numerous direct and indirect methods for measuring evapotranspiration from bare or cropped soil (Baker and Norman, 2002). One simple method for estimating evapotranspiration when the surface is wet and the rate of loss is limited by the external or potential conditions is the *Penman equation* (Penman, 1948). This equation is derived by combining the energy balance equation (5.10) with aerodynamic formulas describing the turbulent transport of water vapor and heat to the atmosphere above the canopy surface.

Aerodynamic Transport Equations   Water vapor and heat are carried away from the immediate vicinity of the soil or the canopy surface by turbulent bulk flow of parcels of air containing quantities of heat and vapor. These parcels are carried upward in a chaotic manner, exchanging their contents with other parcels, which in turn are carried away from the surface. The net result is a transport of heat and water vapor between two heights, z_0 (the soil or the canopy "surface") and an arbitrary elevation z_1. This turbulent flux is proportional to the vapor concentration or temperature difference between z_0 and z_1. Thus, the turbulent flux of sensible heat away from the surface may be written as

$$S = h_H(\rho_a C_a T_0 - \rho_a C_a T_1) = h_H \rho_a C_a (T_0 - T_1) \qquad (5.11)$$

where ρ_a is the density of the air, C_a the specific heat of the air, T the temperature, and h_H the turbulent transfer coefficient for heat between z_0 and z_1. The transfer coefficient depends primarily on wind speed and surface roughness characteristics.

The quantity $\rho_a C_a T$ is the heat concentration or heat content per unit volume. Similarly, the water vapor transport equation may be written as

$$ET = h_v \left(\rho_{v0}^* - \rho_{v1} \right) \tag{5.12}$$

where ρ_v is the water vapor density and h_v is the turbulent transfer coefficient for water vapor between z_0 and z_1. The asterisk denotes vapor saturation, since when using the Penman equation the surface is assumed to be moist. To describe evaporation that may be less than the potential evaporation, the actual water vapor density at the evaporating surface should be used in (5.12) rather than the saturated water vapor density as shown. The saturated vapor density as a function of surface temperature is described by Murray (1967) as follows:

$$\rho_v^* = 1.323 \left\{ \frac{\exp[17.27T/(237.3 + T)]}{T + 273.16} \right\} \tag{5.13}$$

The actual vapor density at a surface can be calculated using the equation

$$\rho_v = \rho_v^* \exp \left[\frac{h}{46.976(T + 273.16)} \right] \tag{5.14}$$

Equation (5.12) may be rewritten in terms of the vapor pressure P_v using the ideal gas law:

$$\rho_v = \frac{M_v P_v}{RT} = \rho_a \frac{M_v P_v}{M_a P_a} \tag{5.15}$$

where M_v is the molecular weight of water vapor, R the universal gas constant, M_a the molecular weight of air, and P_a the air pressure. Thus, (5.12) may be written in terms of the vapor pressure difference:

$$ET = \frac{\rho_a M_v h_v}{M_a P_a} \left(P_{v0}^* - P_{v1} \right) \tag{5.16}$$

Baker and Norman (2002) describe methods for measuring sensible and latent heat fluxes above bare and cropped surfaces. For the special case of sufficiently moist surfaces experiencing potential evaporation, we can use the Penman combination equation to calculate the latent heat flux.

Penman Combination Equation   Equations (5.10), (5.11), and (5.16) are the basis for the Penman equation, which is derived by making the following approximations:

1. The transfer coefficients for heat and water vapor are assumed to be equal:

$$h_v = h_H = h \tag{5.17}$$

This is the *similarity hypothesis*, which is reasonably accurate under most conditions at the surface.

2. The slope ζ of the function $P_v^*(T)$ is approximated by

$$\zeta = \frac{dP_v^*(T)}{dT} \approx \frac{P_{v0}^* - P_{v1}^*}{T_0 - T_1} \tag{5.18}$$

The first step in the derivation is to rewrite (5.16) as

$$\text{ET} = \frac{\rho_a M_v h_v}{M_a P_a}(P_{v0}^* - P_{v1}^* + P_{v1}^* - P_{v1}) \tag{5.19}$$

Equation (5.18) is then inserted into (5.19):

$$\text{ET} = \frac{\rho_a M_v h_v}{M_a P_a}[\zeta(T_0 - T_1) + \Delta P_{v1}] \tag{5.20}$$

where $\Delta P_{v1} = P_{v1}^* - P_{v1}$ is the vapor pressure saturation deficit at z_1. Equation (5.20) is then combined with (5.11) to produce

$$H_v \cdot \text{ET} = \frac{H_v \rho_a M_v h_v}{M_a P_a}\left(\frac{\zeta S}{\rho_a C_a h_H} + \Delta P_{v1}\right) \tag{5.21}$$

If we set $C_a P_a M_a / M_v H_v = \gamma$, called the *psychrometer constant*, and let $h_v = h_H$, (5.21) may be solved for the sensible heat flux S in terms of ET:

$$S = \frac{\gamma}{\zeta} H_v \cdot \text{ET} - \frac{\rho_a C_a h}{\zeta} \Delta P_{v1} \tag{5.22}$$

Finally, (5.22) is inserted into the energy balance equation (5.10). When this equation is solved for the latent heat flux, the result is

$$H_v \cdot \text{ET} = \frac{\zeta}{\zeta + \gamma}\left(R_n - J_H + \frac{\rho_a C_a h}{\zeta}\Delta P_{v1}\right) \tag{5.23}$$

Equation (5.23) is Penman's equation for potential water loss from a bare soil, crop, or canopy. Usually, the group of terms multiplying the saturation deficit is correlated with the wind speed,

$$\frac{\rho_a C_a h}{\zeta} = f(u) = a \cdot u + b \tag{5.24}$$

where a and b are constants and u is horizontal wind speed. Doorenbos and Pruitt (1976) discuss ways of selecting values of a and b to represent a given climate. They also provide tables for calculating net radiation from solar radiation, temperature, and other climate factors. Van Bavel (1966) used measurements of evapotranspiration from a weighing lysimeter to verify that the combination method for estimating potential evaporation was accurate.

5.3   HEAT FLOW IN SOIL

The amount of thermal energy that moves through an area of soil in a unit of time is the soil heat flux or heat flux density. The ability of a soil to conduct heat determines how fast its temperature changes during a day or between seasons. Soil heat flux is important in micrometeorology because it effectively couples energy transfer processes at the surface (surface energy balance) with energy transfer processes in the soil (soil thermal regime). This interaction between surface and subsurface energy transfer processes has led to detailed investigations of soil heat flux for a wide variety of agricultural and nonagricultural systems (Sauer and Horton, 2003).

The magnitude of J_H as a component of the surface energy balance varies with surface cover, soil moisture content, and solar irradiance. Daytime peak hourly values of J_H for a bare, dry soil in midsummer, can be in excess of 300 W m^{-2} (Fuchs and Hadas, 1972). By contrast, hourly J_H for a moist soil beneath a plant canopy, residue layer, or snow cover will often be less than ± 20 W m^{-2}. Surface soil heat flux typically represents 1 to 10% of R_n for growing crops. This percentage can exceed 50% in the fall and spring when R_n is low and the soil is cooling/warming, or in arid climates when there is no vegetation. Most soil heat flux density measurements have been completed using one of four methods (flux plate, calorimetric, gradient, or combination). Sauer (2002) and Sauer and Horton (2003) provide a comprehensive discussion of methods for measuring soil heat flux.

5.3.1   Heat Flux Equation

Heat energy may be transported through soil by a number of different mechanisms, including conduction, radiation, convection of heat by flowing liquid water, convection of heat by moving air, and convection of latent heat. The two most important processes of heat transport in soil under normal conditions are conduction and convection of latent heat. Conduction refers to the transport of heat by molecular collisions. For a pure solid substance, the conductive heat flux J_{Hc} in one dimension is described by Fourier's law as

$$J_{Hc} = \lambda \frac{dT}{dz} \qquad (5.25)$$

where T is temperature and λ is a constant called the *thermal conductivity*. This equation describes the heat flow in rigid bodies whose composition remains unchanged during the transport of heat. Convection of latent heat refers to the transport of the latent heat energy (energy per mass required for vaporization) in water vapor. Since any internal evaporation–condensation phase change in soil will liberate or consume heat energy, transport of water vapor from one location to another constitutes a transport of heat in a latent form. This is illustrated schematically in Fig. 5.8.

Water is evaporated at point A in the system by the addition of heat. The resulting vapor is moved laterally to point B by wind from a fan. At point B it condenses on hitting a cold wall, causing the liberation of heat. The net result is a lateral transport

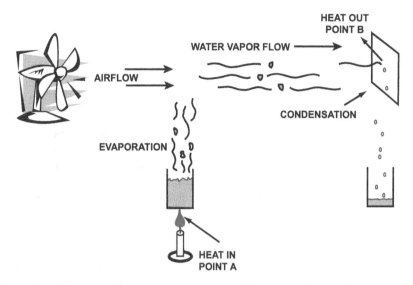

Figure 5.8   Hypothetical illustration of the transport of heat energy from A to B by convection of latent heat carried by water vapor.

of heat from A to B, traveling in latent form as the vapor moves. The equation for the latent heat convection J_{Hv} is simply

$$J_{Hv} = H_v \cdot J_v \qquad (5.26)$$

where J_v (g cm^{-2} s^{-1}) is the water vapor mass flux.

The expression for the net flux of soil heat may thus be written as

$$J_H = -\lambda^* \frac{dT}{dz} + H_v \cdot J_v \qquad (5.27)$$

where λ^* is the instantaneous value of the thermal conductivity of the moist porous medium.

Equation (5.27) is merely a formal expression because as pointed out by de Vries (1958), imposition of a temperature gradient on a moist soil sample will cause liquid water and water vapor to move as well as heat, thus changing the value of λ^*. Therefore, unlike in a rigid solid body, λ^* cannot be measured directly in soil. As shown in Chapter 6, the water vapor flux may be written approximately in one dimension as

$$J_v \sim -D_{Tv} \frac{dT}{dz} \qquad (5.28)$$

where D_{Tv} is the thermal vapor diffusivity. Equation (5.28) is accurate except in very dry soil, where the relative humidity drops significantly below unity. Inserting (5.27) into (5.28), we obtain

$$J_H = -(\lambda^* + D_{Tv}H_v)\frac{dT}{dz} \equiv -\lambda_e\frac{dT}{dz} \tag{5.29}$$

where λ_e is the effective thermal conductivity of the porous medium, including the effects of conduction and convection of latent heat. The parameter λ_e is measurable by any method that measures λ in a solid, provided that (5.27) and (5.28) are valid.

Example 5.3   A steady-state experiment is conducted on a soil sample to measure its thermal conductivity (Fig. 5.9). The sample, of thickness L, is placed between two glass plates of thickness d. The outside edge of each plate is connected to constant-temperature reservoirs. Thermocouples are attached to the inside edge of each plate. The data collected are noted in Table 5.4. Calculate λ_{eff} for the soil layer as a function of the measured temperatures and the known thermal conductivity λ_g of the glass plates.

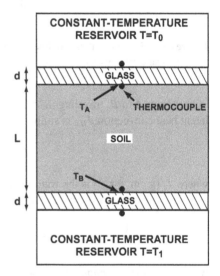

Figure 5.9   Device for the steady-state measurement of soil thermal conductivity λ_e.

TABLE 5.4   Data for Example 5.3

Position	Temperature
$z = 0$	T_0
$z = d$	T_A
$z = d + L$	T_B
$z = 2d + L$	T_1

SOLUTION: If lateral movement of heat is negligible, we may assume that the heat flux through each of the glass plates is equal to the heat flux through the soil. Thus, writing Fourier's law (5.25) through the lower plate, we obtain

$$J_H = -\lambda_g \frac{T_A - T_0}{d} \tag{5.30}$$

which may be calculated if λ_g is known and T_A, T_0, and d are measured. Once J_H is known, Fourier's law may be written across the soil layer between points A and B (assuming that the glass is in good contact with the soil):

$$J_H = -\lambda_e \frac{T_B - T_A}{L} \tag{5.31}$$

After equating (5.30) and (5.31), we solve for λ_e:

$$\lambda_e = \frac{L J_H}{T_B - T_A} = \lambda_g \frac{L}{d} \frac{T_A - T_0}{T_B - T_A} \tag{5.32}$$

Notice how similar the approach to problem solving is for heat flow and for saturated water flow (see Section 3.2). This is because Darcy's law (3.14) and Fourier's law (5.25) have the same mathematical form.

5.3.2 Heat Conservation Equation

The heat conservation equation is derived in exactly the same manner as the water conservation equation in Chapter 3, by conducting a heat energy balance on a small cubic soil volume (see Fig. 3.20). Assuming for simplicity that heat is flowing in the vertical z direction, we may write the differential form of the heat conservation equation as

$$\frac{\partial H}{\partial t} + \frac{\partial J_H}{\partial z} + r_H = 0 \tag{5.33}$$

where H is the heat concentration (heat energy per volume) and r_H is the rate of loss of heat per volume (the heat sink). Note the close correspondence between this equation and the water flow equation (3.67).

The heat sink term r_H should be included in the heat balance whenever a source or sink of heat (e.g., radioactive soil, chemical reactions) generates or consumes nonnegligible quantities of heat. Since we have already included the conversion of latent heat by evaporation in the heat flux term, it should not be placed in the equation as a sink term. Normally, we will assume that $r_H = 0$. The heat content per unit volume H may be written as

$$H = C_{\text{soil}} (T - T_{\text{ref}}) \tag{5.34}$$

where C_{soil} is the soil volumetric heat capacity and T_{ref} is an arbitrary reference temperature at which $H = 0$.

When the heat flux (5.29) and the heat content (5.34) are inserted into the heat conservation equation (5.33), we obtain (assuming that $C_{\text{soil}} = $ const)

$$C_{\text{soil}} \frac{\partial T}{\partial t} = \frac{\partial}{\partial z}\left(\lambda_e \frac{\partial T}{\partial z}\right) \tag{5.35}$$

If the z dependence of λ_e is neglected, (5.35) reduces to

$$\frac{\partial T}{\partial t} = K_T \frac{\partial^2 T}{\partial z^2} \tag{5.36}$$

where $K_T = \lambda_e/C_{\text{soil}}$ is the apparent soil thermal diffusivity, including the effects of conduction and convection of latent heat. Equation (5.36) is called the *heat flow equation*.

5.3.3   Thermal Properties of Soil

Heat Capacity   The volumetric heat capacity of a substance is defined as the quantity of heat required to raise a unit volume of the substance 1 degree of temperature. For a mixture of materials such as soil, the volumetric heat capacity of the composite material is the sum of the heat capacities of the constituents weighted by their volume fractions. Thus, we may express the soil heat capacity as

$$C_{\text{soil}} = X_a C_a + X_w C_w + \sum_{j=1}^{N} X_{sj} C_{sj} \tag{5.37}$$

where X refers to volume fraction, C to volumetric heat capacity, and the subscripts a, w, and sj to air, water, and solid constituent j (out of a total of N different solid materials in the soil). The volumetric heat capacity C of a pure substance may also be expressed as

$$C = \rho c \tag{5.38}$$

where ρ is the density of the substance and c is the specific heat, also called the *heat capacity per unit mass*.

Table 5.5 summarizes specific heat values for the common solid materials found in soil. A study of Table 5.5 reveals that the principal soil minerals differ little in their heat capacity values and that the specific heat of organic matter is higher than that of soil minerals. For this reason, de Vries (1963) recommended using average values of 0.46 and 0.60 cal cm^{-3} °C^{-1} for the volumetric heat capacity of soil minerals and organic matter, respectively, and the value 1.0 cal cm^{-3} °C^{-1} for water. The heat capacity of air is small and may be neglected. With these average values, (5.37) may be expressed as

$$C_{\text{soil}} = \theta + 0.46(\phi - X_o) + 0.6X_o \tag{5.39}$$

where X_o is the volume fraction of organic matter, ϕ the porosity, θ the volumetric water content, and $\phi - X_o$ the volume fraction of all the soil minerals. The soil

TABLE 5.5   Specific Heat (cal g^{-1} °C^{-1}) of Various Soil Constituents

Material	Specific Heat	Material	Specific Heat
Quartz sand	0.190–0.198	Fe$_2$O$_3$	0.163–0.165
Quartz powder	0.189–0.209	Humus	0.443–0.477
Kaolin	0.233–0.244	Calcareous sand	0.249
Feldspar	0.190–0.225	Humus calcareous sand	0.257
Mica	0.206–0.208	Garden soil	0.267
Apatite	0.183–0.220	Clay	0.270
Dolomite	0.222–0.230	Silty clay	0.260
Al$_2$O$_3$	0.217	Silt loam	0.164–0.194

Source: Data from Bowers and Hanks (1962), Kersten (1949), Lang (1878), and Ulrich (1894).

heat capacity can also be measured directly by calorimetry. Equation (5.39) has been shown to correspond well with values of C_{soil} measured directly by this method (de Vries, 1963). Kluitenberg (2002) reviews methods for measuring soil volumetric heat capacity.

Thermal Conductivity   Since soil is a granular medium consisting of solid, liquid, and gaseous phases, the thermal conductivity will depend on the volumetric proportions of these components, the size and arrangement of the solid particles, and the interfacial contact between the solid and liquid phases. The thermal conductivity of quartz is about 20.4 mcal cm^{-1} °C^{-1} s^{-1} (8.5 W m^{-1} K^{-1}), while that of feldspar is 7.0 mcal cm^{-1} °C^{-1} s^{-1} (1.67 W m^{-1} K^{-1}). The corresponding thermal conductivity values for water and air are 1.42 and 0.061 mcal cm^{-1} °C^{-1} s^{-1} (0.59 and 0.026 W m^{-1} K^{-1}), respectively. Thus, the ratio of the thermal conductivities for quartz, water, and air is 333:23:1. Because of this huge difference, it is obvious that the thermal conductivity of a granular soil will depend on the intimacy of the contact of the solid particles and the extent to which air is displaced by water in the pore spaces between the particles.

The relative size of the thermal conductivity of different soils follows the order sand > loam > clay > peat. Since the thermal conductivities of the mineral components of the solid phase are all of the same order of magnitude (Smith and Byers, 1938), thermal conductivity differences between soils at the same dryness are related to their packing and porosity. Thermal conductivity diminishes with decreasing particle size due to reduced surface contact between adjacent particles. Patten (1909) observed that the thermal conductivity of carborundum was lowered about 70% as the size of particles decreased from 450 nm to 6 nm. He also found that the conductivity of quartz particles was only 4 to 7% of the value of a solid quartz block.

Increasing the bulk density of soils lowers the porosity and improves the thermal contact between the solid particles as well as reduces the volume of low-conducting air. The impact of soil porosity on thermal conductivity is shown in Figs. 5.10 and 5.11. Van Rooyen and Winterkorn (1959) evaluated the thermal conductivity measurements reported in Russian investigations on a chernozem soil (Fig. 5.10). As the

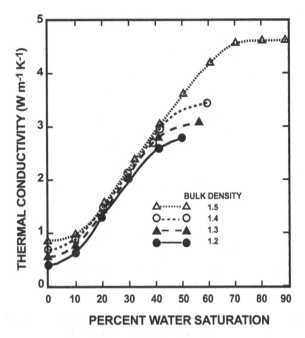

Figure 5.10   Soil thermal conductivity–water content relation as affected by bulk density. (Adapted from data of van Rooyen and Winterkorn, 1959.)

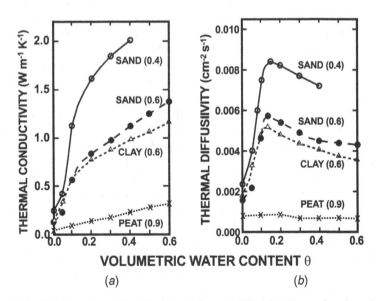

Figure 5.11   Soil thermal conductivity (*a*) and thermal diffusivity (*b*) as a function of water content for various soil types. Numbers refer to porosity. (After van Duin, 1963.)

bulk density of the dry soil increased from 1.1 to 1.5, the thermal conductivity increased from 0.42 W m^{-1} K^{-1} to 0.85 W m^{-1} K^{-1}. The data of van Duin (1963) in Fig. 5.11a showed a similar trend, with a 50% decrease in the porosity of sand, causing a doubling of the thermal conductivity. Similarly, van Duin' s data in Fig. 5.11b show that a 50% increase in porosity caused a large decrease in the thermal diffusivity λ_e / C_{soil} of sand over the entire range of moisture content.

The increase in thermal conductivity as a result of raising the soil bulk density within normal ranges is small compared with the impact of adding water to the soil. The presence of water films at the points of contact between particles not only improves the thermal contact between adjacent solids but also replaces air in the soil pore space with water, which has over 20 times the thermal conductivity of air. The curves in Figs. 5.10 and 5.11 point out quite clearly the rapid increases in thermal conductivity and diffusivity as the percentage of water in the soil pore space rises. The greatest rate of increase in conductivity occurs at the lower moisture contents. If the thermal conductivity of the dry soil (Fig. 5.11) with a bulk density of 1.1 g cm^{-3} is taken as the reference value, the dry soil thermal conductivities at bulk densities of 1.2, 1.3, and 1.5 increase by a factor of 1.3, 1.7, and 2.1, respectively, with respect to the reference value. In contrast, when the pore spaces are 25% filled with water, the thermal conductivities increase by a factor of greater than 4 relative to the reference value. When the water saturation values rise to 50%, the ratios range from 6.7 to 8.6, showing that the effect of bulk density on thermal conductivity is greater at higher moisture contents. The same relationships hold in the curves of van Duin in Fig. 5.11.

To visualize the impact of water on thermal conductivity, consider two dry quartz spheres that are in contact with each other. Most of the heat conduction takes place through a relatively small cross-sectional area at the point of contact. However, with the addition of a small amount of water in the wedges at the point of contact, the surface through which heat is conducted increases greatly because the water flow pathway is small and water conducts heat far more readily than air does. This causes the rapid rise of thermal conductivity at low water contents. As more water is added, the films become thicker, and the effect is less pronounced per unit of water added; thus, the rate of rise of thermal conductivity is smaller in this region.

Although most scientists have described soil thermal conductivity as a function of water content and bulk density, Ochsner et al. (2001) reported that the thermal conductivity of several medium-textured soils was related linearly to volumetric air content a (Fig. 5.12). Notably, the measurements show that each of the four soils in Fig. 5.12 can be accurately described as a decreasing linear function of a ($r^2 = 0.93$). The consistent decrease is logical since the thermal conductivity of air is more than one order of magnitude less than the thermal conductivity of water and two orders of magnitude less than the thermal conductivity of soil minerals. It is worth noting that the thermal conductivity is influenced by both pure conduction and by latent heat transfer across soil pores caused by water vapor movement. An increase in air-filled porosity would be expected to increase the latent heat transfer within the soil, assuming that the relative humidity in the pore space does not change. The data suggest that the influence of the increase in latent heat transfer is overwhelmed by the sharp decrease in pure conduction as a increases.

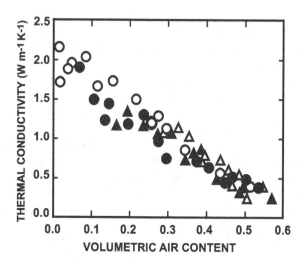

Figure 5.12   Thermal conductivity versus volumetric air content a for four medium-textured soils. (After Ochsner et al., 2001.)

Two common models used to estimate soil thermal conductivity are presented in Campbell (1985) and de Vries (1963). Bristow (2002) provides a detailed description of the use of the models.

Thermal Diffusivity   Thermal diffusivity (K_T) (Fig. 5.11b) increases rapidly with increasing water content to a maximum and then decreases with further increases in water content (Patten, 1909). This is explained by the fact that heat capacity rises linearly with water content, whereas thermal conductivity experiences its most rapid rise at low moisture contents, causing the ratio $\lambda / C_{\text{soil}}$ to have an internal maximum as a function of θ. Despite the change in K_T with θ in the dry range, the thermal diffusivity is relatively constant over a large range of water content. This means that the heat flow equation (5.38) may be treated as though it were linear as a first approximation. This can greatly simplify calculations, as shown in Section 5.3.4.

Determination of Thermal Properties   The steady-state method for determining the thermal conductivity of moist soils (as illustrated by Example 5.3) has two major weaknesses. First, water will redistribute under the influence of a steady-state temperature gradient (Jury and Miller, 1974), creating a nonuniform profile within the column. Second, this method is strictly a laboratory technique and cannot be used in situ. The transient-state cylindrical probe (Bristow, 2002; de Vries and Peck, 1968; Jackson and Taylor, 1965) overcomes these difficulties, although there is some temperature-induced moisture flow. The method consists of a thin metal wire that is heated electrically to serve as the heat source and a thermocouple to measure the temperature rise. These are placed inside a cylindrical tube, which is inserted into the soil. When the wire is connected to a constant power supply, the wire heats up, causing heat to flow radially. The temperature of the thermocouple probe in contact with the soil is given by the equation

$$T - T_0 = \frac{q}{4\pi\lambda}[d + \ln(t - t_c)]$$ (5.40)

where T_0 is the initial temperature, $T - T_0$ the temperature rise, q the power dissipation or heat flow rate per unit length of wire, d a constant, and t_c a correction constant that depends on the dimensions of the probe as well as the thermal properties of both the probe and the soil (Bristow, 2002).

Equation (5.40) is obtained by solving the heat flow equation in cylindrical coordinates for the appropriate initial and boundary conditions (Carslaw and Jaeger, 1959). If $T - T_0$ is plotted against $\ln t$, a straight line is obtained for times $t > t_c$. The thermal conductivity is then calculated by the equation

$$\lambda = \frac{q}{4\pi S}$$ (5.41)

where S is the measured slope of T versus $\ln t$. The value of the power dissipation per unit length $q = i^2\zeta$ is calculated from the current i applied to the wire and ζ, the resistance per unit length of the wire.

The heat pulse probe (Fig. 5.13) has been developed to determine soil thermal properties. The heat pulse method is based on the theory of radial heat conduction of a short-duration heat pulse from a line heat source. In an infinite medium, temperature change as a function of time (Fig. 5.14) at a radial distance from the heat pulse source is given by (Carslaw and Jaeger, 1959)

$$\Delta T(r, t) = \frac{Q}{4\pi K_T}\left\{ \mathrm{Ei}\left[\frac{r^2}{4K_T(t - t_0)}\right] - \mathrm{Ei}\left(\frac{r^2}{4K_T t}\right)\right\}$$ (5.42)

where T (°C) is the temperature change at the sensing needle, t the time (s), t_0 the heat pulse length (s), r the radial distance (m) of the sensing needle, and $-\mathrm{Ei}(-x)$ the exponential integral (Abramowitz and Stegun, 1970). The source strength is defined as $Q = q/C$, where q is the quantity of heat liberated (W m^{-1}). Based on (5.42), soil thermal properties can be expressed analytically as (Bristow et al., 1994; Kluitenberg et al., 1993)

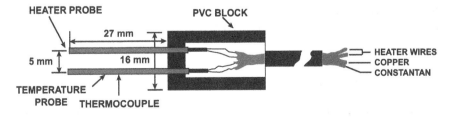

Figure 5.13   Heat pulse probe. (Courtesy of G. J. Kluitenberg.)

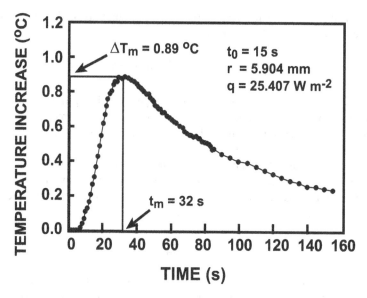

Figure 5.14   Soil temperature increase as a function of time after the input of a heat pulse into a silt loam soil with bulk density of 1.38 g cm^{-3} and a water content of 0.063.

$$K_T = \frac{r^2}{4} \frac{1/(t_m - t_0) - 1/t_m}{\ln [t_m/(t_m - t_0)]} \tag{5.43}$$

$$C = \frac{q}{4\pi K_T \Delta T_m} \left\{ \text{Ei} \left[\frac{r^2}{4 K_T (t - t_0)} \right] - \text{Ei} \left(\frac{r^2}{4 K_T t} \right) \right\} \tag{5.44}$$

$$\lambda = C K_T \tag{5.45}$$

where t_m is the time (s) at which the temperature maximum occurred and T_m (°C) is the maximum temperature change.

When making thermal property measurements, the heat pulse is generated by applying constant current to a heater needle from a power supply. A datalogger controls the heat input and records the heating power and temperature changes at the sensing needle. Once the $\Delta T(r, t)$ data are available, K_T, C, and λ are calculated from (5.43)–(5.45) using the single-point method (Bristow et al., 1994) or by means of nonlinear curve fitting of (5.42) to the data measured (Bristow et al., 1995; Welch et al., 1996).

5.3.4   Applications of the Heat Flow Equation

Steady-State Heat Flow Problems   The steady-state heat flow equation (5.31) is formally identical to Darcy's law [(3.13) and (3.15)] for saturated water flow. In heat flow, the driving force is a temperature gradient instead of a hydraulic head gradient, and the coefficient of proportionality between driving force and flux is the thermal conductivity rather than the hydraulic conductivity. Thus, all of the principles developed for solving flow problems in saturated soil are applicable to steady-state heat flow problems as well. This is illustrated in the next example.

Example 5.4   A soil column contains 50 cm of dry quartz sand ($\lambda_S = 0.5$ mcal cm^{-1} s^{-1} °C^{-1}) over 25 cm of dry loam ($\lambda_L = 0.25$ mcal cm^{-1} s^{-1} °C^{-1}). The top of the column is held at $T = 30$°C and the bottom at $T = 5$°C. Calculate the steady-state heat flux through the two layers and the temperature at the sand–loam interface.

SOLUTION: The first step is to find the equivalent thermal conductivity λ_{eq} of the two layers. By analogy with (3.26), we may write

$$\frac{L_1 + L_2}{\lambda_{eq}} = \frac{L_1}{\lambda_S} + \frac{L_2}{\lambda_L}$$

$$\frac{75}{\lambda_{eq}} = \frac{50}{0.5} + \frac{25}{0.25} = 200 \rightarrow \lambda_{eq} = 0.375 \text{ mcal cm}^{-1} \text{ s}^{-1} \text{ °C}^{-1}$$

Now that the equivalent conductivity has been calculated, the heat flux equation may be written across the entire column using Fourier's law (5.31) expressed between $z_1 = 0$ and $z_2 = L_1 + L_2 = 75$ cm:

$$J_H = -\lambda_{eq}\frac{T_2 - T_1}{z_2 - z_1} = -0.125 \text{ mcal cm}^{-2} \text{ s}^{-1}$$

Since the heat flux is known, we can now write Fourier's law across the lower half of the column between $z_1 = 0$ and $z_2 = L_2 = 25$ cm:

$$J_H = -0.125 = \lambda_L\frac{T - 5}{25} \rightarrow T = 17.5\text{°C}$$

Annual Temperature Changes in Soil   In the field, temperature is constantly changing, so that steady-state models are rarely useful near the surface. Instead, the time-dependent heat flow equation (5.36) must be used to solve for soil temperature. There are a few cases where the time-dependent equation is very easy to use. Figure 5.15 is a graph of the long-term monthly average soil temperature near the surface measured at a research station in Davis, California. It is clear from Fig. 5.15 that we could represent this graph very well by an equation of the form

$$T(t) = T_A + A\sin(\omega t + \phi) \tag{5.46}$$

where T_A is the annual average temperature, A the amplitude of the surface fluctuations, ϕ a phase constant, and $\omega = 2\pi/\tau$ is the angular frequency, where τ is the period of the wave. Equation (5.46) expresses one boundary condition for the heat

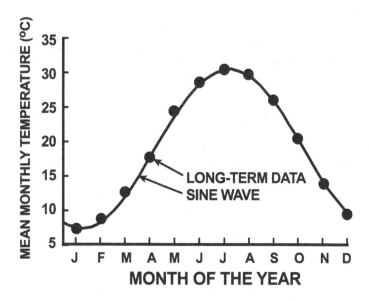

Figure 5.15   Long-term average monthly soil temperatures at 10-cm depth measured at Davis, California. (Data from NOAA, 1986.)

flow equation (5.36). We must also specify what happens deep in the soil as $z \rightarrow -\infty$ in order to solve (5.36) for the temperature. Since the source of heat at the surface is periodic, it is reasonable to assume that far below the surface the soil will stay at the average temperature,

$$\lim_{z \rightarrow -\infty} T(z, t) = T_A \tag{5.47}$$

The solution of (5.36) with the boundary conditions (5.46)–(5.47) may be obtained by several techniques (such as Laplace transforms). However, this is a well-known problem whose solution has been published (Carslaw and Jaeger, 1959):

$$T(z, t) = T_A + A \exp\left(\frac{z}{d}\right) \sin\left(\omega t + \phi + \frac{z}{d}\right) \tag{5.48}$$

where

$$d = \sqrt{\frac{2K_T}{\omega}} = \sqrt{\frac{K_T \tau}{\pi}} \tag{5.49}$$

Equation (5.48) is a sine wave whose amplitude decreases with depth and is phase shifted by an amount that increases with depth (Fig. 5.16). The constant d given in (5.49) is called the *damping depth*. It depends on the apparent thermal diffusivity, K_T, of the soil as well as on the angular frequency ω of the surface temperature. Inspection of (5.48) and Fig. 5.16 reveals several properties of the solution.

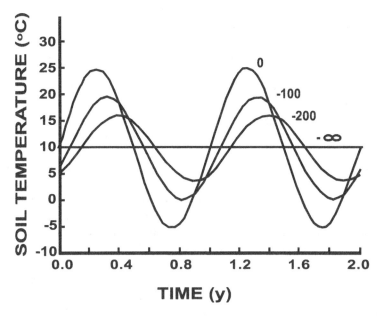

Figure 5.16 Graph of the temperature at the surface and three depths as a function of time using (5.48).

1. The temperature at any depth is a periodic sine wave with amplitude $A \exp (z/d)$.
2. The phase of the wave is retarded with respect to the surface by a time lag $\Delta t = z/\omega d$.
3. All depths have the same period $\tau = 2\pi/\omega$.
4. All depths have the same annual average temperature T_A.

We can insert some realistic values for the soil thermal properties to estimate the size of the predicted phase lag and amplitude attenuation. Table 5.6 shows a

TABLE 5.6   Thermal Profiles for Plainfield Sand

z (cm)	Amplitude (°C)	Time Shift (days)	T_{min}
0	17.0	0	−10.0
−50	13.6	13	−6.4
−100	10.9	26	−3.9
−200	6.8	52	0.2
−300	4.6	78	2.4
−500	1.8	131	5.2
−1000	0.2	262	6.8

calculation of the amplitude and phase lags at several depths for Plainfield sand at Wisconsin, assuming that the soil surface amplitude and average temperature are equal to $A = 17°C$ and $T_A = 7°C$, respectively. Values used for Plainfield sand are $\lambda = 2$ mcal cm^{-1} s^{-1} °C^{-1}, $C = 0.4$ cal cm^{-3} °C^{-1}, $K_T = 430$ cm^2 day^{-1}, and $d = 224$ cm (Jury, 1973).

One striking feature of the data in Table 5.6 is the large phase lag at deeper depths. If the warmest temperatures at the surface occur in July, they will occur in September at 200 cm and in December at 500 cm. Further, we notice that below 200 cm the minimum temperature is above freezing, which is useful to know if we have to bury a water pipe, for example.

Notice that the damping depth completely characterizes the penetration of the thermal wave. Suppose that we were sending down waves with a period of 1 day instead of 1 year. (We have to be more careful here because the daily temperature wave is not as well described by a sine wave, and it would vary more from day to day.) Then $\omega = 2\pi/(1)$ (day^{-1}) and the daily damping depth $d = 11.8$ cm. This means that the wave will damp out before getting very deep at all. For example, a wave of amplitude A at the surface will drop to $0.08A$ at $z = 30$ cm. This effect may be summarized in the following general principle: Rapidly changing surface temperatures will not penetrate as deeply into the soil as slowly changing temperatures of the same amplitude.

Not all climates have soil temperature that varies in a sinusoidal manner. Also, soil surface temperature rather than air temperature should be used as the upper boundary condition because the soil temperature is determined to a great extent by evaporation and soil cover and may differ considerably from the air temperature above it. The sine-wave model provides a means of determining the apparent thermal diffusivity of the soil (Horton, 2002). The procedure is illustrated in the next two examples.

Example 5.5 Two max–min thermometers are buried in the soil for one year and then excavated. The recorded temperatures (°C) are given in Table 5.7. Calculate the apparent thermal diffusivity and damping depth for this soil.

TABLE 5.7   Data for Example 5.5

z (cm)	T_{max}	T_{min}	$T_{max} - T_{min}$
-100	25.0	7.0	18
-200	21.5	10.5	11

SOLUTION: Since the sine function in (5.48) has maximum and minimum values of ± 1, respectively, we may write equations for T_{max} and T_{min} at any z:

$$T_{max} = T_A + A \exp\left(\frac{z}{d}\right) \tag{5.50}$$

$$T_{min} = T_A - A \exp\left(\frac{z}{d}\right) \tag{5.51}$$

Subtracting the second equation from the first, we obtain

$$\Delta T = T_{max} - T_{min} = 2A \exp\left(\frac{z}{d}\right) \tag{5.52}$$

Since the max–min temperature difference has been measured at two depths, both unknowns A and d may be solved for in (5.52). Thus, we have

$$\frac{\Delta T(z_1)}{\Delta T(z_2)} = \frac{A \exp(z_1/d)}{A \exp(z_2/d)} = \exp\left(\frac{z_1 - z_2}{d}\right) \tag{5.53}$$

After taking the natural logarithm of both sides of (5.53), we may solve for d:

$$d = \frac{z_1 - z_2}{\ln[\Delta T(z_1)/\Delta T(z_2)]} \tag{5.54}$$

Inserting the values in the table, we obtain $d = 203$ cm. Finally, solving for K_T in (5.49), we obtain

$$K_T = \frac{\pi d^2}{\tau} = 355 \text{ cm}^2 \text{ day}^{-1}$$

Example 5.6   A similar study is conducted in a different soil, this time using daily thermocouple temperature measurements at two locations to determine the time at which the maximum temperature occurs. From the data in Table 5.8, calculate d and K_T for this soil.

TABLE 5.8   Data for Example 5.6

z (cm)	Day When $T = T_{max}$
−100	August 1
−200	August 31

SOLUTION: The maximum value of the sine function occurs when its argument has the value $\pi/2$. Therefore, the equation describing the maximum value of the temperature reduces to

$$\omega t + \frac{z}{d} = \frac{\pi}{2} \tag{5.55}$$

Thus, the two data points in the table each satisfy (5.55). Subtracting one equation from the other, we obtain

$$\omega(t_2 - t_1) = \frac{z_2 - z_1}{d} \tag{5.56}$$

Solving (5.56) for d, we obtain

$$d = \frac{z_2 - z_1}{\omega(t_2 - t_1)} = \frac{\tau(z_2 - z_1)}{2\pi(t_2 - t_1)} \tag{5.57}$$

Inserting the values in the table, we obtain $d = 194$ cm, and from (5.49), we obtain $K_T = 324$ cm^2 day^{-1}.

5.3.5   Soil Temperature Observations

Diurnal Variations   Early investigations by Wollney (1883) and Bouyoucos (1913) studied the diurnal variations in soils as affected by the nature of the soil, type of surface cover, and incoming radiation. Figure 5.17 shows diurnal soil temperature at four depths in a loam during the summer. In the morning before sunrise, the minimum temperature of the soil was lowest at the surface and increased with depth. For example, at about 4:30 A.M., when the surface temperature was approximately 13°C, the temperature at 20 cm was 24°C. Thus, heat is leaving the soil at this time. Because of the time lag associated with soil heat flow when the surface temperature is changing, the lower depths continue to cool for a period of time. Also evident from Fig. 5.17 is the decreasing amplitude of the diurnal wave at lower depths, which is consistent with the damping depth concept discussed in Section 5.3.4.

Example 5.7   Although the sine-wave model of soil surface temperature does not quantitatively describe the shape of the daily soil surface temperature in most cases, the diurnal damping depth is still useful in analyzing the penetration of waves with a daily frequency. Assuming that the thermal diffusivity of the loam soil of Yakuwa has a constant value of about 4.5×10^{-3} cm^2 s-1 (see Fig. 5.11b), analyze and interpret the data in Fig. 5.17 using the damping depth concept.

SOLUTION: The daily wave has a period of 1 day. Using (5.49) and the value of $K_T = 4.5 \times 10^{-3}$ cm^2 s^{-1} = 389 cm^2 day^{-1}, we obtain a daily damping depth value $d = 11$ cm. Assuming that the mean temperature in Fig. 5.17 is 25°C, the surface amplitude is approximately 20°C. The amplitude predicted at 5 cm is $\exp(-5/d) = 0.63$ as large as at the surface, or 12.7°C. Thus, the maximum temperature at this depth should be about 38°C. Similarly, at 10 and 20 cm, the amplitudes should be 12.1 and 3.3, and the maximum temperatures should be 33 and 28°C, which compare reasonably well with the figure.

Horton et al. (1983) present a harmonic equation for describing surface and shallow soil temperatures as a function of time, and Horton and Wierenga (1983) present a method to estimate J_H that performs a harmonic analysis of soil temperature at one shallow depth if soil thermal diffusivity K_T is known or at two depths if K_T is unknown. A shallow soil temperature can be described with the following equation:

$$T(z \approx 0, t) = T_A + \sum_{N=1}^{M} A_N \sin(N\omega t + \phi_N) \tag{5.58}$$

where T_A is the average soil temperature, M the number of harmonics (usually, one to three harmonics are adequate), A_N and ϕ_N the amplitude and phase angle,

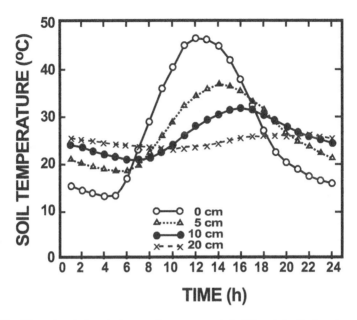

Figure 5.17 Diurnal variations in temperature measured at different depths in a loam soil. (After Yakuwa, 1945.)

respectively, of the Nth harmonic, and $\omega = 2\pi$ day^{-1} for a diurnal cycle. Fitting (5.58) to the observed shallow soil temperature values provides T_A, A_N, and ϕ_N.

Since the soil heat equation is linear, the solution to the heat flow equation for each of the terms of the summation in (5.58) may be added to produce the solution to the sum. Thus, using (5.48), we have

$$T(z,t) = T_A + \sum_{N=1}^{M} A_N \exp\left(\frac{z}{d_N}\right) \sin\left(N\omega t + \frac{z}{d_N} + \phi_N\right) \qquad (5.59)$$

where

$$d_N = \sqrt{\frac{2K_T}{N\omega}} = \sqrt{\frac{K_T \tau}{N\pi}} \qquad (5.60)$$

is the damping depth of the Nth harmonic with angular frequency $N\omega$. We may calculate the soil heat flux from (5.60) using (5.29):

$$J_H = -\lambda \sum_{N=1}^{M} \frac{A_N \sqrt{2}}{d_N} \exp\left(\frac{z}{d_N}\right) \sin\left(N\omega t + \frac{z}{d_N} + \phi_N + \frac{\pi}{4}\right) \qquad (5.61)$$

where we have used the trigonometric identity

$$\cos x + \sin x = \sqrt{2} \sin \left(x + \frac{\pi}{4} \right)$$

Equation (5.61) represents the soil heat flux when the temperature at the surface is described by the Fourier series (5.58). To calculate J_H with (5.61), one has to know the values for A_N and ϕ_N for the temperature at one depth, as well as K_T and λ (or C_{soil}) for the soil. Horton (2002) describes how K_T can be determined from measurements of soil temperature at two depths. Soil heat flux estimated with this harmonic method was in good agreement with J_H measured independently for a clay loam soil in New Mexico (Horton and Wierenga, 1983).

Annual Variations   The seasonal variations in soil temperature with depth are similar in character to the diurnal changes. The summer months (June and July in the northern hemisphere), like midday, represent the peak of the global radiation and the maximum temperatures. The winter months have an effect similar to nocturnal daily temperatures. The California data of Smith (1932) in Fig. 5.18 are typical of the variations observed.

From the middle of May until the first of August, the soil from 15 to 240 cm was warmer than the air except for the 240-cm depth. The deeper layers were warmer during the winter months than the 15-cm depth; therefore, the heat flux was upward

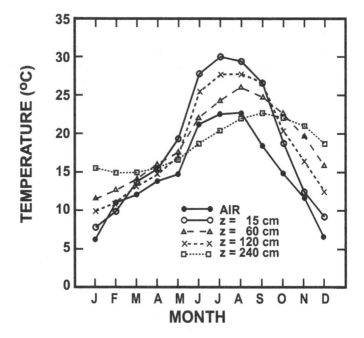

Figure 5.18   Monthly variation of soil temperature in relation to depth. (From data of Smith, 1932.)

during this period. The temperature gradient reversed from about May 1 to the middle of September, so that heat flow was downward during this period. Maximum temperatures at 15 cm and 60, 120, and 240 cm occurred approximately on July 1, July 15, August 1, and September 1, respectively. These seasonal differences were associated with the incoming global radiation and the thermal properties of the soil profile as related to changes in moisture content and temperature gradients.

Example 5.8 The curves in Fig. 5.18 for the soil temperatures yield the values shown in Table 5.9 for $\Delta T = T_{max} - T_{min}$ and $t^* =$ time of maximum temperature. Plot $\ln \Delta T$ and t^* versus z, and use linear regression to estimate d and K_T for this soil.

TABLE 5.9 Data for Example 5.8

Soil Depth (cm)	$T_{max} - T_{min}$ (°C)	t^* (July 1 = 0) (days)
15	22.2	0
60	17.8	14
120	14.2	30
240	7.8	61

SOLUTION: Figure 5.19 shows the plots and the linear regression lines, both of which describe the data well. From (5.52), $\ln \Delta T = \ln 2A + z/d$. Therefore, the inverse slope of the line in Fig. 5.19a is d, which is equal to 217 cm. Also, from (5.55), the time of maximum temperature obeys the equation

$$t^* = \frac{\pi}{2\omega} - \frac{z}{\omega d} = \frac{\tau}{4} - \frac{z\tau}{2\pi d} \tag{5.62}$$

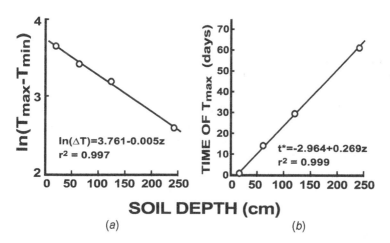

Figure 5.19 Graphical analysis of data in Fig. 5.18 using (a) (5.52) and (b) (5.55).

Therefore, the slope of the line in Fig. 5.19b is equal to $\tau/2\pi d$, from which we calculate $d = 216$ cm.

Obviously, the soil temperature model for the annual wave is very consistent with the data of Smith (1932) in Fig. 5.18.

5.3.6   Using Heat to Determine Soil Physical Properties

Water Fluxes   As we learned in Chapter 4, subsurface water flux is difficult to measure in natural soil. A novel strategy for measuring water flow is to determine it indirectly by the effect it has on temperature profiles under known conditions (Constantz et al., 2003). Heat pulse sensors have been proposed as promising tools for measuring subsurface water fluxes using this procedure. The heat pulse probe of Ren et al. (2000) consists of three 4-cm stainless-steel needles embedded in a waterproof epoxy body. The needles contain resistance heaters and thermocouples. The probes are connected to an external datalogger and power supply and then installed in soil. To measure the water flux, a 15-s heat pulse is generated at the middle needle using the power supply and the resistance heater, and the temperature increases at the needles 6 mm upstream and downstream from the heater are recorded using the thermocouples and datalogger. Wang et al. (2002b) present a simple relationship between water flux and the natural log of the ratio of the temperature increase downstream from the line heat source to the temperature increase upstream from the line heat source for the case of equidistant spacing from the sensors to the heater. The equation is

$$J_H = \frac{\lambda}{x_0 C_w} \ln \frac{T_d}{T_u} \tag{5.63}$$

where x_0 is the spacing from the heater to the sensing needles, T_d the downstream temperature increase, and T_u the upstream temperature increase. The simplicity of this relationship makes heat pulse sensors an attractive option for measuring subsurface water fluxes.

Bulk Density, Air-Filled Porosity, and Degree of Saturation   The thermo-TDR is a tool that enables measurements of thermal and electrical properties of the same volume of soil. The design and construction details of the thermo-TDR probe are given in Ren et al. (1999). Briefly, the probe consists of three parallel stainless steel rods (1.3 mm in diameter and 40 mm in length) with a 6-mm distance from the central rod to the outer rods. Each rod encloses a chromel–constantan thermocouple at the midpoint and a resistance heater made of enameled Evanohm wire. The heater wire and thermocouple are kept in place with high-thermal-conductivity epoxy glue, which also serves to provide a water-resistant and electrically insulated probe. A coaxial cable is connected to the probe by soldering the positive lead to the central rod and the shield to the two outer rods.

The determination of ρ_b using the thermo-TDR technique follows the theory that C_{soil} is the sum of the heat capacities of soil water, solids, and air, (5.37). The contribution of soil air to the soil heat capacity is often negligible and C_{soil} can be approximated as the sum of heat capacities of soil water and solids:

$$C_{soil} \approx \rho_b c_s + \rho_w c_w \theta \qquad (5.64)$$

where c_s is the specific heat of soil solids and c_w is the specific heat of water. The value of c_s can be obtained from the literature or determined experimentally (Ren et al., 2003). Once C_{soil} is determined with the heat-pulse part of the thermal-TDR and θ is measured using the TDR part of the thermo-TDR, ρ_b can be calculated from (5.64) (Ochsner et al., 2001).

PROBLEMS

5.1 Two thermocouples are buried in the soil at $z = -75$ and -150 cm. During the year they record maximum and minimum temperatures (°C) as follows:

z	T_{max}	T_{max}
−75	20.75	−0.75
−150	17.70	2.30

Calculate the damping depth for the annual wave, soil thermal diffusivity K_T, annual average temperature, and amplitude of the wave at the surface. What are T_{max} and T_{min} at $z = -225$ cm?

5.2 A temperature difference of 20°C (0 and 20°C) is established across a soil column containing oven-dry quartz sand of thermal conductivity $\lambda_Q = 20$ mcal cm^{-1} s^{-1} °C^{-1} and oven-dry silt $\lambda_S = 10$ mcal cm^{-1} s^{-1} °C^{-1}. Each layer is 50 cm thick. Calculate the soil heat flux in cal cm^{-2} s^{-1} and the temperature at the interface in the middle of the column (see Fig. 5.20).

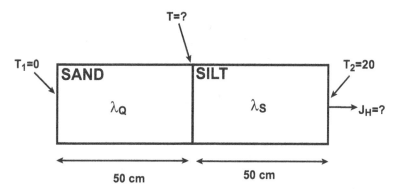

Figure 5.20 Experimental system described in Problem 5.2.

5.3 The climate in Palm Springs, California, is characterized by hot summers and mild winters. A reasonable function to use to represent the annual soil surface temperature is

$$T(t) = 25 + 15 \sin \omega t$$

where $\omega = 2\pi$ yr^{-1}. A person builds a subterranean home in the desert, with a layer of 2.5 m of soil over the roof extending back to the soil surface. The soil has a thermal diffusivity $K_T = 400$ cm^2 day$^{-1} = 146{,}000$ cm^2 yr^{-1}. Calculate the maximum and minimum annual temperatures of the roof of the house and determine how much later the maximum temperature will occur compared to the surface.

5.4 The thermal conductivity–water content data for a soil were fitted to the empirical function

$$\lambda(\theta) = a - b \exp(-c\theta)$$

and the following values were obtained:

$$a = 4.5 \text{ mcal cm}^{-1} \text{ s}^{-1} \text{ }^\circ\text{C}^{-1}$$
$$b = 4.0 \text{ mcal cm}^{-1} \text{ s}^{-1} \text{ }^\circ\text{C}^{-1}$$
$$c = 4.0$$

Use the following relation for soil volumetric heat capacity (in cal cm^{-3} $^\circ$C^{-1}),

$$C_{\text{soil}}(\theta) = 0.46(1 - \phi) + \theta$$

where ϕ is porosity. Calculate and plot $\lambda(\theta)$, $C_{\text{soil}}(\theta)$, and K_T between oven dry and saturation ($\theta_s = \phi = 0.5$). Calculate the average value of K_T between 0.1 and 0.4 approximately. What is the maximum percentage of deviation from the average value in this range?

5.5 An experiment is conducted in which λ_e is measured to be $\lambda_e = 3.0$ mcal cm^{-1} s^{-1} $^\circ$C^{-1} at $T = 20^\circ$C. From a second experiment, the water vapor mass flux J_v is fitted to the following model:

$$J_v = -0.015 \exp\left(\frac{T}{20}\right) \frac{dT}{dz} \text{g cm}^{-1} \text{ day}^{-1}$$

where T is in $^\circ$C. Assume that $H_v = 585$ cal g^{-1} is constant and λ^* is constant.
(a) Calculate λ^* from the information given.
(b) Calculate the fraction of the total λ_e attributable to latent heat flow at each temperature.

5.6 Assume that the soil described in Problem 5.5 is placed in a soil column of length $L = 50$ cm and that a temperature $T = 10^\circ$C is placed at $z = 0$ and

$T = 60°C$ is placed at $z = 50$ cm. Calculate (a) the steady-state heat flux and (b) the temperature $T(z)$ within the column. (*Hint:* λ_e is a function of T.)

5.7 The surface temperature of a soil with $K_T = 300$ cm^2 day^{-1} is measured as a function of time for 1 day when partial cloud cover causes changes to occur about every 30 min, as shown in Fig. 5.21. Sketch a reasonable plot of $T(t)$ at $z = 15$ cm. Quantitatively justify why you sketched it in the way you did.

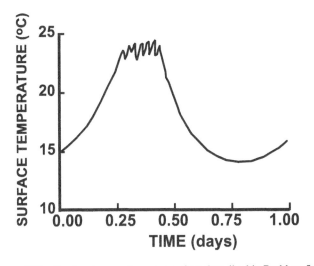

Figure 5.21   Surface temperature versus time described in Problem 5.7.

5.8 Two 20-cm-thick copper plates are used to sandwich a thin 5-cm Styrofoam plate. The left plate outer edge is held at $T = 5°C$ and the right plate outer edge at $T = 25°C$. Calculate the heat flux through the layers and the temperature distribution everywhere in the interior ($\lambda_{copper} \sim 1000$ mcal cm^{-1} s^{-1} °C^{-1}; $\lambda_{Styro} \sim 0.1$ mcal cm^{-1} s^{-1} °C^{-1}).

5.9 The Fourier series formula for the square-wave surface temperature may be expressed as

$$T(0, t) = \begin{cases} T_0 & 0 < t < \tau/2 \\ 0 & \tau/2 < t < \tau \end{cases}$$

$$T(0, t) = \begin{cases} T_0 & \tau < t < 3\tau/2 \\ T_0 & 3\tau/2 < t < 2\tau \end{cases} \quad \text{etc.}$$

where τ is the period of the square wave as given by

$$T(0, t) = \frac{T_0}{2} + \frac{2T_0}{\pi} \sum_{N=1}^{\infty} \frac{1}{2N - 1} \sin \frac{(2N - 1)2\pi t}{\tau}$$

Using the property that the sum of two solutions in the linear heat flow equation (5.36) is a solution of the equation, calculate the soil temperature $T(z, t)$ resulting from such a heat load at the surface. Plot the relative temperature $T(z, t)/T_0$ as a function of time at $z = 10$, 20, and 30 cm assuming that $K_T = 400$ cm^2 day^{-1} and $\tau = 1$ day. [*Hint:* Each term in the series has a different period $\tau_N = \tau/(2N - 1)$.]

6 Soil Aeration

Soil aeration refers to the transport of gases through the soil airspace and to the exchange of gases between the soil and the atmosphere. Two important gases in the soil air are carbon dioxide, which is produced as a by-product of plant root respiration and biological activity, and oxygen, which is consumed in soil by the same processes. CO_2 is a *greenhouse gas*, meaning that it traps outgoing long-wave radiation and prevents it from escaping into space. CO_2 concentrations in the atmosphere have increased significantly over the last century due to fossil fuel burning and are suspected to be at least partially responsible for the observed pattern of global warming (IPCC, 2001). Soils represent a major component of the global carbon cycle, and 10% of atmospheric carbon passes through soils annually (Raich and Tufekcioglu, 2000).

Plant roots require oxygen to function normally. For most plant species, translocation of oxygen from the leaves to the roots is inadequate to supply O_2 at the required rate, and the roots must supplement their supply from the soil air. As the soil O_2 is depleted, it must be replaced by O_2 moving from the atmosphere above the surface into the soil. This transport occurs primarily by gaseous diffusion.

Many soluble chemicals that are added to or accidentally released to the soil partition significantly into the vapor phase, which allows them to volatilize into the atmosphere. Some of these compounds have adverse health effects, and others are potentially hazardous to Earth's climate.

6.1 COMPOSITION OF SOIL AIR

The composition of the gases comprising the soil air depends on the rate of respiration of microorganisms and plant roots, on the solubility of CO_2 and O_2 in water, and on the rate of gaseous exchange with the atmosphere. In general, the CO_2 concentration in soil air will always exceed the small atmospheric value of 0.03%, because CO_2 is produced as a by-product of plant root respiration and microbial breakdown of carbon-based organic compounds in the soil. The extent of buildup of CO_2 gas depends not only on the intensity of these processes but also on the ease with which the gas can escape to the atmosphere. In poorly aerated soils with substantial respiratory and microbial activity, it is not uncommon to observe CO_2 concentrations that are hundreds of times higher than atmospheric levels (Russell, 1973).

The major processes that produce CO_2 in soil also decrease O_2 concentrations. Thus, it is common to observe complementary profiles of these two gases, with CO_2 rising and O_2 falling with distance below the soil surface. The concentrations of

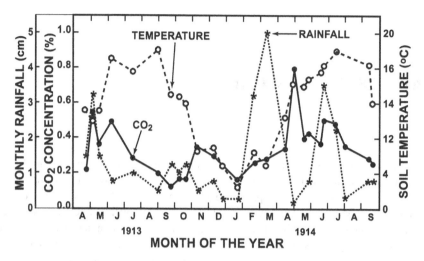

Figure 6.1   Seasonal variation in the CO_2 concentration and temperature in the soil at a 15-cm depth. (Data from Russell and Appleyard, 1915.)

both gases are likely to vary at a given location at different times of the year (Fig. 6.1). These seasonal differences reflect not only changes in respiratory and microbial activity but also varying diffusive resistance in the soil caused by differences in water content.

6.2   GAS REACTIONS IN SOIL

6.2.1 CO_2 Production in Soil

Production of CO_2 gas in soil will be highest when microbial and plant activity is at a maximum, which will depend on the crop and climate. The metabolic activity of microorganisms tends to increase with temperature up to a maximum and then decrease as temperature is raised further. In addition, CO_2 concentrations will build up when the soil is moist and transport of gas to the atmosphere is impeded. Figure 6.2 shows O_2 and CO_2 profiles over time for a well-drained sandy loam soil and a poorly drained silty clay from the same geographical area. The clay soil shows much more pronounced O_2 depletion and CO_2 buildup than the well-aerated soil (Russell, 1973).

Wide ranges of CO_2 evolution rates in soil have been observed under different conditions. Monteith et al. (1964) reported CO_2 fluxes of 1.5 g m^{-2} day^{-1} in winter and 6.7 g m^{-2} day^{-1} in summer for a clay soil without vegetation. Currie (1970) measured values of 1.2 g m^{-2} day^{-1} in winter and 16 g m^{-2} day^{-1} in summer in bare soil and corresponding values of 3.0 and 35 g m^{-2} day^{-1} in soil cropped with kale.

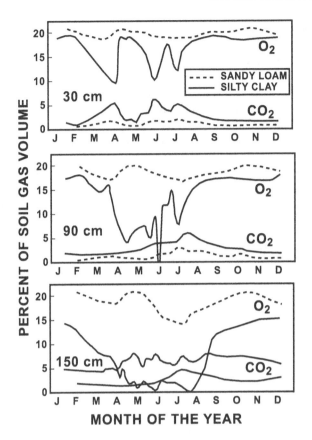

Figure 6.2 Oxygen and carbon dioxide content of the soil air at three depths in a sandy loam and a silty clay apple orchard. (After Russell, 1973. Reprinted with permission of Longman.)

6.2.2 O_2 Consumption in Soil

When the soil is aerobic, the volumes of CO_2 evolved and O_2 depleted tend to be comparable. However, O_2 depletion has frequently been found to be less than the commensurate CO_2 evolution, which implies that anaerobic respiration is occurring in part of the soil volume. Clark and Kemper (1967) reported mean values of O_2 depletion between 2.5 and 5.0 m^{-2} day^{-1} in bare soil and values approximately twice as high under cropped conditions. Greenwood (1971) measured a mean consumption rate of about 8 m^{-2} day^{-1} in a crop at full cover. Jacinthe et al. (2002) measured values of 2.79 g m^{-2} day^{-1} in winter and 2.45 g m^{-2} day^{-1} in summer from a wheat straw–amended Ohio soil. In the study by Currie (1970), the O_2 consumption rates were between 60 and 75% of the CO_2 production rates, reaching a maximum of 24 m^{-2} day^{-1} under cropped kale in the summer.

Even in bare soil, the rate of oxygen consumption by native microorganisms would rapidly remove all oxygen from the soil air if not replaced from the soil atmosphere. This is illustrated in Example 6.1.

Example 6.1   Assuming that the soil air is at atmospheric O_2 concentration, calculate how long it would take to deplete the O_2 level of the top 0.5 m of a soil that is at a volumetric air content $a = 0.2$ if the O_2 consumption rate by microorganisms is 5.0 g m^{-2} day^{-1}. Assume that no gas enters the soil from the atmosphere.

SOLUTION: The atmospheric O_2 concentration is approximately 300 g m^{-3}. The top 0.5 m of soil contains 0.1 m^3 m^{-3} of soil air. Thus, there is an oxygen supply of about 30 g m^{-2} in the top 0.5 m, which would last about 6 days if not replenished from above.

As Example 6.1 illustrates, biological activity in the soil requires frequent replenishment of oxygen from the atmosphere above the surface. Because it is evolved as oxygen is consumed, carbon dioxide gas similarly would reach very high levels in soil if the gas did not escape to the atmosphere. Romell (1922) estimated that the CO_2 concentration at 20 cm in the soil could double in 1.5 h and increase by a factor of 10 in 14 h just from bacterial production if gaseous exchange with the atmosphere were prevented.

6.3   GAS TRANSPORT THROUGH SOIL

6.3.1   Gas Conservation Equation

In this section the mathematical framework for describing gas transport through soil will be developed out of the more general framework for describing the transport and reaction of a particular chemical of interest. The starting point for the analysis, just as it was with water and heat flow, is a statement of conservation for the chemical of interest, which is subject to transport and transformation processes in the soil. To simplify the analysis, we will assume that the chemical is moving in the z direction only.

As we did with water flow in soil, we imagine that there is a small cubical volume of dimensions Δx, Δy, Δz located inside the soil (see Fig. 3.13). The chemical we wish to characterize is flowing into and out of the volume in the z direction, reacting in the soil, and changing its mass concentration within the soil volume. Thus, the differential statement of mass conservation for the chemical in the volume V is

$$\frac{\partial C_T}{\partial t} + \frac{\partial J_s}{\partial z} + r_s = 0 \tag{6.1}$$

where the mass of chemical per unit soil volume C_T is called the *total chemical concentration*, the flow of chemical per unit area per unit time J_s is called the *soil chemical flux*, and the loss of mass per unit soil volume per unit time r_s is called the *chemical reaction loss rate*. It is sometimes called a *sink term*. Notice the similarity of this equation to the water conservation and heat conservation equations (3.65) and (5.33), respectively. In this chapter, we adapt this general equation (6.1) to the case of an insoluble gas which is only present in the soil air phase. In this case, the total chemical concentration is equal to

$$C_T = aC_g \tag{6.2}$$

where a is the volumetric air content (volume of soil air per volume of soil) and C_g is the mass of chemical vapor per volume of soil air space. This is the concentration we measure when we remove a soil air sample.

Since the chemical is only present in the gas phase, we may write the chemical mass flux as

$$J_s = J_g \tag{6.3}$$

where J_g is the chemical vapor flux, or the mass of chemical moving in the vapor phase per unit area and per unit time.

Finally, we formally replace the chemical sink term with

$$r_s = r_g \tag{6.4}$$

where r_g is the rate of loss of chemical per volume occurring by reaction in the vapor phase. With these changes, (6.1) becomes

$$\frac{\partial(aC_g)}{\partial t} + \frac{\partial J_g}{\partial z} + r_g = 0 \tag{6.5}$$

Equation (6.5) is the *gas conservation equation* for an insoluble gas. Before (6.5) can be applied to solve gas transport problems, a gas flux law must be specified, relating J_g to the gas concentration C_g.

6.3.2 Gas Convection in Soil

Gases in soil can be transported by bulk movement of the soil air phase in response to differences in air pressure. The process is similar to the movement of water through soil in response to differences in liquid water pressure. Therefore, the process can be described by Darcy's law as follows:

$$J_c = -K_a \frac{dP}{dz} \tag{6.6}$$

where J_c is air flux density (m s^{-1}), K_a is air conductivity (m s^{-1}), P is air pressure in head units (m), and z is distance (m). The air conductivity is a function of soil pore structure, air content, air density, and air viscosity. Ball and Schjonning (2002) describe various laboratory and field methods for measuring K_a, many of which are formally similar to the measurement of hydraulic conductivity for the water phase. In

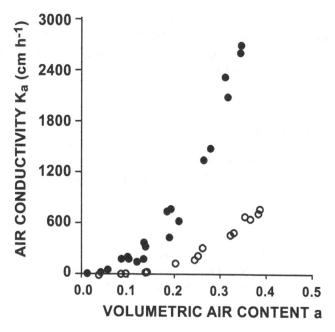

Figure 6.3 Measured air conductivity K_a as a function of volumetric air content a for Oso Flaco fine sand (solid circles) and Columbia fine sandy loam (open circles). (After Tuli and Hopmans, 2003.)

the laboratory, the most common procedure is to measure the airflow rate in response to a known imposed pressure difference across a column.

Figure 6.3 shows measured air conductivity as a function of air content for two soils. For a particular soil, the maximum value of air content is in dry soil, while a zero value occurs at or near water saturation. In Figure 6.3 the sand soil has larger air conductivity than the sandy loam soil, which is a consequence of its larger larger pathways for air to move through. Moldrup et al. (1998, 2001) discuss methods for predicting air conductivity, but these are subject to great uncertainty because of the complexity of the soil geometry and do not serve as an adequate substitute for measurement.

Within the airspace of the soil, natural pressure changes may be induced by soil temperature changes, barometric pressure fluctuations in the atmosphere above the surface, wind blowing over the soil surface, and infiltration. In addition to naturally occurring pressure changes, human-induced pressure changes are sometimes used to encourage the removal of volatile pollutants from soil. These effects are described in what follows.

Temperature Effects Temperature may influence the renewal of soil air in two ways. First, there may be temperature differences between various parts of the soil. It is possible that the contraction and expansion of the air within the pore spaces as well as the tendency for warm air to move upward may cause some exchange between soil regions with different temperatures and perhaps with the atmosphere. Second, the soil and the atmosphere usually have different temperatures. This temperature differential could cause an exchange of air volume between the atmosphere and the soil air in the immediate surface layer.

It is difficult to estimate the significance of temperature effects on gas exchange in soils. Romell (1922) suggested that the contribution to normal soil aeration caused by daily variations of temperature within the soil was less than 0.13% and that caused by temperature differences between soil and atmosphere could be responsible for no more than 0.5%. Thus, it appeared that temperature was a minor factor in soil aeration. However, more recent measurements highlight our current lack of understanding of the contributions of thermally induced convection to water vapor transfer out of soil. Cahill and Parlange (1998) and Parlange et al. (1998) reported that diffusion-based theory was inadequate to describe the vapor fluxes and evaporation rates observed in their field study. As a result of their findings, they proposed that diurnal heating of the soil caused convective transport of soil air due to temperature-driven expansion of the soil air. They also proposed that this convective transport (soil "breathing") explained the fluxes of water vapor observed. Thermally induced convection is a topic in need of further evaluation.

Barometric Pressure Effects   Theoretically, according to Boyle's law, any increase in the barometric pressure of the atmosphere should cause a compression and subsequent decrease in the volume of the soil air, thereby allowing air from the atmosphere to penetrate the soil pores. On the other hand, any decrease in barometric pressure should produce an expansion of the soil air, allowing part of it to enter the atmosphere above the soil. Thus, gas exchange between soil and atmosphere caused by pressure fluctuations should occur from time to time, provided that changes in the barometric pressure of the atmosphere are not damped out before they influence the air pressure within the soil pores.

Buckingham (1904) calculated the magnitude of the soil–atmosphere gas exchange due to barometric pressure changes and showed that atmospheric air would penetrate a permeable soil column 3 m deep only to a depth of about 3.0 to 5.6 mm, depending on the magnitude of the barometric pressure change. Thus, it seems that fluctuations in atmospheric pressure have little influence on soil aeration, even where there is good contact between the soil air and atmosphere. Romell (1922) estimated that no more than about 1% of the normal aeration of soils can be attributed to variations in barometric pressure.

Wind Effects   One might expect that the pressure and suction effects of high winds at the soil surface would exert some influence on the renewal of the soil air. Romell (1922) also analyzed this contribution to soil aeration and concluded that wind action could not be responsible for more than about 0.1% of the normal aeration on vegetated soils. More recent experiments on the effects of air turbulence on the transfer of vapor in the soil suggest that bulk flow of air in response to pressure fluctuations may not be negligible in porous, bare soils (Farrell et al., 1966). For example, air blowing across a bare surface with a wind speed of 15 mph can penetrate coarse sand and mulches to a depth of several centimeters. Even though there is no net mass flow of soil gases, the fluctuations in air pressure at the soil surface result in a mixing of air within the surface that enhances gas transport beyond that due to diffusion alone (Scotter et al., 1967). Takle et al. (2003) report high-frequency pressure variations in the vicinity of surface structures.

Windbreaks and shelterbelts are designed to influence wind profiles. The most frequent uses relate to wind erosion reduction or snow capture. The impacts on wind also affect surface pressure fluctuations in time and space, and thus influence convective soil and atmospheric gas exchange (Takle, 2003; Wang et al., 2001).

Rainfall Effects   The infiltration of rainfall into the soil may cause a renewal of the soil air in two ways. First, as the water infiltrates, it displaces soil air from the pores. During the subsequent redistribution of the infiltrating front, the pores are refilled with air from above. Second, soil air is enriched by dissolved O_2 that is carried into the soil by infiltrating water and exchanged later with the soil air phase. We might at first expect that a large rainfall could completely renew the soil air, especially if the water is able to displace the major portion of the air within the pores. In many instances, however, a considerable amount of entrapped air remains, which is not forced out of the soil by infiltrated water. Romell (1922) estimated that rainfall accounts for only about 7 to 9% of the normal aeration, depending on the soil type and climate.

Mass Flow of Gases into Buildings   Concern over high levels of radon gas in certain homes has prompted intense research recently on the mechanisms by which this gas moves from the soil (Wang and Ward, 2002). Radon gas is a vapor-phase reaction product formed during the radioactive decay of radium, a natural constituent of many soils. Experimental monitoring studies (Nero and Nazeroff, 1985) have verified that substantial air pressure differences exist between the soil air and the interior of houses in contact with the soil air. The lower pressure of the air within the house is a consequence of ventilation, circulation of air in the heating and cooling systems, and the geometry of the home and can permit substantial entry of soil gases into the house by mass flow. Once in the home, the radon gas can pose a substantial health hazard since it is a potent carcinogen. It appears to be possible to minimize the air entry by altering the points of contact between the soil and the house (Nazeroff et al., 1987).

Remediation of Volatile Pollutants   In recent years there have been a number of engineering developments, such as air and steam flushing (Sleep and McClure, 2001), for pumping or withdrawing gas in soil to remove volatile pollutants. In some cases air is pumped through natural systems, while in other cases, *pneumatic fracturing* is used. In the fracturing technique, a large volume of air at large pressure is injected into the ground. The rapid air input causes fractures to develop, which provide additional access to the polluted soil volume.

A common procedure for pumping air through soil is air sparging, which uses injected air to remove volatile or biodegradable contaminants from saturated soil. For volatiles, such as solvents or gasoline, air is injected into the soil so that the contaminants can be removed by air stripping. For semivolatiles, such as diesel and jet fuels, pumped air increases the dissolved oxygen, which increases the biodegradation rate of the contaminants. Recent investigations of air sparging use have been reported by Bass et al. (2000), Benner et al. (2002), Braida and Ong (2000, 2001), Fields et al. (2002), Rogers and Ong (2000), and Thomson and Johnson (2000). Although air sparging effectively removes gaseous-phase pollutants from convectively flowing channels in the soil, investigators make it clear that diffusion of pollutants from zones adjacent to channels is often the rate-limiting step for remediation.

6.3.3   Gas Diffusion

Since convection is usually negligible except under the preceding special circumstances, the major mechanism of gas transport is diffusion of vapor within the soil air space. In free air, the gas diffusion flux J_g is expressed by *Fick's law of diffusion*:

$$J_g = -D_g^a \frac{\partial C_g}{\partial z} \tag{6.7}$$

where D_g^a (cm^2 s^{-1}) is the binary gaseous diffusion coefficient in free air. Its value depends on the chemical diffusing in the air and on the temperature and air pressure (Bird et al., 2001). At atmospheric pressure and $T = 25°C$, it varies between about 0.15 and 0.25 cm^2 s^{-1} for gases of low molecular weight and decreases with increasing size of the diffusing molecule. Jury et al. (1983) recommended using an average value of about 0.05 cm^2 s^{-1} for pesticides of intermediate molecular weight (e.g., 100 to 300 g mol^{-1}). For small molecules like CO_2, the value is higher ($D = 0.14$ cm^2 s^{-1}; Lide, 2002). Equation (6.7) will overestimate the flux of gas through soil, both because gas must diffuse through a longer path length to get from one point to the other and because the cross-sectional area available for flow is reduced by solid and liquid barriers. Consequently, the gas flux in soil is calculated by modifying the diffusion flux in air (6.7) by a gas tortuosity factor $\xi_g < 1$, producing the new flux equation

$$J_g = -\xi_g D_g^a \frac{\partial C_g}{\partial z} \equiv -D_g^s \frac{\partial C_g}{\partial z} \tag{6.8}$$

where $D_g^s = \xi_g D_g^a$ is the soil gas diffusion coefficient.

A number of studies have been conducted to determine the influence of the porous medium on gas diffusion through soil. In early work, such studies were confined to porous media with very high air content a, so that the principal factor affecting the tortuosity ξ_g was the solid matrix. Buckingham (1904) proposed an equation of the form $\xi_g = \epsilon a$, where ϵ is a constant. Penman (1940a,b) studied the diffusion of carbon dioxide through packed soil cores with a range of $0.195 < a < 0.676$ and recommended 0.66 as an average value for ϵ. Thus, the Penman tortuosity model is given by

$$\xi_g = 0.66a \tag{6.9}$$

Flegg (1953), working with soil aggregates in the range of $0.35 < a < 0.75$, obtained values of ϵ between 0.53 and 0.89. Using soil with a porosity of 0.355, van Bavel (1952) measured a value of ϵ of 0.58. Marshall (1959) proposed the nonlinear relation $\xi_g = \epsilon^{4/3}$ for air-dry soils. This model was found to give good agreement with observation in sieved and repacked media (Moldrup et al., 2000).

When soil is not air dry, the relation between ξ_g and air content becomes more complex, particularly in soils that are undisturbed from their natural state. Substantial effort has been put forth in the last decade to develop tortuosity models with parameters that represent measurable properties of the soil pore size. Equation (6.10) from Moldrup et al. (2000) is a versatile model for describing tortuosity in undisturbed soils:

$$\xi_g = (2a_{100}^3 + 0.04a_{100}) \left(\frac{a}{a_{100}} \right)^{2+3/b} \tag{6.10}$$

where a_{100} is the volumetric air content when the matric potential is $h = -100$ cm and b is the PSD index of Campbell (1974), defined as the negative of the slope of the $h(\theta)$ function plotted on a log-log scale:

$$b = -\frac{d \ln[-h(\theta)]}{d \ln \theta} \tag{6.11}$$

Equation (6.10) was able to describe the gas tortuosity factor well over a range of water contents in 24 soils in their natural state (Moldrup et al., 2000). Other models have been developed relating gas tortuosity to soil structural or pore-size characteristics. These are reviewed in Rolston and Moldrup (2002) and Moldrup et al. (2001). In a recent effort, Hunt and Ewing (2003) demonstrated that the current gas diffusion model equations are consistent with the predictions of continuum percolation theory.

6.3.4   Measurement of Gas Diffusion Coefficients in Soil

The standard method for measuring gas diffusion coefficients in soil in the laboratory is the diffusion chamber (Rolston and Moldrup, 2002). In this method, a soil column that is devoid of gas and is open at one end is suddenly brought into contact with a closed air chamber containing a concentration of the gas of interest. Over time, the gas diffuses from the chamber into the soil at a rate that depends in a known way on the soil gas diffusion coefficient. Thus if the chamber concentration is monitored over time, the diffusion coefficient can be calculated from the theory describing the migration from the chamber into the soil.

Measuring Gas Flux in the Field   The gas flux is difficult to measure in the field because the gas diffusion coefficient is normally not known. Therefore, (6.8) cannot be used directly to evaluate J_g. The most common methods for estimating gas flux involve flux chambers at the soil surface (Hutchinson and Livingston, 2002). These chambers either allow gas to accumulate in a box placed over the surface (Matthias et al., 1980) or trap the gas at the surface while maintaining a low concentration (Ryden et al., 1978). In either method, only the gas escaping at the surface is measured, which might be very different from the rate of production (Jury et al., 1982a). Errors involved in the operation and interpretation of these devices are discussed in Hutchinson and Livingston (2002).

6.3.5   Gas Transport Equation

The gas transport equation is formed by inserting the gas flux equation (6.8) into the gas conservation equation (6.5). If we assume that D_g^s and a are constant, we obtain

$$a\frac{\partial C_g}{\partial t} = D_g^s \frac{\partial^2 C_g}{\partial z^2} + r_g \tag{6.12}$$

This equation describes the changes in gas concentration as a function of space and time in soil.

6.4   GAS TRANSPORT MODELING IN SOIL

The gas transport equation (6.12) is identical mathematically to the heat transport equation (5.35) when the gas reaction term r_g is absent. For this reason, the heat transport examples discussed in Chapter 5 apply equally well to inert gas transport problems with the same boundary conditions. This is illustrated in the next example.

Example 6.2   An inert gas ($r_g = 0$) diffuses through a soil column whose inlet end $z = 0$ is at a uniform concentration C_0 while the outlet end $z = L$ of the column is maintained free of gas, $C_L = 0$. Calculate the steady-state gas concentration within the column.

SOLUTION: In steady state, $\partial C_g / \partial t = 0$. Thus, the gas transport equation (6.12) reduces to

$$\frac{d^2 C_g}{dz^2} = 0 \tag{6.13}$$

This may be integrated once so that

$$\frac{dC_g}{dz} = \phi_1 \tag{6.14}$$

where ϕ_1 is an unknown constant of integration. Equation (6.14) may be integrated a second time, with the result

$$C_g = \phi_1 z + \phi_2 \tag{6.15}$$

where ϕ_2 is a constant of integration. The two constants may be evaluated using the boundary conditions

$$C_g(0) = C_0 = \phi_2 \tag{6.16}$$

$$C_g(L) = 0 = \phi_1 L + \phi_2 \tag{6.17}$$

Thus, $\phi_1 = -C_0/L$, and the gas concentration profile is

$$C_g(z) = C_0 \left(1 - \frac{z}{L}\right) \tag{6.18}$$

6.4.1 Steady-State O_2 Transport and Consumption

Most transport processes occurring in the natural environment are time dependent. However, steady-state solutions to certain problems whose parameters do not change appreciably over time may provide reasonable approximations to the actual situation. One such problem is oxygen consumption in the surface zone of soil. In a fully developed crop root zone, it is reasonable to treat the oxygen consumption rate as constant in space and time. Since the O_2 concentration in the atmosphere is uniform, a steady-state flow of gas from the atmosphere to the soil is possible, provided that gas cannot escape below the root zone. In steady state, the gas conservation equation (6.12) reduces to

$$\frac{dJ_g}{dz} + r_g = 0 \tag{6.19}$$

If we assume that no gas flows below the root zone at $z = -L$, the lower boundary condition may be expressed as

$$J_g(-L) = 0 \tag{6.20}$$

Equation (6.19) may be rearranged and integrated from $z = -L$ to z:

$$\int_0^{J_g} dJ_g = -r_g \int_{-L}^{z} dz \tag{6.21}$$

This trivial integration results in

$$J_g(z) = -r_g(z + L) \tag{6.22}$$

Thus, the gas flux decreases linearly from a maximum value of $-r_g L$ at the surface to a low of zero at the bottom of the root zone. It is negative because the direction of flow is downward.

The gas concentration in the profile may be calculated by substituting the gas flux equation (6.8) into (6.22):

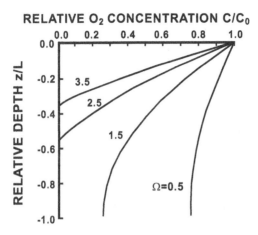

Figure 6.4 Steady-state O_2 profiles in a root zone as a function a for various values of the dimensionless parameter $\Omega = r_g L^2 / D_g^s$, calculated using (6.26).

$$J_g = -D_g^s \frac{dC_g}{dz} = -r_g(z + L) \tag{6.23}$$

Equation (6.23) is a first-order differential equation, which may be integrated after all factors depending explicitly on z are placed on the same side:

$$dC_g = \frac{r_g}{D_g^s}(z + L)\, dz \tag{6.24}$$

Equation (6.24) may now be integrated from $z = 0$ (where $C_g = C_0$) to z:

$$\int_{C_0}^{C_g} dC_g = \frac{r_g}{D_g^s} \int_0^z (z + L)\, dz \tag{6.25}$$

The integrals in (6.25) are straightforward, and we obtain

$$C_g(z) = C_0 + \frac{r_g}{2D_g^s}\left(z^2 + 2Lz\right) \tag{6.26}$$

Equation (6.26) is plotted in Fig. 6.4 for various values of $\Omega = r_g L^2 / D_g^s$. Note that for large values of Ω the gas concentration drops to zero within the root zone. This behavior is explored further in the next example.

Example 6.3   Calculate the maximum value of soil air content a below which some part of the root zone will be completely depleted of oxygen. Assume that the tortuosity model (6.8) is valid.

SOLUTION: Since the gas concentration decreases uniformly with depth, we can define the condition at which the soil begins to suffer from oxygen depletion by setting the concentration (6.26) equal to zero at the bottom of the root zone:

$$C_g(-L) = 0 = C_0 - \frac{r_g L^2}{2 D_g^s} \tag{6.27}$$

Thus, the critical value of the soil gas diffusion coefficient that satisfies (6.27) is

$$D_g^s = \xi_g D_g^a = \frac{r_g L^2}{2 C_0} \tag{6.28}$$

If we substitute the tortuosity model (6.10) into (6.28), we can solve for the value of air content that satisfies (6.28):

$$a = a_{100} \left[\frac{r_g L^2}{2 C_0 D_g^a (2 a_{100}^3 + 0.04 a_{100})} \right]^{1/(2 + 3/b)} \tag{6.29}$$

This is the smallest value of air content for which the entire root zone receives oxygen. Above this value we can expect symptoms of oxygen deficiency to develop.

6.4.2 Steady-State and Transient CO_2 Transport and Evolution

The steady-state model used for O_2 depletion in a root zone is less applicable to the problem of CO_2 evolution because the evolving gas can diffuse downward as well as upward unless a barrier to downward diffusion, such as a water table, is present near the surface. This special case is discussed in the next example.

Example 6.4 Calculate the steady-state CO_2 profile in soil, assuming that the gas is produced at a constant rate $rg = -R$ (negative because of production rather than consumption of gas) and that the CO_2 flux is zero at $z = -L$.

SOLUTION: The gas conservation equation (6.5) reduces to

$$\frac{dC_g}{dz} - R = 0 \tag{6.30}$$

for the case of steady flow with uniform evolution of CO_2. Equation (6.30) may be easily integrated from $z = -L$ to z, producing

$$J_g = R(z + L) \tag{6.31}$$

Thus, the flux has a maximum value of RL at the surface. When the gas flux equation (6.8) is inserted into (6.31), the result is a first-order differential equation for the concentration:

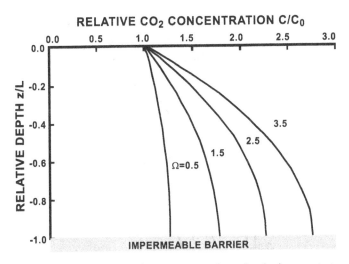

Figure 6.5 Steady-state CO_2 profiles in a root zone when a barrier is present at $z = -L$ as a function of a for various values of the dimensionless parameter $\Omega = RL^2/D_g^s$, calculated using (6.33).

$$J_g = R(z + L) = -D_g^s \frac{dC_g}{dz} \qquad (6.32)$$

Equation (6.32) may be integrated after placing all factors that depend on z on one side and integrating from $z = 0$ (where $C = C_0$) to z, producing the concentration profile

$$C_g(z) = C_0 - \frac{R}{2D_g^s}\left(z^2 + 2zL\right) \qquad (6.33)$$

Equation (6.34) is plotted in Fig. 6.5 as a function of $\Omega = RL^2/D_g^s$. Large values of this parameter, such as when the air content is low and diffusion resistance is high, result in substantial CO_2 evolution in the soil.

It is interesting to contrast the steady-state solution in Fig. 6.5 with the solution to the time-dependent problem described by (6.12) for the case where there is no barrier to downward gas diffusion in the soil. We may solve this problem by assuming that the gas production term is equal to $r_g = -R$ for $-L < z < 0$ and $r_g = 0$ for $z < -L$ and that the air content is uniform. Equation (6.12) with this value of the source term r_g may be solved by the method of Laplace transforms (Arfken, 2000). The solution is shown in Fig. 6.6 for the case $\Omega = RL^2/D_g^s = 1$. Note that concentration always lies below the steady concentration calculated for the case of no flow below $z = -L$ because now gas may diffuse downward as well as upward. Steady state can be reached only when the soil air below the root zone reaches the steady-state value

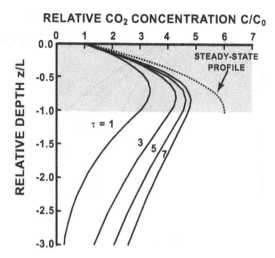

Figure 6.6   Transient CO_2 profiles in a root zone when no barrier is present at $z = -L$ for $\Omega = RL^2/D_g^s = 1$ at various values of the scaled time $\tau = D_g^s t/aL^2$. The shaded region corresponds to the zone where CO_2 is evolving at a rate R.

at $z = -L$, or when a barrier restricts downward movement of gas. This situation is unlikely ever to occur in soils with deep water tables.

6.4.3 O_2 Depletion at the Plant–Root Interface

The O_2 concentration in the soil air space at a given depth in the soil is regulated by the diffusive resistance of the soil and by the rate of demand of the plant roots and soil biota. For a given O_2 level, the rate of entry of gas to the root is regulated further by diffusion from the air space into the root (Currie, 1962; Lemon, 1962). Thus, the amount of O_2 that reaches the root surface is controlled both by the rate of gaseous exchange between the soil air and the atmosphere and by the rate of transfer of O_2 from the soil pores to the root. The latter process takes place through water films that exist around the plant roots and the soil particles; therefore, the final O_2 diffusion to the root must take place through the liquid phase. Wiegand and Lemon (1958) proposed the following equation to explain the O_2 supply to plant roots:

$$C_R = C_p - \frac{q R^2}{2D_l^s} \ln \frac{R_e}{R} \tag{6.34}$$

where C_R is the oxygen concentration dissolved in water at the root surface (g m^{-3}), C_p the dissolved O_2 concentration at the liquid–gas interface (g m^{-3}), q the rate of O_2 consumption within the root (g m^{-3} s^{-1}), R the root radius (m), R_e the radius of the root plus the moisture film (m), and D_l^s the diffusion coefficient in the liquid–solid soil matrix around the root (m^2 s^{-1}). Equation (6.34) is derived by solving the steady-state diffusion equation in cylindrical coordinates.

The liquid diffusion coefficient in soil, D_l^s, may be represented by a tortuosity model

$$D_l^s = \xi_l(\theta) D_l^w \tag{6.35}$$

where D_l^w is the diffusion coefficient of O_2 in pure water and $\xi_l(\theta)$ is the liquid diffusion tortuosity factor.

The water film around the root represents an enormous barrier to diffusion. The ratio of D_l^w to D_g^a is approximately 0.0001 for gases that also dissolve. This means that a 1-mm water film offers about the same diffusion flow resistance as 10 m of air with the same concentration difference across it. As a result, even relatively high values of O_2 in the soil air may not be sufficient to prevent aeration problems in plant roots surrounded by thick water films (Kristensen and Lemon, 1964).

6.5 FLOW OF WATER VAPOR THROUGH SOIL

6.5.1 Water Vapor Flux Equation

Because Fick's law of diffusion in soil (6.8) describes well the transport of gases of low solubility in soil, the first researchers studying water vapor movement assumed that it would behave similarly to other gases. Thus, they wrote the vapor flux equation as

$$J_{wv} = -\xi_g D_{wv}^a \frac{\partial \rho_v}{\partial z} \tag{6.36}$$

where ξ_g is the gas tortuosity factor, which might be represented by a model like (6.9). When the solute potential of the soil water is negligible (i.e., nonsaline soil), the water vapor density or vapor concentration ρ_v can be expressed in terms of the temperature and matric potential head h as

$$\rho_v = \rho_v^* \cdot RH = \rho_v^* \exp\left(\frac{hM_v}{\rho_w RT}\right) \tag{6.37}$$

where RH is the relative humidity, M_v the molecular weight of water, and ρ_v^* the saturated vapor density. For saline soils the h term in (6.37) must be replaced by the sum of the matric potential and the solute potential. Unless the soil is saline or very dry (i.e., $h < -20{,}000$ cm) the vapor density may be approximated by its saturated value, which is a function only of temperature. Thus, we may rewrite (6.37) in the region where the vapor approximation is valid as

$$J_{wv} = -\xi_g D_{wv}^a \frac{d\rho_v^*}{dT} \frac{dT}{dx} \qquad RH \sim 1 \tag{6.38}$$

Note that the slope of the saturated vapor density-temperature curve is a known property of water that has been compiled in numerous handbooks.

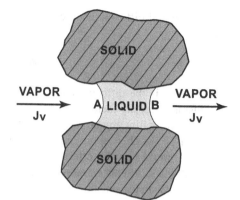

Figure 6.7 Liquid film between two soil grains forming a gas–liquid–gas pathway for vapor transport. Water condenses on side A of the film and evaporates from side B.

It was thought that (6.38) should be able to describe water vapor movement in soil. However, in laboratory tests of (6.38), when a temperature gradient was placed across a soil column, the measured vapor flux was up to 10 times larger than the value predicted (Gurr et al., 1952). The reason for this discrepancy eluded researchers for many years but was finally resolved satisfactorily in a theoretical paper written by Philip and de Vries in 1957.

The model of Philip and de Vries contains two theoretical assumptions causing it to differ from the simple theory described by (6.38). Their first observation was that water vapor may condense on one side of a liquid barrier and evaporate across the other side, thereby short-circuiting part of the liquid-filled pore space (Fig. 6.7).

Thus, until the porous medium is very wet and the liquid-filled pore space is continuous, the available pore space for diffusion to be used in the tortuosity factor in (6.38) is not the air content a but $a + \theta = \phi$.

The second enhancement mechanism proposed by Philip and de Vries (1957) relates to the use of an average temperature gradient in (6.42). The thermal conductivity of the solid phase is greater than that of the liquid phase, which in turn is very much greater than the thermal conductivity of the air phase. Since water vapor moves primarily through the air spaces, the mean temperature gradient measured by sensors that average over all three phases should be much smaller than the temperature gradient across the vapor-filled space, which drives the water vapor. Thus, (6.38) should be written as

$$J_{wv} = -\xi_g(\phi)D_{wv}^a \frac{d\rho_v^*}{dT}\omega\frac{dT}{dx} \tag{6.39}$$

where if we use the Penman tortuosity model (6.9),

$$\xi_g(\phi) = 0.66\phi \tag{6.40}$$

and

$$\omega = \frac{(\partial T/\partial x)_v}{(\partial T/\partial x)_{\text{soil}}} \tag{6.41}$$

By using a theoretical model of the soil thermal conductivity (de Vries, 1963), the authors were able to calculate values for ω that ranged from 2 to 3. With the two enhancement mechanisms, the modified theory of Philip and de Vries given by (6.39)–(6.41) removed most of the discrepancies between theory and experiment.

Cass et al. (1984) reported values of enhancement factor for several soils. They proposed an empirical model that has the enhancement factor increasing as a function of water content. The model parameters depend on soil texture.

6.5.2 Approximate Water Vapor Flux Law

A detailed analysis of a modified form of the Philip and de Vries (1957) model showed that a very simple approximate flux law may be used to describe water vapor movement in soil. This flux law is given by

$$J_{wv} \approx -2F(T)\frac{dT}{dx} \tag{6.42}$$

where $F(T)$ is a known function of the temperature that may be calculated without doing any experiments. Details of this approximate flux law are given in Jury and Letey (1979). Equation (6.42) was shown to be in reasonable agreement with existing experimental data while having the advantage of circumventing the extremely difficult problem of measuring water vapor movement directly.

6.5.3 Osmotic Effects on Water Vapor Flux

As the solute or osmotic potential decreases, the associated vapor pressure in the soil atmosphere also decreases. Thus, for equal water contents and temperatures, vapor fluxes in a saline soil will be less than in an equivalent nonsaline soil. Figure 6.8 presents two water distribution curves for soil columns that have been exposed to the same thermal gradient. Both soil columns had the same initial water content, but one soil column contained almost pure water, while the other column had salinized soil solution. The curves in Figure 6.8 indicate that greater water redistribution in response to the thermal gradient occurred in the soil column that was not salinized.

In another study, Nassar and Horton (1999a) reported that evaporation from salinized soils ranged from 78 to 95% of the evaporation from nonsalinized soils. Since most natural soils contain some dissolved salts, it is sometimes important to account for soil solution concentrations when estimating water vapor transfer. To aid in making an account, Nassar and Horton (1997) provide a theoretical description of coupled heat, water, and salt tranport in unsaturated soil.

6.5.4 Soil Wettability Effects on Water Vapor Flux

Most studies of water vapor fluxes have been done using wettable (hydrophilic) soils. For wettable soils the thermal water vapor fluxes have been enhanced. Bachmann et al. (2002) measured isothermal and nonisothermal (heated from below) evaporation from columns containing wettable soil and columns containing hydrophobic soil.

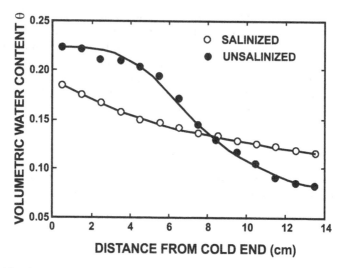

Figure 6.8   Observed steady-state water content distributions within moist salinized and solute-free soil columns. (From Nassar and Horton, 1989.)

Evaporation from the hydrophobic soil was about 75 and 50% of the evaporation from the wettable soil under isothermal and nonisothermal conditions, respectively. When the Philip and deVries model was used to evaluate thermally driven water vapor transport in the nonisothermal experiment, the wettable soil was found to have an enhancement factor of about 3, while the enhancement factor of the hydrophobic soil was only about 0.3. This means that water vapor transfer was reduced rather than enhanced in the hydrophobic soil. One reason for the differences is that liquid water films covered the wettable soil particles but not the hydrophobic soil particles. The differences in the wetted areas undoudtedly contributed to different water vapor transfer rates.

6.6   MEASUREMENT OF O_2 DIFFUSION AND CONSUMPTION IN SOIL

One of the earliest methods used to determine the O_2 diffusion coefficient in the soil pore space was the technique described by Raney (1949). It consisted of allowing O_2 to diffuse into a chamber that was inserted into the soil. From analyses of the gas concentration in the chamber made at 10-min intervals, the O_2 diffusion rate was estimated from a solution to the gas flow equation (6.12) applied to the geometry of the chamber.

Willey and Tanner (1963) determined the O_2 concentration in the soil air with a polarographic electrode, which is enclosed in a plastic membrane that is permeable to O_2. When a small voltage is applied between the polarographic cathode and the nonpolar anode, almost all of the O_2 that diffuses to the cathode is reduced and the concentration of O_2 at the electrode surface is very close to zero. The resulting current is proportional to the rate of reduction of O_2, which in turn is limited by the rate of diffusion of O_2 to the electrode. The measurement is applicable to O_2 diffusion in situ.

The platinum electrode was introduced by Lemon and Erickson (1952) to measure the oxygen diffusion rate (ODR) in soils. The method is based on the principle that the electric current resulting from the reduction of O_2 at the platinum surface is governed (within limits) solely by the rate at which O_2 diffuses to the electrode surface. By placing a known voltage between the platinum electrode and a saturated calomel cell, the authors were able to calculate the O_2 diffusion rate from the observed current. A review of the platinum electrode and its use in soils is presented by Farrell et al. (2002).

According to Fick's law of diffusion (6.8), the measured ODR for an electrode placed in soil should depend on both the O_2 concentration gradient at the electrode surface and the O_2 diffusion coefficient in the water film around the electrode. Thus, the measured value of ODR represents the integrated effects of the O_2 concentration in solution, the O_2 diffusion coefficient, the thickness of the moisture film around the electrode, the length of the diffusion path to the electrode, and the electrode radius (Letey and Stolzy, 1964; Stolzy and Letey, 1964). The method fails in relatively dry soils, since the electrode must be wetted so that the entire electrode surface is covered by a moisture film. For this reason, it works better in finer-textured soils whose pores remain filled at lower matric potentials than in coarse-textured soils. The ODR decreases as the moisture content increases, because the water film around the electrode becomes thicker, increasing the gas diffusion resistance.

Temperature affects the ODR by increasing the rate of plant respiration and the diffusion coefficient of O_2 in water. Also, the solubility of O_2 in the liquid phase is decreased. These combined effects result in about a 1.8% increase in the ODR per degree Celsius (Stolzy and Letey, 1964). Because so many factors can affect the ODR measurement, it is only qualitatively related to the aeration status of soil (McIntyre, 1971). It is important therefore that standard techniques be used so that results from different investigators will be comparable.

Bertrand and Kohnke (1957) obtained a good correlation between the ODR values measured at the 20-cm depth with the platinum electrode and the gas tortuosity factor ξ_g determined at the same depth with an oxygen analyzer. However, the correlation was not as good at the 40-cm depth. Wiersum (1960) found that visual observations of the depth of root penetration correlated very well with measurements by the platinum electrode.

Because of the similarity between the electrode shape and a plant root, ODR measurements can give relative estimates of soil aeration status when they are taken at different depths. Figure 6.9 shows measured ODR values with depth as a function of O_2 concentration at the soil surface. The values decrease with depth and with O_2 concentration, showing the need to take measurements at several locations to obtain a representative value for the root zone. Absolute values of ODR have meaning only when correlated with observed plant performance. Stolzy and Letey (1964) indicate that root initiation is reduced or stopped in many plants at ODR values between 30 and 38 μg m^{-2} s^{-1}.

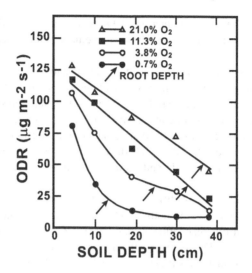

Figure 6.9 Oxygen diffusion rate as a function of depth for various oxygen concentrations at the soil surface. (After Stolzy and Letey, 1964.)

The ODR for optimum top growth appears to be greater than about 67 μg m^{-2} s^{-1}. The reader is referred to the comprehensive review of Stolzy and Letey (1964) for detailed evaluations of the impact of the ODR on plant responses.

The importance of drainage on the ODR is illustrated in Fig. 6.10. These data are self-explanatory and emphasize the necessity of lowering the water table to provide

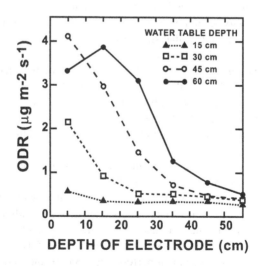

Figure 6.10 Oxygen diffusion rates at a given soil depth as a function of depth of water table. (After Williamson and van Schilfgaarde, 1965.)

the proper aeration status for plant roots. The data indicate that aeration falls below optimum levels 20 to 25 cm above the water table, irrespective of its depth.

PROBLEMS

6.1 The steady-state gas concentration in spherical coordinates may be written for radially symmetric flow as

$$\frac{1}{r^2}\frac{d}{dr}(r^2 J_g) + \Omega_g = 0$$

where J_g is the gas flux in the radial direction and Ω_g is the gas consumption rate.

(a) For the special case of constant gas consumption rate, calculate the gas flux $J_g(r)$ inside a spherical aggregate of radius r. Assume that the flux vanishes at $r = 0$.

(b) Using the gas flux equation

$$J_g = -D_g^s \frac{dC_g}{dr}$$

calculate the gas concentration $C_g(r)$ inside the aggregate assuming that it remains at the constant value $C_g(R) = C_0$ at $r = R$.

6.2 Using the result calculated from part **(b)** of Problem 6.1, calculate the critical aggregate radius R_c above which the inner part of the aggregate becomes anaerobic. Using the gas tortuosity model (6.9), calculate a relation between R_c and a.

6.3 Calculate a relation between the b parameter in (6.11) and the parameters of the van Genuchten model (3.82) in the range where $\alpha(-h)^N \gg 1$. Assume that θ_r is negligible.

6.4 How would the predicted CO_2 profiles in Fig. 6.6 be changed if a water table or some other gas barrier was present 1 m below the bottom of the root zone?

6.5 Calculate the ratio of the water vapor flux predicted by the Philip and de Vries model (6.39) and from the simple theory of water vapor flux in soil [equation (6.8)]. Assuming that $\omega = 2.5$ and $\phi = 0.5$, calculate the size of the ratio as a function of the air content a.

6.6 Carbon dioxide gas is moderately soluble in water. Therefore, some of the mass of CO_2 per soil volume resides in the dissolved phase, and (6.2) should be written as

$$C_T = aC_g + \theta C_l$$

Furthermore, the flow of CO_2 mass per area per time is equal to (assuming that the water phase is stationary)

$$J_s = J_s + J_l = -D_g^s \frac{\partial C_g}{\partial z} - D_l^s \frac{\partial C_l}{\partial z}$$

where D_l^s is the liquid diffusion coefficient for CO_2 in soil.

(a) Write the new CO_2 transport equation to be used instead of (6.5).

(b) If we assume that $C_g = K_H C_l$ at all times where K_H is the dimensionless Henrys coefficient, show that the new equation may be written as (assuming constant a)

$$\frac{\partial C_g}{\partial t} = D_E \frac{\partial^2 C_g}{\partial z^2}$$

where

$$D_E = \frac{K_H D_g^s + D_l^s}{a K_H + \theta}$$

is the generalized effective liquid–gas diffusion coefficient.

7 Chemical Transport in Soil

Chemicals are ubiquitous in our modern technological society. They provide innumerable services, ranging from energy to food production to virtually all manufacturing processes. But many of them pose health problems if injested, so that contamination of surface and underground water supplies has become a major problem worldwide. Agricultural use of pesticides and fertilizers poses a particularly difficult problem, because these chemicals must be applied to soil to raise crop yields, but they also must be kept out of surface and ground water. For this reason, a significant amount of modern soil physics has been devoted to agricultural chemical mangement. The design of optimum pesticide and fertilizer application rates and timing involves developing procedures to maximize the effectiveness of these chemicals within the root zone while minimizing their movement below it. The reclamation of saline or sodic soils by application of water and chemicals may involve complex chemical reactions in the soil solution as well as displacement of dissolved chemicals from the surface soil by leaching.

Increased public awareness of significant contamination of groundwater by industrial, municipal, and agricultural chemicals (Pye et al., 1983) has focused much attention on chemical movement, creating a burst of experimental and theoretical research in this area (van Genuchten and Jury, 1987). In this chapter we develop the description of chemical transport and reactions in soil using mass balance and flow principles similar to those used in describing water and gas transport in previous chapters.

7.1 CHEMICAL CONSERVATION EQUATION

In Chapter 3 we derived the water conservation equation by performing a water mass balance on a unit volume of soil. The same procedure may be used on a chemical of interest to derive the following chemical conservation equation in one dimension, with the result

$$\frac{\partial C_T}{\partial t} + \frac{\partial J_c}{\partial z} + r_c = 0 \tag{7.1}$$

where C_T is the total concentration of chemical in all forms expressed as mass of chemical per soil volume, J_c the chemical mass flux (mass of chemical flowing per unit area per unit time), and r_c the rate of loss of mass per unit volume by reactions or other loss mechanisms (e.g., plant uptake). If chemical is created, $r_c < 0$. For the special case where the chemical is insoluble, (7.1) reduces to the gas conservation equation (6.5).

7.1.1   Chemical Storage in Soil

In general, chemicals that are soluble in water and have a nonnegligible vapor pressure can exist in three phases in soil[1]: as a gas in the soil air, as a solute (dissolved chemical) in the soil water, and as a stationary phase adsorbed to soil organic matter or mineral surfaces. Mathematically, we may express the total chemical concentration C_T in terms of the phase contributions as

$$C_T = \rho_b C_a + \theta C_l + a C_g \tag{7.2}$$

where C_a is adsorbed chemical concentration, expressed as mass of sorbant per mass of dry soil, C_l is dissolved chemical concentration, expressed as mass of solute per volume of soil solution, and C_g is gaseous chemical concentration, expressed as mass of chemical vapor per volume of soil air. The units in which the chemical phases are expressed correspond to the way in which they are measured. For example, if a soil air sample of volume V_a is withdrawn and a mass M_g of chemical vapor is detected in it, $C_g = M_g / V_a$. The bulk density ρ_b, volumetric water content θ_v, and volumetric air content a convert each of these phase concentrations to mass per soil volume, which is the unit in which C_T is expressed.

7.1.2   Flux of Chemical in Soil

When chemical is present in more than one phase, it can move through soil either as a vapor or as a solute within the liquid phase[2]:

$$J_c = J_l + J_v \tag{7.3}$$

where J_c is the chemical mass flux (mass of chemical flowing per unit area per unit time), J_l the solute flux, and J_g the flux of chemical vapor. As discussed in Chapter 6, gas is transported primarily by diffusion, so that the gas flux is given by Fick's law (6.8), repeated here as

$$J_g = -\xi_g(a) D_g^a \frac{\partial C_g}{\partial z} = -D_g^s \frac{\partial C_g}{\partial z} \tag{7.4}$$

where D_g^s is the soil gas diffusion coefficient.

The solute flux consists of two terms: the bulk transport or convection of solute moving with flowing soil solution and the diffusive flux of solute moving by

---

[1] Certain chemicals may also exist in a separate or nonaqueous phase (see Section 7.4).
[2] The sorbed phase is assumed to be stationary.

molecular diffusion within the soil solution. The bulk flow or convection of solute J_{lc} may be written formally as

$$J_{lc} = J_w C_l \tag{7.5}$$

where J_w is the water flux. In porous media, (7.5) does not represent all of the transport by flowing solution because the soil water flux J_w is an approximate quantity that has been averaged over many soil pores. It does not represent the actual water flow paths, which must curve around solid obstacles and air space. Consequently, (7.5) does not describe the solute convection completely; it fails to describe the tortuous convective motion of solute relative to the average convective motion. This extra motion is called *hydrodynamic dispersion* (Bear, 1972), and the total solute convection is described by

$$\text{total convection} = J_{lc} + J_{lh} \tag{7.6}$$

where J_{lh} is the *hydrodynamic dispersion flux*.

The other term contributing to chemical transport in the dissolved phase is the *liquid diffusion flux* J_{ld}, which is written formally as

$$J_{ld} = -\xi_l(\theta) D_l^w \frac{\partial C_l}{\partial z} = -D_l^s \frac{\partial C_l}{\partial z} \tag{7.7}$$

where D_l^w is the binary diffusion coefficient of the solute in water, D_l^s the soil liquid diffusion coefficient, and $\xi_l(\theta)$ the liquid tortuosity factor to account for the increased path length and decreased cross-sectional area of the diffusing solute in soil. Since air as well as solid particles form barriers to liquid diffusion, $\xi_l(\theta)$ is a strongly increasing function of water content θ. Since the liquid tortuosity modifies liquid diffusion in the same way that gas tortuosity modifies gas diffusion, any of the gas tortuosity models [e.g., (6.10)] may be modified by substituting θ for a to represent $\xi_l(\theta)$:

$$\xi_l(\theta) = \xi_g(a \rightarrow \theta) \tag{7.8}$$

7.1.3 Multiphase Chemical Reactions in Soil

The general reaction term r_c in (7.1) is by definition the rate of loss of chemical mass per unit soil volume from all reactions. For modeling purposes, it is often useful to express this quantity formally in terms of the reactions occurring in each phase. Thus,

$$r_c = \rho_b r_a + \theta r_l + a r_g \tag{7.9}$$

where the sorbed, dissolved, and vapor phase reactions r_a, r_l, and r_g are expressed per unit mass of soil, volume of water, and volume of air, respectively.

7.1.4   Convection–Dispersion Model of Hydrodynamic Dispersion

The erratic transport of solute by hydrodynamic dispersion has been studied theoretically and experimentally for many years. A special case of all possible flows, *convective–dispersive transport*, occurs in porous media when the following two conditions are met:

1. The porous medium is homogeneous through the volume in which solute transport occurs.
2. The time required for solutes in stream tubes of different velocity to mix (by diffusion or transverse dispersion) along a direction normal to the direction of mean convection is short compared to the time required for solutes to move through the volume by mean convection (Taylor, 1953).

For example, in a cylindrical soil column, this condition requires that solutes mix completely in the radial direction before they reach the outflow end of the column in the axial direction.

For this important case, the solute hydrodynamic dispersion flux has a form that is mathematically identical to the diffusion flux (Bear, 1972):

$$J_{lh} = -D_{lh}\frac{\partial C_l}{\partial z} \tag{7.10}$$

where D_{lh} ($L^2 T^{-1}$) is the *hydrodynamic dispersion coefficient*. This coefficient has frequently been observed to be proportional to the pore water velocity $V = J_w/\theta$ (Anderson, 1979; Bear, 1972; Biggar and Nielsen, 1967):

$$D_{lh} = \lambda V \tag{7.11}$$

where $\lambda(L)$ is the dispersivity. The size of the dispersivity depends on the scale over which the water flux and solute convection are averaged. Typical values of λ are 0.5 to 2 cm in packed laboratory columns and 5 to 20 cm in the field, although they can be considerably larger in regional groundwater transport (Fried, 1975; Gelhar et al., 1985). The total solute flux in the convection–dispersion model is thus

$$J_l = J_w C_l - D_{lh}\frac{\partial C_l}{\partial z} - D_l^s\frac{\partial C_l}{\partial z} \tag{7.12}$$

which is commonly written as

$$J_l = -D_e\frac{\partial C_l}{\partial z} + J_w C_l \tag{7.13}$$

where D_e is the effective diffusion–dispersion coefficient. Unless water is flowing very slowly through repacked soil, D_e is normally dominated by the hydrodynamic dispersion term, as illustrated in the next example.

Example 7.1   In a repacked sandy soil column of porosity $\phi = 0.5$, the measured dispersivity λ is 1 cm. Assuming that $D_l^w = 1$ cm^2 day^{-1}, calculate D_e, D_{lh}, and $\xi_l(\theta)D_l^w$ at applied water fluxes of 0.2, 1.0, 2.0, and 5.0 cm day^{-1}, which create average water contents of 0.25, 0.30, 0.35, and 0.40, respectively. Use the Millington–Quirk (1961) tortuosity model

$$\xi_l(\theta) = \frac{\theta^{10/3}}{\phi^2} \qquad (7.14)$$

SOLUTION: Since $V = J_w/\theta$, the water velocities are 0.80, 3.33, 5.71, and 12.50 cm day^{-1}, respectively, and the corresponding tortuosity factors using (7.14) are 0.039, 0.072, 0.121, and 0.189. Thus, the final values of the various dispersion and diffusion terms are given in Table 7.1.

TABLE 7.1   **Values of Diffusion and Dispersion Parameters in Example 7.1**

θ	V	D_{lh}	$\xi_l(\theta)$	D_l^s	D_e	D_{lh}/D_e
0.25	0.80	0.80	0.04	0.04	0.84	0.95
0.30	3.33	3.33	0.07	0.07	3.37	0.98
0.35	5.51	5.51	0.12	0.12	5.82	0.98
0.40	12.50	12.50	0.19	0.19	12.69	0.99

7.2   CHEMICAL TRANSPORT EQUATION

For the most general case where chemical is present in the dissolved, vapor, and sorbed phases, the chemical transport equation is obtained by combining the conservation equation (7.1), the storage equation (7.2), the vapor flux equation (6.8), the reaction term (7.9), and the solute flux equation (7.13), to produce

$$\frac{\partial}{\partial t}(\rho_b C_a + \theta C_l + a C_g) = \frac{\partial}{\partial z}\left(D_g^s \frac{\partial C_g}{\partial z}\right) + \frac{\partial}{\partial z}\left(D_e \frac{\partial C_l}{\partial z}\right)$$
$$- \frac{\partial}{\partial z}(J_w C_l) - (\rho_b r_a + \theta r_l + a r_g) \qquad (7.15)$$

Special cases of (7.15) are discussed in the next section.

7.3   CONVECTION–DISPERSION EQUATION

For those chemicals whose vapor phase is negligible, the chemical transport equation (7.15) may be simplified by setting C_g and r_g to zero. This produces

$$\frac{\partial}{\partial t}(\rho_b C_a + \theta C_l) = \frac{\partial}{\partial z}\left(D_e \frac{\partial C_l}{\partial z}\right) - \frac{\partial}{\partial z}(J_w C_l) - (\rho_b r_a + \theta r_l) \qquad (7.16)$$

Equation (7.16) is called the *convection–dispersion equation* (CDE) (Biggar and Nielsen, 1967).

7.3.1   Transport of Inert, Nonadsorbing Solutes

The simplest chemical transport processes in soil are those that involve nonvolatile ($C_g = 0$) dissolved chemicals that neither react ($r_a = r_l = 0$) nor adsorb ($C_a = 0$) to soil solids. Examples of these kinds of tracers are chloride or bromide ions flowing through soil that has only negatively charged or neutral minerals. These ions do not react chemically with the kinds of compounds normally found in soil solution and are not attracted to clay or organic matter surfaces (see Section 1.1.4). Thus, they may act as tracers of the water flow pathways.

Imagine an experiment where water is flowing uniformly at steady state through a homogeneous soil column of length L that is at a constant water content. Equation (7.16) reduces to

$$\frac{\partial C_l}{\partial t} = D\frac{\partial^2 C_l}{\partial z^2} - V\frac{\partial C_l}{\partial z} \qquad \text{where } D = D_e/\theta \qquad (7.17)$$

At t = 0, we instantaneously switch the water inlet valve of the soil column from its initial tracer-free source over to a chloride solution at a concentration C_0, which continues to flow at the same flux rate J_w through the column (see Fig. 7.1) From $t = 0$ onward, we begin to monitor the chloride concentration at the outflow end $z = L$ of the column. The plot of outflow concentration versus time (Fig. 7.2) is called an outflow curve, or sometimes a "breakthrough" curve (representing the solute breaking through the outflow end).

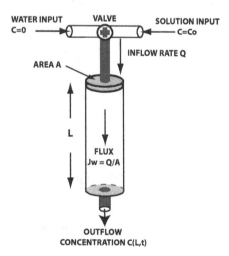

Figure 7.1   Soil column experiment in which solute is suddenly added to the inlet end at $t = 0$.

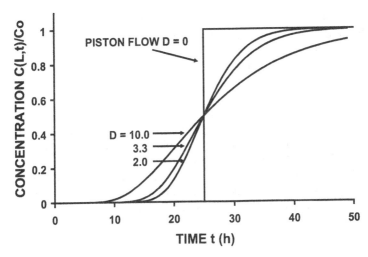

Figure 7.2 Outflow concentration versus time for a step change in solute input at $t = 0$. Column length is 50 cm and water velocity $V = 2$ cm h^{-1}. Curves are for different values of D in cm^2 h^{-1}.

The curves shown in Fig. 7.2 represent mathematical solutions to (7.17) corresponding to the initial and boundary conditions describing the experiment. The outflow concentration[3] is expressed as (Jury and Roth, 1990)

$$C(L, t) = \frac{C_0}{2}\left[\text{erfc}\left(\frac{L - Vt}{\sqrt{4Dt}}\right) + \exp\left(\frac{VL}{D}\right)\text{erfc}\left[\frac{L + Vt}{\sqrt{4Dt}}\right]\right] \qquad (7.18)$$

where $\text{erfc}(x)$ is the complementary error function (Abramowitz and Stegun, 1970), defined as

$$\text{erfc}(x) = \frac{2}{\sqrt{\pi}}\int_x^{\infty} \exp(-\xi^2)\,d\xi \qquad (7.19)$$

There are a number of important features illustrated in Fig. 7.2, which display various characteristics of convective–dispersive flow.

Breakthrough Time   The center of each of the solute fronts, drawn for different values of the dispersion coefficient D, arrive at the outflow end of the column at the same time $t_b = L/V$, called the *breakthrough time*. When dispersion is neglected (the $D = 0$ curve), all the solutes move at the same identical velocity V, and the front arrives as one discontinuous jump to the final concentration C_0 at $t = t_b$. This model, in which dispersion is neglected, is called the *piston flow model* of solute movement, so named because the solute is displaced through the soil like a piston.

[3] Technically, this is the *flux concentration*, defined as the ratio of the solute flux to the water flux (Jury and Roth, 1990).

Effect of Dispersion   As seen in Fig. 7.2, the effect of dispersion on the break-through curve is to cause some early and late arrival of solute with respect to the breakthrough time. This deviation is due to diffusion and small-scale convection ahead of and behind the front moving at velocity V and becomes more pronounced as D becomes larger.

Drainage Breakthrough Curves   Instead of plotting outflow concentration as a function of time, concentration can be plotted against the cumulative water drainage d_w passing through the outflow end of the column. If the water is flowing at steady state, $d_w = J_w t$, and the outflow concentrations in Fig. 7.2 are identical when plotted against d_w except for a change in scale for the abscissa. The drainage evolved at the breakthrough time $d_{wb} = J_w t_b$ is by definition at steady state equal to the amount of water that must be added to the column to move solute from the inflow port to the outflow end:

$$d_{wb} = J_w t_b = \frac{J_w L}{V} = L\theta \qquad (7.20)$$

which is the volume of water per unit area held in the wetted soil pores of the column during transport. For this reason $d_{wb} = L\theta$ is called a *pore volume*. Thus, it requires approximately one pore volume of water to move a mobile solute through a soil volume of water content θ and length L.

Equation (7.20) provides additional insight into the piston flow model ($D = 0$) and its assumptions. According to the piston flow model, the incoming solution replaces the water initially present in the soil, or, equivalently, pushes it ahead of the front like a piston. Thus, one may calculate solute transport with the piston flow model by estimating how long it will take to replace the water between the point of entry and the final location. This is illustrated in the next example.

Example 7.2   Calculate the amount of time required to transport nitrate (a mobile anion) from the bottom of the root zone to groundwater 10 m below it if the average water content of the soil is 0.15 and the average annual drainage rate is 0.5 m yr^{-1}.

SOLUTION: The total amount of water in the soil profile is $L\theta = 10 \times 0.15 = 1.5$ m, which would require 3 yr to add at 0.5 m yr^{-1}. Thus, $t_b = 3$ yr. Equivalently, we may calculate the nitrate velocity $V = J_w/\theta = 10/3$ m yr^{-1}, and a corresponding breakthrough time $t_b = L/V = 3$ yr.

Transport of Pulses through Soil   In some soil column outflow experiments, a narrow pulse of solute, rather than a front, might be added to the inlet end at $t = 0$. Figure 7.3 shows the corresponding outflow curves for a narrow pulse input of mass M_0 that is added at $t = 0$, calculated with the mathematical solution to (7.17) and the appropriate boundary conditions. This solution is given by (Jury and Sposito, 1985)

$$C(l, t) = \frac{M_0 L}{2 J_w \sqrt{\pi D t^3}} \exp\left[-\frac{(L - Vt)^2}{2Dt} \right] \qquad (7.21)$$

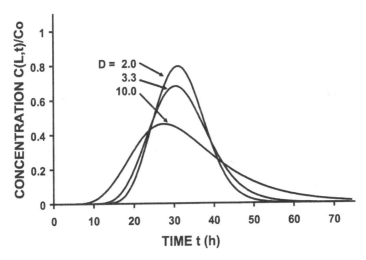

Figure 7.3 Outflow concentration versus time for a narrow pulse of solute input at $t = 0$. Column length is 50 cm and water velocity $V = 2$ cm h^{-1}. Curves are for different values of D in cm^2 h^{-1}.

Note that as D becomes larger, the pulse spreads out more and the pulse height decreases.

7.3.2 Transport of Nondecaying Adsorbing Chemicals

Certain chemicals, although they do not react chemically or biologically in soil, adsorb to stationary soil solids. For this case, the transport equation (7.16) may be written (assuming homogeneous soil and steady-state water flux)

$$\frac{\rho_b}{\theta}\frac{\partial C_a}{\partial t} + \frac{\partial C_l}{\partial t} = D\frac{\partial^2 C_l}{\partial z^2} - V\frac{\partial C_l}{\partial z} \tag{7.22}$$

Adsorption Isotherms   Equation (7.22) differs from (7.17) only by the addition of the first term, which represents the rate of change of mass storage in the sorbed phase. To solve (7.22), we must develop a relationship between the adsorbed concentration C_a and the dissolved concentration C_l. This relationship at equilibrium is called an *adsorption isotherm* (see Section 1.1.6). Figure 7.4 shows different isotherm shapes found for various compounds in soil.

Linear Equilibrium Sorption During Transport   A special form of the isotherm, the *linear adsorption isotherm*, may be expressed as

$$C_a = K_d C_l \tag{7.23}$$

where K_d, the slope of the isotherm, is called the *distribution coefficient*. It has SI units of (m^3 kg^{-1}) and cgs units of (cm^3 g^{-1}).

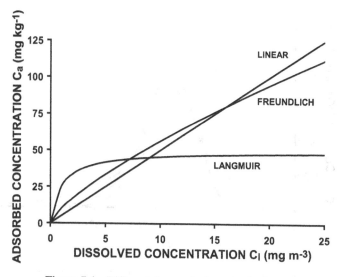

Figure 7.4   Different shapes of adsorption isotherm.

If we assume that the linear isotherm is valid at all times in soil (i.e., the dissolved and adsorbed concentrations are in instantaneous equilibrium), the time derivative of C_a may be written as

$$\frac{\partial C_a}{\partial t} = K_d \frac{\partial C_l}{\partial t} \tag{7.24}$$

Using (7.24), (7.22) may be written as

$$\left(1 + \frac{\rho_b K_d}{\theta}\right) \frac{\partial C_l}{\partial t} = R \frac{\partial C_l}{\partial t} = D \frac{\partial^2 C_l}{\partial z^2} - V \frac{\partial C_l}{\partial z} \tag{7.25}$$

where

$$R = 1 + \frac{\rho_b K_d}{\theta} \tag{7.26}$$

is the retardation factor. If we then divide each side of (7.25) by R, we obtain

$$\frac{\partial C_l}{\partial t} = D_R \frac{\partial^2 C_l}{\partial z^2} - V_R \frac{\partial C_l}{\partial z} \tag{7.27}$$

where $D_R = D/R$ and $V_R = V/R$ are the retarded dispersion coefficient and velocity, respectively.

Sorbing Chemical Breakthrough Time   Equation (7.27) now has the same mathematical form as (7.17), and therefore it has the same solutions as those shown in Figs.

7.2 and 7.3 for the same value of the parameters. Thus, the breakthrough time for a chemical with retardation factor R in a soil whose water velocity is V is given by

$$t_{bR} = \frac{L}{V_R} = \frac{RL}{V} = Rt_b \qquad (7.28)$$

where $t_b = L/V$ is the breakthrough time for a mobile chemical that does not sorb to soil solids. Thus, in steady flow a sorbing chemical would take R times as long to reach the outlet end as a mobile one. By the same reasoning, it will require R times as much drainage to move a sorbing chemical to the outlet end as a mobile one. Since it requires $d_b = L\theta = 1$ pore volume to move a mobile chemical to the outlet end, it will require $d_{bR} = RL\theta = R$ pore volumes to move a sorbing one to the outlet.

Example 7.3   For the soil in Example 7.2, calculate how long it will take to move a pesticide with a distribution coefficient $K_d = 2$ cm^3 g^{-1} to the groundwater. Assume $\rho_b = 1.5$ g cm^{-3} and $\theta = 0.15$.

SOLUTION: We obtain a retardation factor $R = 21$ using (7.26). Thus, by (7.28) and Example 7.2, the breakthrough time of the pesticide is $21 \times 3 = 63$ yr, requiring 21 pore volumes or 11.5 m of water to pass through the profile.

Effect of Dispersion   Although the dispersion coefficient in (7.27) is reduced by a factor R, the travel time is increased, so that the amount of solute spreading as a function of time observed at the outflow end actually appears to be greater than that of a mobile chemical. This is illustrated in Fig. 7.5, which shows outflow curves for chemicals with different R. The breakthrough times increase as R increases, and the apparent amount of solute spreading of the outflow curves increases with R as well.

Transport and Nonequilibrium Sorption   When the time of contact is short compared to the time required for sorption, the assumption of instantaneous equilibrium between the adsorbed and dissolved chemical concentrations is not valid. In this case,[4] (7.23) is not always satisfied within a unit volume of soil. Instead, we can substitute a rate-limited sorption equation such as

$$\rho_b \frac{\partial C_a}{\partial t} = -\alpha(K_d C_l - C_a) \qquad (7.29)$$

where α is a rate parameter that expresses the rate of mass transfer of chemical between the phases. Note that the rate of change of mass in the sorbed phase is assumed to be proportional to the deviation from equilibrium (7.23). Equation (7.29) may be solved simultaneously with (7.22) if the initial and boundary conditions are known.[5] Figure 7.6 shows the outflow curve for a sorbing chemical as a function of the rate parameter α.

---

[4] Here we are assuming a linear isotherm for simplicity.
[5] Mathematical solution methods and equations for rate-limited transport and other advanced topics may be found in Jury and Roth (1990).

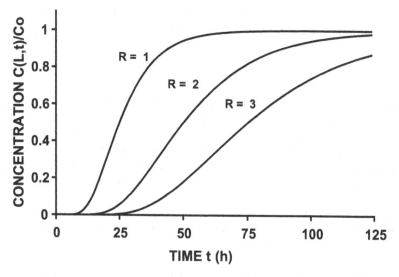

Figure 7.5   Outflow concentration versus time of chemicals with different retardation factors for a step change of solute input at $t = 0$. Column length is 50 cm and water velocity $V = 2$ cm h^{-1}.

For large values of α, the sorption process reaches equilibrium during the time of contact, and the equilibrium sorption model (7.27) may be used. When the rate parameter approaches zero, sorption is largely blocked, and the chemical moves like a mobile tracer described by (7.17). For intermediate values of α, breakthrough begins earlier than with equilibrium sorption, and the outflow curve has a long tail.

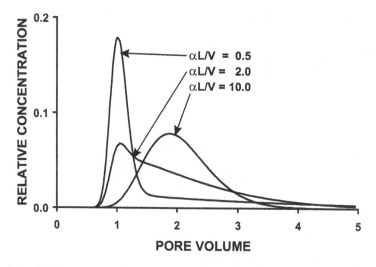

Figure 7.6   Outflow concentration versus pore volume of drainage calculated with (7.22) and (7.29) for a chemical with a retardation factor $R = 2$ as a function of the rate parameter α.

Transport and Nonlinear Sorption   Not all sorbing chemicals may be described by a linear isotherm. A number of organic compounds for example, obey a Freundlich isotherm,

$$C_a = K_F C_l^{\beta} \tag{7.30}$$

generally with $\beta < 1$ (Hamaker and Thompson, 1972). The transport equation (7.22) must be solved numerically in this case, but it is possible to deduce some of the qualitative features of the process by looking at the Freundlich isotherm shape in Fig. 7.4. At low concentration, the slope of the isotherm is steeper than at higher concentration when $\beta < 1$, which implies that more sorption (and hence more retardation) will occur. This effect is shown in the outflow curves of Fig. 7.7 for the case of a front of applied solute at $t = 0$. The upper figure shows the effect of β for three compounds applied at $C_0 = 10$ mmol m^{-3} whose sorption concentrations are equal at the input concentration when $C_l = 10$ mmol m^{-3}. All compounds reach half of the input concentration at about the same time. Also, the dispersion or spreading of the front decreases as β decreases, because the sorption values were set to coincide

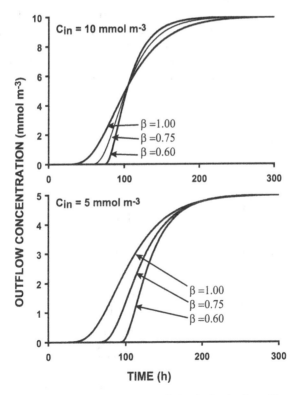

Figure 7.7   Outflow concentration versus time of chemicals obeying a Freundlich sorption isotherm (7.30) for different values of β. All chemicals have the same sorption concentration C_a when C_l is 5.0 mmol m^{-3}.

at the maximum concentration $C_l = C_0$. The bottom graph in Fig. 7.7 shows the outflow for the same chemicals, except this time with $C_0 = 5$ mmol m^{-3}. Now the chemicals arrive at the outflow end at different times, because the effective retardation increases as β decreases for all $C_l < 10$ mmol m^{-3}, where all compounds have the same sorption. This figure illustrates a significant feature of nonlinear sorption— solute velocity is a function of concentration during steady flow.

7.3.3 Effect of Soil Structure on Transport

As we saw in Chapter 4, soil structure has a pronounced effect on water movement, and hence can substantially influence solute transport as well. Aggregated soils have a continuous network of large pores through which water can flow rapidly, and a set of finer pores within the aggregates where water is present but largely stagnant. Soils with a massive structure may have fracture planes that channel virtually all of the water in the medium. In contrast, soils that have been packed uniformly have a distribution of pore sizes, and most of the water is either moving or in close contact with moving water. For such media, the uniform CDE (7.17) generally provides a good description of the transport process.

Figure 7.8 shows breakthrough curves for chloride pulses added to 1-m-long soil columns containing both undisturbed and repacked sandy loam from a field site. The undisturbed soil columns were formed by pushing cylinders into the soil surface and returning the intact soil to the laboratory, while the repacked soil from the same site was sieved, broken down, and carefully packed to a uniform bulk density (Khan and Jury, 1990).

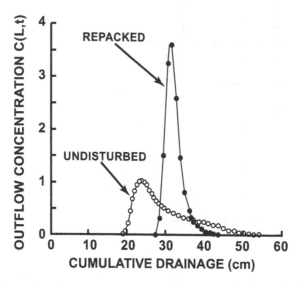

Figure 7.8 Comparison of breakthrough curves in repacked and undisturbed soil. (After Khan, 1988.)

There are several obvious differences between the two breakthrough curves. Chloride arrives after less drainage in the undisturbed soil column than in the repacked one and has substantial delay in the arrival of the final portion of the applied pulse at the outflow end. Because of their finer pore sizes, the aggregates hold a substantial volume of water at normal soil matric potentials, but this water is relatively stagnant compared to that flowing in the large channels between aggregates. For soils of this type, some of the solute can be transported much more rapidly than one would expect based on the total volumetric water content and the piston flow model (7.20).

The delayed transport of part of the solute mass to the outflow end in the structured soil compared to the repacked soil can be attributed to removal of solute from the flow channels by transverse diffusion into stagnant regions of the soil while the pulse is passing through the column, and gradual release of the solute to the flow channels by diffusion after the pulse has passed through the system. Fracture planes or large channels in the soil can create extreme differences in the flow characteristics of the medium, as shown in the next example.

Example 7.4 Calculate the breakthrough time for a front of solute moving through a vertical 100-cm soil column of saturated hydraulic conductivity $K_s = 5$ cm h^{-1} and $\theta_s = 0.5$ that has 10 cm of water ponded on it. Further assume that somewhere in the 100-cm^2 cross section of the column is a 1-mm-diameter wormhole that extends through the column and through which water flows according to Poiseuille's law (3.10), while it flows according to Darcy's law (3.14) in the soil.

SOLUTION: Within the soil matrix, the steady water flux rate may be calculated with Darcy's law (3.14):

$$J_w = -K_s \frac{L+d}{L} = -5.5 \text{ cm h}^{-1}$$

Therefore, within the matrix the solute moves with a downward velocity equal to

$$V = \frac{J_w}{\theta_s} = 11 \text{ cm h}^{-1}$$

Thus, solute breakthrough in the soil occurs at a time

$$t_b = \frac{L}{V} \approx 9.1 \text{ h}$$

In the wormhole, the water volume flow rate is given by Poiseuille's law (3.10):

$$Q = \frac{\rho_w g \pi R^4 (L+d)}{8Lv} = 0.265 \text{ cm}^3 \text{ s}^{-1}$$

and the water velocity in the wormhole is

$$V_h = \frac{Q}{\pi R^2} = \frac{\rho_w g R^2 (L+d)}{8Lv} = 33.7 \text{ cm}^3 \text{ s}^{-1}$$

Thus, the breakthrough time through the wormhole is

$$t_b = \frac{L}{V_h} = 2.97 \text{ s}$$

Clearly, structural voids can have a dominant effect on water and solute flow if they are filled with water that is flowing.

Mobile–Immobile Water Model   The mobile–immobile water model (MIM) was developed to include the effects of structure on solute transport. It was first described by Coats and Smith (1956) and later applied to transport in soil columns by van Genuchten and co-workers (van Genuchten and Wierenga, 1976, 1977; van Genuchten et al., 1977). This model divides the wetted pore space into two regions: a mobile water content θ_m through which water is flowing and an immobile water content $\theta_{im} = \theta - \theta_m$ which contains stagnant water. For this system, the solute concentration is divided into an average concentration C_m in the mobile region and a second concentration C_{im} in the immobile region. The solute is transported by a convection–dispersion process in the mobile region while mass is exchanged with the immobile region by a rate-limited diffusion process. For an inert, nonreactive solute the conservation equation (7.1) may be written as (assuming constant J_w, θ_m, and θ_{im})

$$\theta_m \frac{\partial C_m}{\partial t} + \theta_{im} \frac{\partial C_{im}}{\partial t} = D_e \frac{\partial^2 C_m}{\partial z^2} - J_w \frac{\partial C_m}{\partial} \qquad (7.31)$$

where in this case the total solute mass per soil volume is $C_T = \theta_m C_m + \theta_{im} C_{im}$. The rate-limited mass transfer between the mobile and stagnant regions is expressed in the model as

$$\theta_{im} \frac{\partial C_{im}}{\partial t} = \alpha(C_m - C_{im}) \qquad (7.32)$$

where α is the rate coefficient. Equation (7.32) states that the rate of change of solute stored in the immobile region is proportional to the difference in concentration between the two regions. Thus, when solute is added to the soil as a pulse, it will first diffuse from the mobile to the stagnant region and then return to the mobile region after the pulse has passed through the system.

Figure 7.9 illustrates solutions to the mobile–immobile water model (7.31) and (7.32) for different values of the model parameters. The information has been summarized in dimensionless variables (see Section 7.3.7) for compactness. These figures show how versatile this model is for representing shapes of breakthrough curves. For example, in Fig. 7.9a the mobile water fraction $\beta = \theta_m/\theta$ is varied from 0.5 to 1.0 while keeping the other parameters constant. The effect of lowering β on the outflow pulse is to move the breakthrough curve farther to the left as more and more of the wetted pore space is excluded from an active role in transport. The limiting case $\beta = 1$ reduces to the simpler one-region convection–dispersion model (7.17) discussed earlier.

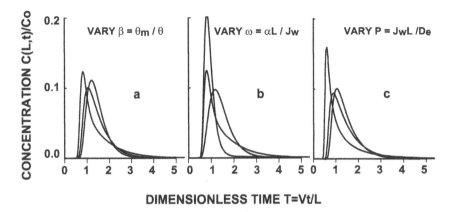

Figure 7.9 Sensitivity of the model parameters for the MIM model (7.31)–(7.32).

In Fig. 7.9*b*, the dimensionless rate parameter $\omega = \alpha L/J_w$ is varied over a large range, from 0.1 to 10.0. In the limit as $\alpha \to \infty$ the two concentrations merge ($C_m = C_{im}$) because they mix instantaneously [see (7.32)], and the MIM again reduces to the convection–dispersion model (7.17). At the other extreme of zero mixing when $\alpha \to 0$, then $\partial C_{im}/\partial t = 0$, and (7.17) reduces to a convection–dispersion model (7.17) but with a total water content θ_m, since the immobile region is blocked off completely. For this case the solute moves at a velocity $V_m = J_w/\theta_m$ and breaks through sooner than if $\alpha = \infty$. At intermediate values of α, the solute arrives at a time between the arrival times of the two limiting cases, but now has a long effluent tail, corresponding to rate-limited diffusion from the stagnant regions.

Figure 7.9*c* illustrates the effect of varying $P = J_w L/D_e$ on the outflow shape for with the other parameters held constant. Here the influence of dispersion is far less pronounced than in the simple convection–dispersion model because much of the mixing and pulse spreading has been described instead by rate-limited diffusion between mobile and stagnant regions. The remaining spreading within the mobile region is less significant for this reason.

The MIM includes three significant model parameters: mobile water content θ_m, mass exchange coefficient α, and dispersion coefficient D_e, to describe nonsorbing, conservative solute transport. Determining these parameters is not easy, especially in the field, and only a few indirect methods have been developed to estimate them independently. The model can be fit by simultaneous optimization methods, but the parameter uncertainty is often high.

Clothier et al. (1992) proposed an approximate method to estimate θ_m in surface soil. Their approach involved infiltration of a single conservative tracer solution of concentration C_0 via a tension infiltrometer followed by rapid sampling of the soil immediately under the disk. They assumed that during the time of the experiment solute remained completely in the mobile domain, so that by measuring the final concentration C of a completely mixed extract and the total water content θ they could determine θ_m from

$$\theta_m = \theta \frac{C}{C_0} \tag{7.33}$$

Jaynes et al. (1995) provided a simple procedure to estimate the θ_{im} and α parameters of the MIM without the need for extensive breakthrough experiments. Their method involved the application of a sequence of conservative tracers having similar transport characteristics (benzoic acid tracers) at different times during a single infiltration experiment. Assuming piston flow of chemicals in the mobile domain of the soil, and constant concentration C_0 in the mobile zone, Jaynes et al. (1995) derived the following equation:

$$\ln\left(1 - \frac{C}{C_0}\right) = \ln \frac{\theta_{im}}{\theta} - \frac{\alpha t^*}{\theta_{im}} \tag{7.34}$$

where C is the chemical mass per volume of soil solution, C_0 is the chemical concentration of the input solution, θ is the total water content, $t^* = t - \tau$ is the elapsed time since the front containing the tracer of interest reached the location where the final sample was obtained, t is the time since the tracer was applied, and τ is the travel time of the tracer from the entry to the location of sampling. As shown in 7.34, plotting

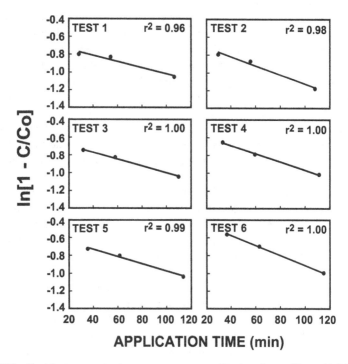

Figure 7.10   Resident concentrations versus tracer application times. (From Al-Jabri et al., 2002a.)

$\ln(1 - C/C_0)$ versus t^* results in a straight line with a negative slope. The immobile water fraction can be determined from the resulting intercept, and α can be computed from the slope.

Figure 7.10 shows the results of this method in a laboratory experiment conducted by Al-Jabri et al. (2002a) and their fitted regression lines produced by the point source method for all observation sites. Differences between the measured and predicted transport behavior are attributed to some experimental limitations, such as difficulty in measuring the small differences in tracer concentration (Jaynes and Horton, 1998). Linear regression lines for all tests produced good fits of the measured data, suggesting that the multiple application of tracers from the point source method produced a good linearity and can be well described using 7.34.

The Jaynes et al. (1995) method was tested in the field (Casey et al., 1997, 1998) and in the laboratory (Lee et al., 2000b), and it was reported that the method provided MIM parameters representative of the soil. Al-Jabri et al. (2002a) extended the method to be used with a dripper-supplied ponded infiltration setup. Al-Jabri et al. (2002b) used the dripper setup in the field to determine surface properties at 50 locations simultaneously. Lee et al. (2000a) showed how to use the Jaynes et al. (1995) method with TDR-collected relative concentration data as a function of time. The use of TDR means that a simple salt solution can be used as a single tracer instead of the suite of tracers used by Jaynes et al. (1995). Lee et al. (2002) extended the TDR method so that all of the MIM parameters, including D_e, could be determined. Al-Jabri (2001) combined the TDR method with a dripper-supplied ponded infiltration setup in order to measure all three MIM transport properties simultaneously at several surface locations. The setup is depicted in Fig. 7.11.

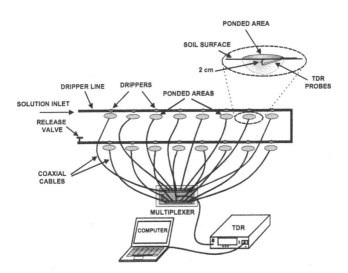

Figure 7.11 Experimental TDR dripper-supplied ponded infiltration setup for two transects. (Courtesy of S. Al-Jabri.)

7.3.4   Reactions of Chemicals in Soil

First-Order Decay   Many organic chemicals in soil are broken down by microbial or chemical reactions. Although this reaction can depend in a complex manner on temperature, pH, microbial population density, carbon content, and other factors, under optimum conditions it may be described as a first-order decay process (Hamaker, 1972). A mass of chemical $M(t)$ undergoing first-order decay loses material at a rate proportional to its mass. This loss rate is expressed mathematically as

$$\frac{dM}{dt} = -\mu M \tag{7.35}$$

where μ is the first-order decay rate constant.

Example 7.5   Calculate the mass remaining in the soil as a function of time for a chemical obeying (7.35). Assume that $M = M_0$ at $t = 0$.

SOLUTION: Equation (7.35) may be integrated if all factors that depend on M are placed on the same side. Thus, we rearrange the equation to

$$\frac{dM}{M} = -\mu \, dt$$

This may be integrated from $t = 0$ to t, with the result

$$\int_{M_0}^{M(t)} \frac{dM}{M} = \ln \frac{M}{M_0} = -\mu \int_0^t dt = -\mu t$$

Taking the exponential of both sides, we obtain finally

$$M(t) = M_0 \exp(-\mu t)$$

Thus, a chemical undergoing first-order decay loses its mass at an exponentially declining rate over time. This simple result will help us understand what happens when a moving mass of chemical is experiencing first-order decay.

Convection–Dispersion Equation and First-Order Decay   A mobile chemical obeying the CDE and undergoing first-order decay must obey (7.16) with $C_a = 0$, $r_a = 0$, and $r_l = \mu C_l$. Therefore, the transport equation may be written as (assuming steady flow)

$$\theta \frac{\partial C_l}{\partial t} = D_e \frac{\partial^2 C_l}{\partial z^2} - J_w \frac{\partial C_l}{\partial z} - \mu \theta C_l \tag{7.36}$$

The breakthrough curve at $z = L$ for this chemical is shown in Fig. 7.12 for various values of μ. The effect of first-order decay on the outflow curve is to remove the long tail of the breakthrough curve preferentially, since this portion of the curve contains

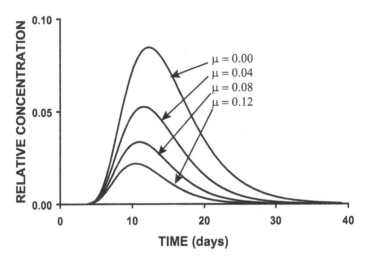

Figure 7.12 Outflow concentration versus time calculated for a mobile chemical as a function of the value of the first-order decay constant μ.

solutes that spend the longest time in the column. As a result, the average travel time of the pulse decreases as decay increases, and it appears to speed up.

Piston Flow Model and First-Order Decay   The piston flow model approximation ($D = 0$) for first-order decay is quite simple. The chemical moves with a velocity $V = J_w/\theta$ and reaches a distance $z = L$ in a time $t_b = L\theta/J_w$. If the chemical was added as a front of concentration C_0 at time $t = 0$, it will have a concentration $C_0 \exp(-\mu t_b) = \exp(-\mu L/V)$ when it arrives at $z = L$. If it is added as a pulse of mass M_0, the mass of the pulse that passes $z = L$ at time $t = L/V$ will be $M_0 \exp(-\mu L/V)$.

7.3.5 Transport of Volatile Organic Compounds through Soil

Volatile organic compounds, including many pesticides, display quite complex behavior in soil because they may move in the vapor phase as well as within solution. They adsorb to stationary soil solid material and can be degraded by biological and chemical reactions. To describe all of these processes, the chemical mass balance equation (7.1) must be combined with the full chemical storage expression (7.2) and both vapor (7.4) and solute (7.12) fluxes, producing

$$\frac{\partial}{\partial t}(\rho_b C_a + \theta C_l + a C_g) = \frac{\partial}{\partial z}\left(D_s^g \frac{\partial C_g}{\partial z}\right) + \frac{\partial}{\partial z}\left(D_e \frac{\partial C_l}{\partial z}\right)$$

$$- \frac{\partial}{\partial z}(J_w C_l) - r_c \tag{7.37}$$

This equation has three unknowns (C_a, C_l, C_g). Therefore, it cannot be solved until two additional equations are specified defining the relationship between the phases. These are sometimes called *phase partitioning laws.*

Vapor–Liquid Partitioning   We have already discussed equilibrium and nonequilibrium laws relating the dissolved and adsorbed phases. The relationship most often used to describe liquid–vapor partitioning is the linear equilibrium formulation known as *Henry's law.* Henry's law was originally formulated as a relationship between the solute mole fraction X in solution and the vapor pressure P_v of the vapor in equilibrium with the solution

$$P_v = k_H X \tag{7.38}$$

where k_H (pressure per mole fraction) is Henry's constant. We may write it as a relation between C_g and C_l:

$$C_g = K_H C_l \tag{7.39}$$

where K_H is the dimensionless Henry's constant. It is straightforward to demonstrate with the ideal gas law (see Problem 7.8) that

$$K_H = \frac{M_w}{\rho_w RT} k_H \tag{7.40}$$

where M_w is the molecular weight of water.

Linear Partitioning between Phases   If a compound satisfies both linear equilibrium adsorption (7.23) and the equilibrium form of Henry's law (7.39), the total concentration of chemical C_T in (7.2) may be expressed in terms of any one of the phases as follows (Jury et al., 1983):

$$\begin{aligned} C_T &= \rho_b C_a + \theta C_l + a C_g \\ &= \rho_b K_d C_l + \theta C_l + a K_H C_l \\ &= (\rho_b K_d + \theta + a K_H) C_l \equiv R_l C_l \end{aligned} \tag{7.41}$$

where

$$R_l = \frac{C_T}{C_l} = \rho_b K_d + \theta + a K_H \tag{7.42}$$

is the liquid-phase partition coefficient. Similarly,

$$R_a = \frac{C_T}{C_a} = \rho_b + \frac{\theta}{K_d} + \frac{a K_H}{K_d} \tag{7.43}$$

$$R_g = \frac{C_T}{C_g} = \frac{\rho_b K_d}{K_H} + \frac{\theta}{K_H} + a \tag{7.44}$$

are the adsorbed-phase and gas-phase partition coefficients, respectively.

We may also define phase mass fractions, which express the fraction of the total chemical mass residing in a given phase:

$$f_a = \frac{\rho_b C_a}{C_T} = \frac{\rho_b}{R_a} = \frac{\rho_b}{\rho_b + (\theta + aK_H)/K_d} = \frac{\zeta}{1 + \zeta + \xi} \qquad (7.45)$$

$$f_l = \frac{\theta C_l}{C_T} = \frac{\theta}{R_l} = \frac{\theta}{\theta + \rho_b K_d + aK_H} = \frac{1}{1 + \zeta + \xi} \qquad (7.46)$$

$$f_g = \frac{aC_g}{C_T} = \frac{a}{R_g} = \frac{a}{a + (\rho K_d + \theta)/K_H} = \frac{\xi}{1 + \zeta + \xi} \qquad (7.47)$$

where

$$\zeta = \frac{\rho K_d}{\theta} \qquad \xi = \frac{aK_H}{\theta} \qquad (7.48)$$

With these equations we can determine where mass resides as a function of soil and chemical properties. This is illustrated in the next example.

Example 7.6 Table 7.2 presents chemical phase properties K_d and K_H for six pesticides in a loamy soil. The phase mass fractions for these compounds are given in Table 7.3 assuming that $\theta = 0.25$, $a = 0.25$, and $\rho_b = 1.5$ g cm^{-3}.

TABLE 7.2   Distribution Coefficients K_d (cm^3 g^{-1}) and Henry's Constant Values for Six Pesticides in a Loamy Soil

Compound	K_d	K_H
Atrazine	1.6	2.5×10^{-7}
Bromacil	0.7	3.7×10^{-8}
DBCP	1.6	8.3×10^{-3}
DDT	2400.0	2.0×10^{-3}
Lindane	13.0	1.3×10^{-4}
Phorate	6.6	3.1×10^{-4}

Source: Jury et al. (1984b).

TABLE 7.3   Calculated Percent Mass in Each Phase for the Six Compounds in Table 7.2

Compound	$100 f_a$	$100 f_l$	$100 f_g$
Atrazine	90.570	9.430	2.4×10^{-6}
Bromacil	81.200	18.800	7.0×10^{-8}
DBCP	88.550	11.350	9.0×10^{-2}
DDT	99.993	0.007	1.4×10^{-5}
Lindane	98.730	1.270	1.7×10^{-4}
Phorate	97.540	2.460	7.6×10^{-4}

Example 7.6 illustrates several features of pesticide partitioning. When the compound is attracted to soil solids, much of the mass resides in that phase. Such compounds might be expected to resist leaching to groundwater but may attach to soil particles that are carried off agricultural fields by erosion. Chemicals with substantial mass in the dissolved phase might be prone to leaching to groundwater. Finally, very little of the mass resides in the vapor phase. However, this does not mean that vapor-phase transport is negligible, as will be seen in the next section.

Effective Liquid–Vapor Diffusion   If we ignore liquid dispersion, which might be reasonable if the convective velocity of the compound is low in soil, then using (7.4) and (7.7), the combined liquid and vapor diffusion of the compound may be written as

$$J_v + J_{ld} = -\xi_g(a)D_g^a \frac{\partial C_g}{\partial z} - \xi_l(\theta)D_l^w \frac{\partial C_l}{\partial z} \tag{7.49}$$

Assuming equilibrium linear partitioning, we may rewrite (7.49) in terms of the partition coefficients as

$$J_v + J_{ld} = -\left[\frac{\xi_g(a)D_g^a}{R_g} + \frac{\xi_l(\theta)D_l^w}{R_l} \right] \frac{\partial C_T}{\partial z} \tag{7.50}$$

Equation (7.50) may be written compactly as

$$J_g + J_{ld} = -D_E \frac{\partial C_T}{\partial z} \tag{7.51}$$

where

$$D_E = \frac{\xi_g(a)D_g^a}{R_g} + \frac{\xi_l(\theta)D_l^w}{R_l} \tag{7.52}$$

is the effective vapor and liquid diffusion coefficient (Jury et al., 1983). Whether this coefficient represents predominantly liquid or vapor diffusion for a given compound depends on the value of its vapor and liquid partitioning coefficients and on how much water is in the soil.

Example 7.7   For the six pesticides in Example 7.6, calculate and plot D_E as a function of water content, assuming that the porosity $\phi = 0.5$, $\rho_b = 1.5$ g cm^{-3}, $D_l^w = 0.432$ cm^2 day^{-1}, and $D_g^a = 4320$ cm^2 day^{-1} for all pesticides (Jury et al., 1983). Assume for illustration purposes that each tortuosity factor can be expressed by the Millington–Quirk (1961) model:

$$\xi_g(a) = \frac{a^{10/3}}{\phi^2} \qquad \xi_l(\theta) = \frac{\theta^{10/3}}{\phi^2} \tag{7.53}$$

SOLUTION: The effective diffusion coefficient D_E in (7.52) may be written as

$$D_E = \frac{(\phi - \theta)^{10/3} K_H D_g^a + \theta^{10/3} D_l^w}{[\rho_b K_d + \theta + (\phi - \theta)K_H]\phi^2} \tag{7.54}$$

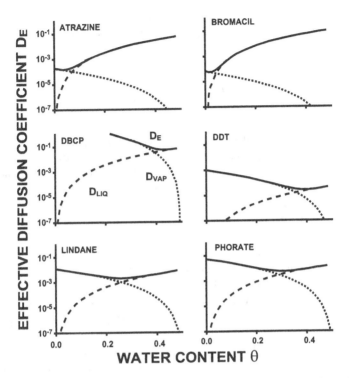

Figure 7.13 Effective diffusion coefficient D_E (7.54) and the contribution from the vapor and dissolved phases (7.31)–(7.32).

as a function of water content. Figure 7.13 shows a plot of D_E, along with the contribution from the vapor and liquid terms, for the six compounds as a function of θ.

The total diffusion coefficient is vapor dominated for compounds with large K_H (DBCP, DDT) and liquid-dominated for the low-K_H compounds (atrazine, bromacil). Lindane and phorate are vapor-dominated at low θ and liquid dominated at high θ. As a rough rule, compounds with $K_H \gg 2.5 \times 10^{-5}$ are vapor dominated and those with $K_H \ll 2.5 \times 10^{-5}$ are liquid dominated at 50% water saturation (Jury et al., 1984a; Letey and Farmer, 1973).

Total Chemical Flux   If we neglect solute dispersion, the total chemical flux is the sum of the chemical convection and diffusion terms. Thus, using (7.41) and (7.51) yields

$$J_c = J_w C_l - D_E \frac{\partial C_T}{\partial z} = V_E C_T - D_E \frac{\partial C_T}{\partial z} \tag{7.55}$$

where $V_E = J_w / R_l$ is the chemical velocity.

If the flux equation (7.51) is plugged into the transport equation (7.37) and first-order decay is assumed, the new equation may be written as (assuming uniform soil properties and constant J_w)

$$\frac{\partial C_T}{\partial t} = D_E \frac{\partial^2 C_T}{\partial z^2} - V_E \frac{\partial C_T}{\partial z} - \mu_E C_T \qquad (7.56)$$

where μ_E is the effective first-order degradation coefficient.

Equation (7.56) is now identical to the convection–dispersion equation (7.36) for a compound undergoing first-order decay. It may be solved exactly for many boundary conditions of interest, yielding different behavior for compounds with distinct chemical and adsorption properties (K_d, K_H, μ_E). Thus, it makes an ideal screening model for predicting the relative behavior of different chemicals in the same environment (Jury et al., 1983, 1984a–c).

Volatilization of Chemicals from Soil   Chemical volatilization through the soil surface is an important problem that may be studied by solving (7.56) with the appropriate boundary condition. Jury et al. (1983) formulated the upper boundary condition for this problem by assuming that the chemical flux at the surface had to pass through an air boundary layer by vapor diffusion before entering the well-mixed region of the atmosphere.

Figure 7.14 illustrates the boundary layer region above the soil surface. It is conceptualized as an air layer of thickness d through which vapor diffuses according to Fick's law (6.7), written in finite-difference form as

$$J_{cb} = -D_g^a \frac{C_g(0, t) - C_g^a}{d} \qquad (7.57)$$

where $C_g(0, t)$ is the vapor concentration at the soil surface, J_{cb} the chemical vapor flux through the boundary layer, and C_g^a the chemical concentration in the atmosphere above the air layer, which Jury et al. (1983) assumed was zero. They also estimated the value of d from measured water evaporation rates and recommended $d = 0.5$ cm as an average value for a bare surface. The transport equation (7.56) was solved by these authors, together with the upper boundary condition

$$J_c(0, t) = V_E C_T - D_E \frac{\partial C_T}{\partial z}\bigg|_{z=0} = J_{cb} = -D_g^a \frac{C_g(0, t) - C_g^a}{d} \qquad (7.58)$$

and the initial condition

$$C_T(z, 0) = \begin{cases} C_0 & \text{for } -L < z < 0 \\ 0 & \text{for } z < -L \end{cases} \qquad (7.59)$$

where L is the depth of incorporation of the compound.

The model calculations were conducted under standard conditions consisting of uniform upward water flux (evaporation), zero water flux, and uniform downward

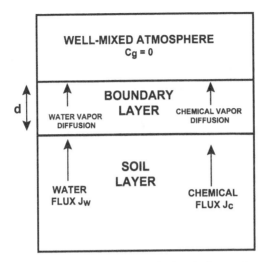

Figure 7.14 Stagnant boundary layer in the organic chemical screening model of Jury et al. (1983).

water flux (leaching). The behavior of the chemicals can be summarized in various ways, most compactly in terms of the percentage of the initial mass $M = C_0L$ that has been volatilized, has degraded, has leached below a specified layer, or remains in the soil. Figure 7.15 summarizes the volatilization flux as a function of time for the six chemicals in Table 7.2 that were incorporated to a depth of 10 cm under the same soil and climate conditions ($\rho_b = 1.5$ g cm^{-3}, $\theta = 0.25$, $f_{oc} = 0.01$).

The compounds behave differently over time and react differently to the presence or absence of water flux. Atrazine and bromacil both have insignificant volatilization fluxes when water is not flowing, but their volatilization fluxes increase with time when water is flowing upward and can eventually become significant. In contrast, the other compounds all have volatilization rates that decrease as a function of time under all water flow regimes. The volatilization rates of DBCP, lindane, and phorate are all enhanced significantly by evaporation, whereas DDT, which is virtually immobile in soil in the liquid phase, has the same volatilization–time curve irrespective of the rate and direction of water flow.

Jury et al. (1984a) explained this volatilization behavior by the influence of the boundary layer. Compounds with a low K_H value, such as atrazine and bromacil, normally have a low vapor density in soil, and the boundary layer represents a much greater resistance to flow than the soil. As a result, when water is flowing upward, the chemical accumulates at the interface and the volatilization rate increases with time as the vapor density at the surface increases.

Compounds with a large K_H value, such as the other four pesticides in Fig. 7.15, have a higher vapor density in soil, and the boundary layer offers comparatively little resistance. As a result, the chemical is carried off as rapidly as it is transported to the surface and does not build up with time. The authors classified compounds into

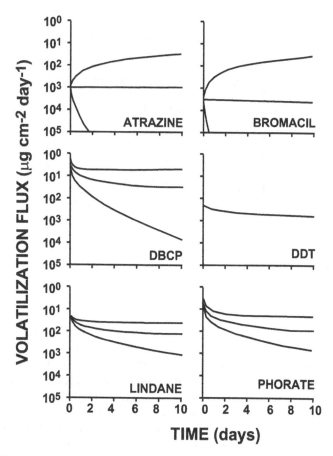

Figure 7.15 Predicted volatilization flux rate versus time for six pesticides. Upper, middle, and lower curves on each diagram are for uniform water evaporation at 0.5 cm day^{-1}, zero evaporation, and uniform downward flow at 0.5 cm day^{-1}, respectively.

different categories based on whether $K_H > 2.5 \times 10^{-5}$ (category 1) or $K_H < 2.5 \times 10^{-5}$ (category 3) (Jury et al., 1984a). Their predictions were verified experimentally for prometon (category 3) and lindane (category 1) (Spencer et al., 1988). In this study, prometon accumulated over time at the surface, reaching values in the top few millimeters of soil that were over 20 times higher than the initial concentration. In contrast, lindane had a very low surface concentration at the end of the study.

The environmental fate summary for the six pesticides during the 10-day simulations is given in Table 7.4. Notable in Table 7.4 are the significant losses of DBCP and phorate under water evaporation and the high degradation of atrazine and phorate.

Isothermal conditions have been assumed for the equations we have presented. In reality nonisothermal conditions usually occur in shallow field soil. See Nassar and Horton (1999b) and Nassar et al. (1999) for equations that can be used to describe transport of volatile organic compounds under nonisothermal soil conditions.

TABLE 7.4   Environmental Fate Summary for Six Pesticides after 10 Days of Volatilization with and without Water Evaporation

Compound	Percent Volatilized $E = 0.5$	$E = 0$	Percent Degraded $E = 0.5$	$E = 0$	Percent Remaining $E = 0.5$	$E = 0$
Atrazine	2.08	0.07	9.23	9.30	88.69	90.63
Bromacil	1.32	0.02	1.95	1.98	96.73	98.00
DBCP	24.28	6.68	0.60	0.77	75.12	92.55
DDT	0.22	0.21	0.18	0.18	99.60	99.61
Lindane	2.88	1.26	2.53	2.56	94.59	96.17
Phorate	5.74	2.61	7.84	7.99	86.41	89.40

7.3.6   Solute Transport during Transient Water Flow

All of the preceding examples of solute transport involved water flowing at a constant flux rate throughout the medium. In the more general case when water flow is transient, (7.11) must be solved simultaneously with the water flow equation (3.70) by numerical methods. Two examples using the HYDRUS-1D code (Simunek et al., 1997) will help to illustrate the versatility of numerical methods.

Downward Transport of a Solute Pulse   In Chapter 3 (see Fig. 3.21) we showed that downward water flow becomes relatively constant at sufficient depth below the point of application. It is useful to explore the corresponding case of downward chemical transport when water flow into the soil varies as a function of time. Figure 7.16a shows the $z = 30$ cm breakthrough curve (solid curve) of a chemical pulse added at $C_0 = 5$ during $0 < t < 3$ days to soil initially at uniform matric potential ($h = -40$ cm). The soil receives water at the rate shown as the dashed curve for $t > 0$. The breakthrough curve as a function of time has a complex shape, reflecting the changing speed with which it is moved through the soil. However, if the outflow concentration is plotted (Fig. 7.16b) as a function of the cumulative drainage past $z = 30$ cm, the breakthrough (solid circles) appears quite regular and can be fitted quite well to the linear CDE model with a constant V and D. Thus, for net downward flow processes such as chemical movement to groundwater under an irrigated field, the variable water flux rate may be replaced with a constant flux without loss of accuracy, provided that an appropriate value of D is used.

Upward Transport of a Solute Pulse   In Chapter 3 we noted that finer-textured soils could transport water and chemicals upward over greater distances above a water table than could coarser-textured soils. As a result, the former posed a greater risk for salinization of surface layers by salt that remained after water evaporated at the surface. Figure 7.17 shows the result of a 200-day simulation of water and solute movement upward from a water table located at $z = -100$ cm. The soil surface is held at a low matric potential and zero solute flux, and the water table is held at $h = 0$ and $C_0 = 5$ (arbitrary units). The profile is assumed to be at $h = -50$ cm initially, causing a high upward flux at short times due to the sharp potential gradient

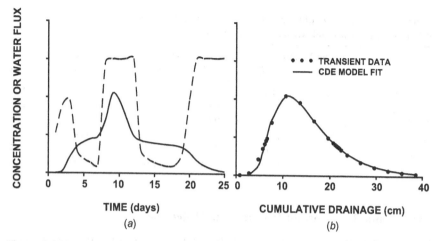

Figure 7.16 (a) Calculated breakthrough curve at $z = 30$ cm (solid line) of a solute pulse added at $C_0 = 5$ (arbitrary units) to a soil initially at $h = -40$ cm while the surface receives water at a rate shown by the dashed line; (b) replot of the solute concentration against cumulative drainage past $z = 30$ cm (solid circles). The solid line is a sum of squares fit to the curve with the linear convection–dispersion equation (CDE) (7.17). All calculations were performed with the HYDRUS-1 code of Simunek et al. (1997).

near the water table. This high flux declines over time to a final steady value as the matric potential gradient smoothes out. The finer-textured silt soil is able to move water upward over the distance, reaching a final water flux value of 0.11 cm day^{-1}. The loamy sand, on the other hand, can support only an insignificant final water flux, causing the solute profile to remain suspended below the surface after the initially high water flux dies out.

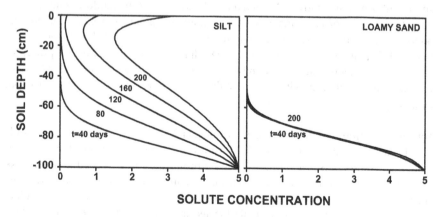

Figure 7.17 Upward flow of solute from a water table ($h = 0$) at $z = -100$ cm in two soils initially at $h = -50$ cm. Surface is held at a low $h = -500$ cm matric potential. All calculations were performed with the HYDRUS-1 code of Simunek et al. (1997).

7.3.7 Dimensional Analysis

The models discussed in this chapter contain many parameters. Depending on the values of rate parameters, the models may or may not behave similarly to equilibrium limits of the models. The task of deciding which form of the model to use, and what assumptions to make, can be simplified by the use of dimensional analysis. For transport problems, this requires defining a characteristic length and time scale. If we are looking at one-dimensional outflow experiments under steady water flow, a logical length scale is the distance L from the inlet to the outlet. Similarly, a convenient time scale is the breakthrough time $t_b = L/V$ for a mobile solute. Thus we define new dimensionless distances and times as

$$Y = \frac{z}{L} \tag{7.60}$$

$$T = \frac{tV}{L} = \frac{J_w t\theta}{L\theta} \tag{7.61}$$

Dimensionless CDE   If we insert the expressions (7.60)–(7.61) into the CDE (7.17) and rearrange, the result is

$$\frac{\partial C}{\partial T} = \frac{1}{P}\frac{\partial^2 C}{\partial Y^2} - \frac{\partial C}{\partial Y} \tag{7.62}$$

where

$$P = \frac{LV}{D} \tag{7.63}$$

is the *Peclet number*,[6] the ratio of the convective time scale L/V (the time to move a distance L by convection) to the diffusive time scale L^2/D (the time to move a distance L by diffusion). Note also that $T = 1$ corresponds to one pore volume of drainage, or to the breakthrough time. Thus, the value of P may be used to characterize the nature of the process. In addition, plots of $C(Y, T)$ versus T for different P values are valid for all values of D, L, and V that have the same P.

Dimensionless MIM   As shown in Problem 7.10, substitution of (7.60)–(7.61) into the MIM (7.31)–(7.32), yields the dimensionless equations

$$B\frac{\partial C_m}{\partial T} + (1 - B)\frac{\partial C_{im}}{\partial T} = \frac{1}{P}\frac{\partial^2 C_m}{\partial Y^2} - \frac{\partial C}{\partial Y} \tag{7.64}$$

$$(1 - B)\frac{\partial C_{im}}{\partial T} = W(C_m - C_{im}) \tag{7.65}$$

---

[6]This is often called the *Brenner number* in engineering to distinguish it from the Peclet number, which uses a microscopic length scale.

where

$$P = \frac{L J_w}{D_e} \qquad B = \frac{\theta_m}{\theta} \qquad W = \frac{\alpha L}{J_w} \tag{7.66}$$

The parameter B expresses the extent of immobile water in the system, and the parameter W is the ratio of the convective time scale $L/V = L\theta_w$ to the mass transfer time scale α^{-1}.

7.4   TRANSFER FUNCTION MODEL OF SOLUTE TRANSPORT THROUGH SOIL

The convection–dispersion equation (CDE) of solute transport is based on a set of assumptions that are valid only under certain conditions in soil. The CDE makes the hypothesis that the dispersion process is formally equivalent to diffusion even though dispersion is a convective transport process. In order for the dispersive mixing to behave analogously to diffusion, solute transported by convection along the direction of mean transport at a rate other than the average convective velocity must have time to mix with other regions of the wetted porous medium before the outlet end is reached; otherwise, the CDE cannot describe the transport with a constant dispersion coefficient (Gelhar and Axness, 1983; Taylor, 1953).

The simple hypothetical example shown in Fig. 7.18 illustrates the criteria for the validity of the CDE. In this situation the transport medium receiving a constant water flux consists of two parallel regions with water contents θ_1 and $\theta_2 = 2\theta_1$, so that the water velocity through each region is V_1 and $V_2 = V_1/2$. A narrow pulse of solute is added to the inlet end at $t = 0$ and proceeds down the column as shown in the figure. At an early time t_1, the solute front has progressed much farther down the column in the fast region than in the slow region, and transfer of solute from the fast to slow regions (by diffusion or transverse convection) has just begun. Later, at time t_2, the fronts are closer together because more transfer between the regions has occurred. Finally, at t_3 the mixing between the regions has had time to reach

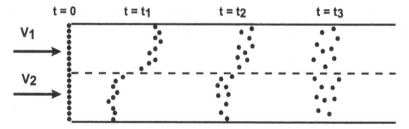

Figure 7.18   Solute pulse moving through a medium with $V = V_1$ in the upper region and $V = V_2 = V_1/2$ in the lower region. Solute mixes between the two fronts as it moves through the medium.

completion and there is only one front, which moves at the average[7] velocity $3V_1/4$. For times $t > t_3$, the CDE will describe transport and longitudinal mixing of solute through the system.

Because the CDE is valid only after sufficient time has elapsed to smooth out transport by convective velocity differences along the direction of motion, it cannot be valid at early times and requires longer to be valid in large, structured systems (such as natural fields) than in small, uniform systems (such as repacked soil columns). For this reason, Jury (1982) developed an alternative formulation for solute transport, called the transfer function model, which does not require the restrictive assumptions of the CDE and can be applied to problems at different scales.

7.4.1 Solute Transport as a Transfer Function

If we add a large number of solute molecules at $t = 0$ to the inlet end of a solute transport volume like a soil column, we can interpret the outflow concentration as a measurement of the distribution of travel times for the chemical from the inlet to the outlet end. For example, suppose we add a total of N molecules at $t = 0$, and capture the effluent in a series of time intervals $\Delta t, 2\Delta t, 3\Delta t, \ldots$, measuring $n_1, n_2, n_3, \ldots$ molecules in each interval. Then we can make the statement that the probability of arriving at the outflow end in a time $t \leq \Delta t$ is $P_1 = n_1/N$. Similarly, the probability of arriving at the outflow end in a time $t \leq 2\Delta t$ is $P_2 = (n_1 + n_2)/N$, and so forth. If we plot P versus t as in Fig. 7.19, we obtain an experimental estimate of the cumulative probability distribution $P(t)$ for travel time (see the Appendix). By the same reasoning, we may interpret the ratio n_K/N of solute molecules n_K arriving in any time interval $\Delta t_K : (K - 1)\Delta t < t < K\Delta t$ to the total number N added as the probability of having *a* travel time $(K - 1)\Delta t < t < K\Delta t$. This function, shown in Fig. 7.20, is called the *travel-time probability density function* (pdf) $f(t)$, and is the derivative of the cumulative probability function.

$$f(t) = \frac{dP(t)}{dt} \rightarrow P(t) = \int_0^t f(t')\,dt' \tag{7.67}$$

The probability density function is normalized to have unit area under its curve when plotted as a function of time. This is equivalent [see (7.67)] to the statement that $P(\infty) = 1$, i.e. that the probability of having less than an infinite travel time is 1.

Of course, we don't have to count molecules to measure $f(t)$ or $P(t)$; we only have to measure solute concentration at the outflow end. By definition, if we add a step change of solute concentration C_0 to the inflow end of a medium at $t = 0$ that has been receiving concentration-free water previously, then

$$P(t) = \frac{C(L, t)}{C_0} \tag{7.68}$$

---

[7] This is the average velocity calculated by examining the system at a fixed time. The average velocity calculated by observing solute passing a fixed distance over time would be $2V_1/3$.

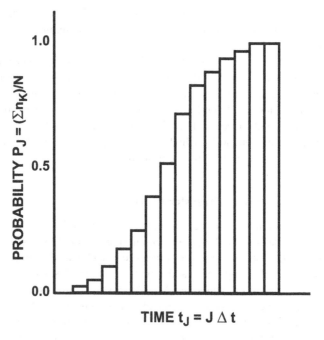

Figure 7.19 Experimental measurement of the cumulative travel-time probability distribution.

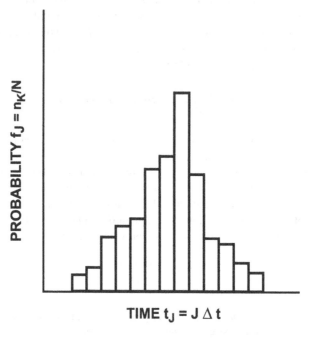

Figure 7.20 Experimental measurement of the travel-time probability density function.

where L is the distance to the outflow end. Similarly, if we add a narrow pulse of solute to the inflow end at $t = 0$, then

$$f(t) = \frac{C(L, t)}{\int_0^\infty C(L, t') \, dt'} \tag{7.69}$$

7.4.2 Transfer Function Equation

If we make the following two assumptions about our system:

1. The solute transport process is linear.
2. The solute travel time probabilities do not change over time.

then it is easy to show that the solute concentration at the outflow end of a solute transport volume that is receiving a time-dependent input of solute $C_0(t)$ is given by

$$C_{\text{out}}(t) = \int_0^t C_{\text{in}}(t - t') f(t') \, dt' \tag{7.70}$$

Equation (7.70) is called the *transfer function equation*. The outflow at time t consists of the superposition of solute added at all times less than t, weighted by its travel-time probability. In its present form it is suitable for describing systems with outlets that can be sampled, such as soil columns or tile drains. Note that the function only describes outflow concentration, and says nothing about solute characteristics between the inflow and outflow surfaces.

In its present form, (7.70) is not very useful, because it only applies to systems that meet the strong assumption that the travel-time probabilities do not change over time. The only circumstance where this would be true is if the water flux through the system is constant and has the same value as it had when $f(t)$ was measured. Recognizing this extreme limitation, Jury (1982) expressed the travel-time pdf as a function of cumulative net applied water I, where

$$I = \int_0^t J_w(t') \, dt' \tag{7.71}$$

He then made the additional assumption that $f(I)$ is a unique function of the solute transport volume, which is equivalent to assuming that solute outflow expressed as a function of cumulative drainage does not depend on the water flux. With this assumption, we may rewrite (7.70) as

$$C_{\text{out}}(I) = \int_0^{I(t)} C_{\text{in}}(I - I') f(I') \, dI' \tag{7.72}$$

The assumption that solute outflow will be a unique function of the drainage or net applied water volume has been shown to be accurate for repacked soil columns subjected to different water flow rates in the laboratory (Khan, 1988). It has also been used successfully to describe area-averaged solute concentrations in a large field study (Butters et al., 1989; Jury et al., 1982b, 1990) and has been shown to be accurate by computer simulation of solute transport in a sandy soil (Wierenga, 1977). The assumption is less likely to be met in structured soils, where preferential flow channels become active at high water flux rates (Dyson and White, 1986).

Example 7.8   The travel-time pdf $f(I)$ was measured for a tile-drained field and could be described well by the function[8]

$$f(I) = \alpha \exp(-\alpha I) \tag{7.73}$$

Calculate the nitrate outflow concentration resulting from a single nitrate input of duration I_s and concentration C_N.

SOLUTION: The nitrate input concentration is thus $C_{in} = C_N$ for $0 < I < I_s$ and $C_{in} = 0$ for $I > I_s$. If we insert this function and (7.73) into (7.72), we obtain

$$C_{out}(I) = \int_0^I C_{in}(I - I')\alpha \exp(-\alpha I')\, dI' = \int_0^I C_{in}(I')\alpha \exp\left[-\alpha(I - I')\right] dI'$$

$$= \int_0^I \alpha \exp\left[-\alpha(I - I')\right] dI' = 1 - \exp(-\alpha I) \qquad \text{if } I < I_s$$

$$= \int_0^{I_s} \alpha \exp\left(-\alpha\left[I - I'\right]\right) dI' = \exp\left[-\alpha(I - I_s)\right] - \exp(-\alpha I) \quad \text{if } I > I_s$$

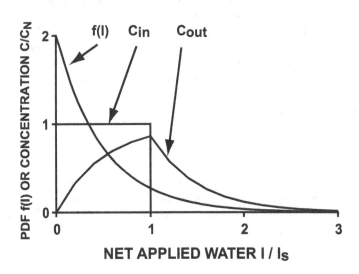

Figure 7.21   Exponential pdf (7.73) and input and output concentrations for Example 7.8. The value of α is $2/I_s$.

[8] Note that $\int_0^\infty f(I)\, dI = 1$ as required for a pdf.

Figure 7.21 shows a plot of the pdf and the input and output concentrations for a particular value of α. Note how the square input pulse is deformed by the distribution of travel times in the system.

Once the travel-time pdf has been obtained in an outflow experiment such as those described in (7.68) or (7.69) and is converted to net applied water, the transport volume fraction can be calculated from either the median $\widehat{I}$ or mean $\overline{I}$ water input displacement, where the mean value is calculated as

$$\overline{I} = \int_0^\infty I' f(I') \, dI' \tag{7.74}$$

Thus, $\overline{I}$ is equal to the first moment of the pdf $f(I)$ (Himmelblau, 1970). The median value $\widehat{I}$ is the value at the middle of the population. It is calculated from

$$P(\widehat{I}) = \int_0^{\widehat{I}} f(I') \, dI' = 0.5 \tag{7.75}$$

The median value $\widehat{I}$ is measured most easily by adding a solute front of concentration C_0 and determining the value of I at which the outflow concentration rises to $0.5C_0$ (White et al., 1986). If solute is moving in one dimension[9] from an inlet surface at $z = 0$ to an outflow surface at $z = L$, we may determine mean and median water displacement fractions or water contents, as

$$\widehat{\theta} = \frac{\widehat{I}}{L} \qquad \overline{\theta} = \frac{\overline{I}}{L} \tag{7.76}$$

where L is the effective thickness of the transport volume and $\widehat{\theta}$ corresponds roughly to the mobile water content θ_m in the MIM equations (7.31) and (7.32) or to the mobile zone of Addiscot (1977). For some applications, the transport volume fraction $\widehat{\theta}$ or $\overline{\theta}$ may be the only measurement of the soil properties required to make rough predictions of solute leaching under field conditions (White, 1987).

7.4.3 Model Distribution Functions

The travel-time pdf is normalized to unit area and is defined for all nonnegative values of t or I. Therefore, it can be represented by parametric frequency distributions that have only positive values of the independent variable I. Two such distributions are the Fickian pdf[10]

$$f(I) = \frac{L}{2\sqrt{\pi D I^3}} \exp\left[-\frac{(L - VI)^2}{4DI} \right] \tag{7.77}$$

---

[9] The movement could be macroscopically one-dimensional, as for a field application from the surface to groundwater at depth L.

[10] Normally, the parameters V and D are used in the CDE when the independent variable is time rather than net applied water I.

and the lognormal distribution

$$f(I) = \frac{1}{\sqrt{2\pi}\,\sigma I} \exp\left[-\frac{(\ln I - \mu)^2}{2\sigma^2}\right] \tag{7.78}$$

where D, V, μ, and σ are model parameters and L is the distance from the outflow surface to the inflow surface (Jury and Sposito, 1985; Simmons, 1982).

The Fickian pdf is the concentration solution at $z = L$ to the CDE (7.17) for a narrow-pulse input of solute, after normalization as in (7.69). Thus, the CDE can be represented as a transfer function equation in the form of (7.70). In fact, any linear solute transport model that obeys the solute conservation equation (7.1) can be represented as a transfer function equation (Jury et al., 1986). Thus, it is a more general formulation than any specific model and can be used to represent any model hypothesis that obeys mass balance and is linear in the concentration.

7.4.4 Stochastic–Convective Transfer Function Model

The foundation of the transfer function model is the pdf $f(t)$ or $f(I)$. Since it is measured by an outflow experiment in which solute moves through the entire transport volume, any model function of $f(I)$ representing its shape well can be used in (7.72) to predict outflow concentrations for different inputs. However, no predictions can be made about the solute concentration anywhere except at the outflow location unless additional assumptions are built into the model (Jury and Roth, 1990). The CDE (7.17) is an example of a process model that can make predictions at every location z. However, its assumptions represent only one possibility for a solute transport model. Jury (1982) proposed a new model that postulated the behavior of the solute concentration at depths $z \neq L$ in the following way: The cumulative distribution function $P_L(I)$ for solute arriving at $z = L$ is by definition equal to the probability that solute will reach the outflow surface at or before an amount of water I has been added. The stochastic–convective hypothesis (Jury, 1982) postulates that in the same system at $z \neq L$, the cumulative probability distribution $P_z(I)$ is related to the cdf at $z = L$ by

$$P_z(I) = P_L\left(\frac{IL}{z}\right) \tag{7.79}$$

Equation (7.79) states that the probability that a solute tracer will reach a depth z when an amount of water I has flowed through the system is equal to the probability that a solute tracer will arrive at $z = L$ (called the *reference depth*) when an amount of water IL/z has been added. For example, if 5% of a solute pulse reaches $z = 30\,\text{cm}$ after $I = 4\,\text{cm}$ of water has been added, then 5% of the pulse will reach $z = 60\,\text{cm}$ after $I = 8\,\text{cm}$ of water has been added, and so on. This model may be interpreted physically as a representation of solute traveling at different velocities in isolated stream tubes whose contents do not have time to mix with adjacent stream tubes before reaching the depth where the observation is made. Thus, it may be thought of as the early time limit of solute transport (see Fig. 7.18), just as the CDE, which assumes complete mixing of adjacent stream tubes, may be thought of as the long-time limit of possible solute behavior.[11]

The pdf $f_z(I)$ at any depth z is by definition equal to the derivative of the cdf. Thus,

$$f_z(I) = \frac{dP_z}{dI} = \frac{L}{z} f_L\left(\frac{IL}{z}\right) \tag{7.80}$$

Therefore, for a stochastic–convective process, solute arrival at any depth z may be expressed in terms of the pdf at the reference depth (Jury, 1982):

$$C(z, I) = \int_0^I C(0, I - I') \frac{L}{z} f_L\left(\frac{I'L}{z}\right) dI' \tag{7.81}$$

A process that obeys (7.80) is said to be stochastic–convective (Jury, 1982; Simmons, 1982). Thus, for a transport volume obeying the stochastic–convective assumption (7.80), (7.81) can be used to predict $C(z, I)$ using only the single experiment calibrating $f_L(I)$.

In contrast, the pdf for the CDE at $z \neq L$ is given by the solution of (7.17) at z for a narrow-pulse input of solute, which is simply equal to the Fickian pdf (7.77) with z substituted for L (Jury and Sposito, 1985):

$$f_z(I) = \frac{z}{2\sqrt{\pi D I^3}} \exp\left[-\frac{(z - VI)^2}{4DI}\right] \tag{7.82}$$

Convective Lognormal Transfer Function Model   The stochastic–convective assumption (7.80), when applied to the lognormal pdf (7.78), yields the simple result

$$f_z(I) = \frac{1}{\sqrt{2\pi}\,\sigma I} \exp\left\{-\frac{[\ln(IL/z) - \mu]^2}{2\sigma I}\right\} \tag{7.83}$$

The transfer function employing this pdf is called the *convective lognormal transfer* (CLT) *function model* (Jury, 1982).

Figure 7.22 plots the CLT model pdf (7.83) and the CDE pdf (7.82) for three values of z when the parameters of the two models have been adjusted for maximum agreement at $z = 30$ cm. Two important features of these curves should be noted. First, at outflow locations other than 30 cm the concentrations predicted by the models do not agree. The CLT model pulse has spread much more at depths $z > 30$ cm than the CDE pulse. Second, at $z = 30$, the model pdfs are almost identical, showing how closely the shapes of the two mathematical functions can be made to agree. Thus, for any soil column experiment in which outflow from a single exit point is modeled, the two approaches would be judged equivalent even though they are based on completely different assumptions about the dispersion process. Consequently, a

---

[11] Note that the CDE does not satisfy (7.79).

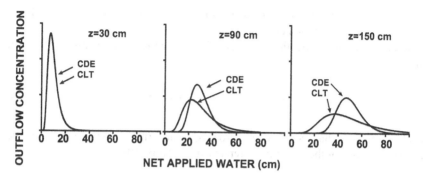

Figure 7.22 Outflow concentrations of the CDE and CLT models after calibration at $z = 30$ cm.

model assumption about dispersion can only be tested by observing solute movement to several different distances from the solute entry point (Khan and Jury, 1990).

7.4.5   Model Parameter Estimates

When parametric models are used to represent the frequency distribution $f_L(I)$, the model parameters must be estimated by optimizing the agreement between the model and the sample measurement of $f_L(I)$. Jury and Sposito (1985) discussed three methods for solute parameter estimation: sum of squares, method of moments, and maximum likelihood. Only the method of moments is discussed here, because it lends itself most easily to a discussion of the differences between model hypotheses. The STANMOD package of parametric models (Simunek et al., 1999) allows data to be fitted by sum of squares optimization.

Method of Moments   The method of moments procedure for estimating model parameters from outflow experiments is used for narrow solute input pulse experiments, in which a set of concentrations $C(z, I_J)$, $J = 1, \ldots, M$ are measured at a fixed distance z from the inlet end at $I_1, I_2, \ldots, I_M$ [or equivalently, at $t_1, t_2, \ldots, t_M$ if $C(t)$ is measured]. The Nth moment T_N of I is defined theoretically as

$$T_N(z) = \int_0^\infty I^N C(z, I) \, dI \tag{7.84}$$

where $T_0 = 1$ if $C(z, I)$ has been normalized [i.e., $C(z, I) = f_z(I)$]. Equation (7.84) may be evaluated in terms of model parameters if the theoretical expression for $f_z(I)$ for a given model can be integrated exactly. For example, it can be shown (Jury and Sposito, 1985; Valocchi, 1985) that for the Fickian pdf (7.77),

$$T_0 = 1 \tag{7.85}$$

$$T_1 = \frac{z}{V} \tag{7.86}$$

$$T_2 = \frac{z^2}{V^2} + \frac{2Dz}{V^3} \tag{7.87}$$

When the data are measured, (7.84) can be evaluated numerically to yield sample measurements T_N^* of T_N. Once these have been obtained, the model parameters may be evaluated. For example, the CDE model parameters are

$$V = \frac{z}{\widehat{T}^1} \tag{7.88}$$

$$D = \frac{V^3}{2z}\left(\widehat{T}_2 - \widehat{T}_1^2\right) \tag{7.89}$$

$$\lambda = \frac{D}{V} = \frac{V^2}{2}\left(\widehat{T}_2 - \widehat{T}_1^2\right) \tag{7.90}$$

where λ is the dispersivity and $\widehat{T}_N = T_N^*/T_0^*$ are the normalized sample moments of the travel-time pdf $f_z(I)$. Since the CDE requires that these parameters be constant at different z, evaluation of T_N at several depths can be used to test this model's validity.

The expression $\widehat{T}_2 - \widehat{T}_1^2$ in (7.90) has a special name (see the Appendix) the *variance of the distribution* $f_z(I)$:

$$\mathrm{Var}_z[I] = \int_0^\infty (I - \mathrm{E}_z[I])^2 f_z(I)\,dI \tag{7.91}$$

where

$$\mathrm{E}_z[I] = \int_0^\infty I f_z(I)\,dI \tag{7.92}$$

is called the *expectation* or *mean value* of I. By (7.91) we see that the variance is the mean-square deviation from the average value. We may use (7.88)–(7.92) to develop a fundamental definition of the dispersivity of the travel-time pdf of a particular solute transport model, as

$$\text{dispersivity of } f_z(I) = \frac{z}{2}\frac{\mathrm{Var}_z[I]}{\mathrm{E}_z^2[I]} = \frac{z}{2}\mathrm{CV}_z^2[I] \tag{7.93}$$

where

$$\mathrm{CV}_z[I] = \frac{\sqrt{\mathrm{Var}_z[I]}}{\mathrm{E}_z[I]} \tag{7.94}$$

is the coefficient of variation of the travel-time pdf.

The general definition (7.94) allows us to calculate the dispersivity of various solute transport models in terms of their model parameters and z. Table 7.5 summarizes $\mathrm{E}_z[\mathrm{I}]$, $\mathrm{Var}_z[\mathrm{I}]$, and the dispersivity (7.93) of the travel-time pdf of three solute transport models we have studied thus far: the CDE (7.17), the MIM (7.31)–(7.32), and the CLT (7.83).

Several features are notable in Table 7.5. First, both the CDE and MIM have constant dispersion as a function of z when their parameters are constant. In contrast,

TABLE 7.5   Mean, Variance, and Dispersivity of Three Solute Transport Models

Model	Mean $E_z[I]$	Variance $Var_z[I]$	Dispersivity [Eq. (7.93)]
CDE	$\dfrac{z}{V}$	$\dfrac{2Dz}{V^3}$	$\dfrac{D}{V}$
MIM	$\dfrac{\theta z}{J_w}$	$\dfrac{2z D_m \phi_m \theta^3}{J_w^3} + \dfrac{2z\theta^2(1-\phi_m)^2}{J_w \alpha}$	$\dfrac{D_m \theta_m}{J_w} + \dfrac{J_w(1-\phi_m)^2}{\alpha}$
CLT	$\dfrac{z\exp(\mu+\sigma^2)}{L}$	$\dfrac{z^2\exp(2\mu+\sigma^2)[\exp(\sigma^2)-1]}{L^2}$	$\dfrac{z[\exp(\sigma^2)-1]}{2}$

Source: Adapted from Jury and Sposito (1985).

the CLT has a linearly increasing dispersion with z when its parameters are constant. This explains the greater spreading observed in Fig. 7.9 compared to the CDE.

Second, by comparing the CDE and MIM, we see that part of the dispersivity or spreading in the MIM is caused by the rate-limited diffusion process between the mobile and stagnant regions of the wetted pore space (Valocchi, 1985). Finally, all three models have a mean $E_z[I]$ that increases proportionally to z. Thus, the center of mass of a pulse is predicted to move at the same velocity by all three models; only the spreading or dispersion differs. For this reason the piston flow model will describe mean transport equally well for all three processes but does not describe the dispersion or spreading process.

Rate Processes and Equilibrium   The first and second moments (or the mean and variance) of the travel-time pdf contain much of the essential information about a solute transport model. To illustrate this point, Fig. 7.23 shows a plot of the CDE and

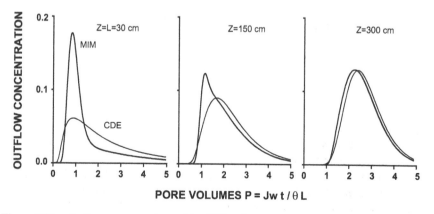

Figure 7.23   Outflow concentrations of the CDE and MIM models at different distances from the inlet surface. The models have the same travel-time mean and variance.

MIM outflow concentrations from a narrow pulse application at different distances from the inlet end. The models have been calibrated to have the same mean and variance. As seen in the figure, the models become more and more similar at larger distances from the inlet end, because the rate-limited diffusion has more time to reach equilibrium during the time of transport. As a result, the simpler CDE model would be adequate to describe all of the transport features at sufficiently long times after application.

7.5   SOLUTE MANAGEMENT IN FIELD SOIL

7.5.1   Salinization of Crop Root Zones

When poor-quality water is used for irrigation, salts can build up in the crop root zone as water is removed by plant root extraction. Management of irrigation to minimize the salt burden of the irrigation drainage water while avoiding plant stress from salinity has long been an active area of research in soil science. Some simple calculations with the water and solute transport equations developed in this chapter will help to illustrate the principles of salinization in the root zone.

Example 7.9   Calculate the steady-state solute concentration as a function of depth in a crop root zone of depth L receiving high-frequency irrigation of concentration $C = C_0$ at a rate $J_w = -i_0$ in which water is extracted at a spatially uniform rate r_w in $-L < z < 0$. Assume no solute reactions, no solute dispersion, and no plant uptake of solute.

SOLUTION: As shown in Chapter 3 (see Example 3.11), the steady-state water flux equation in one dimension for a spatially uniform water uptake rate is given by

$$J_w = -i_0 - \text{ET}\frac{z}{L} \tag{7.95}$$

where L is the thickness of the root zone, i_0 the steady irrigation flux, and ET the uniform evapotranspiration rate.

   When dispersion is neglected and solutes do not precipitate, dissolve, or enter plant roots, the steady-state solute flux is given by the piston flow model

$$J_c = \text{constant} = J_w(z)C(z) = -i_0C_0 \tag{7.96}$$

where C_0 is the solute concentration in the irrigation water. Thus, combining (7.95) and (7.96), we obtain

$$C(z) = \frac{i_0C_0}{i_0 + \text{ET} \cdot z/L} = \frac{C_0}{1 + (1 - f_L)z/L} \tag{7.97}$$

where

$$f_L = \frac{\text{drainage flux}}{\text{irrigation flux}} = \frac{i_0 - \text{ET}}{i_0} \qquad (7.98)$$

is the leaching fraction. Figure 7.24 shows a plot of C/C_0 versus z/L for various values of the leaching fraction. The maximum solute concentration occurs at the bottom of the root zone, where

$$C_{\max} = C(-L) = \frac{C_0}{f_L} \qquad (7.99)$$

Scientists have long searched for a reliable index of the effect of soil salinity on plant growth and yield. For a fully salinized root zone such as the one described by Fig. 7.24, Raats (1975) presented the following analysis. He defined the average salinity $< C >$ of the root zone as the flow-averaged concentration of the water extracted by the plant:

$$< C > = \frac{\int_{-L}^{0} r_w(z) C(z) \, dz}{\int_{-L}^{0} r_w(z) \, dz} \qquad (7.100)$$

where the denominator has the value ET. In steady state, $C(z) = -C_0 i_0 / J_w(z)$ by (7.96) and $r_w = -dJ_w/dz$ by (3.58). Thus, (7.100) may be written as

$$< C > = \frac{i_0 C_0}{\text{ET}} \int_{-L}^{0} \frac{dJ_w}{dz} \frac{dz}{J_w} = \frac{i_0 C_0}{\text{ET}} \int_{f_L i_0}^{i_0} \frac{dJ_w}{J_w} = \frac{i_0 C_0}{\text{ET}} \ln \frac{1}{f_L} \qquad (7.101)$$

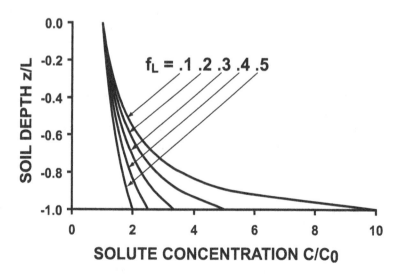

Figure 7.24   Steady-state solute concentration in a root zone with uniform water uptake as a function of the leaching fraction f_L.

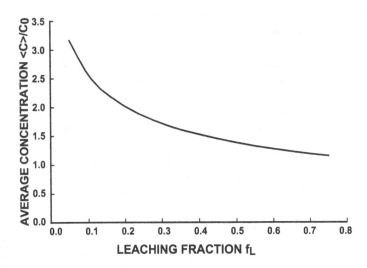

Figure 7.25   Average root zone concentration $< C >$ calculated from (7.102) as a function of the leaching fraction f_L.

Finally, using (7.98), we may rewrite (7.101) as

$$\frac{<C>}{C_0} = \frac{1}{1 - f_L} \ln \frac{1}{f_L} \tag{7.102}$$

Figure 7.25 shows a plot of the average root zone concentration (7.102) seen by the plant as a function of the leaching fraction. Note that this result is independent of the shape of the water uptake distribution and depends only on f_L and C_0.

More frequent and intensive monitoring of groundwater quality and improvements in the analytical detection of chemicals dissolved in water have greatly increased research on the transport of contaminants from the soil surface to groundwater.

Nitrate Pollution   Nitrates can reach groundwater from a variety of sources. Nitrogen is a primary component of organic and inorganic fertilizers (Scarsbrook, 1965) and transforms rapidly to nitrate under normal soil conditions (Alexander, 1965). Septic tanks, which serve about 30% of the households in the United States, can be major sources of nitrate pollution in groundwater, particularly if they are poorly maintained (Novotny and Chesters, 1981; Pye et al., 1983). Dairy and poultry feedlots can form substantial sources of nitrate pollution, especially if a site is abandoned and subsequently exposed to rainfall or irrigation (Novotny and Chesters, 1981). In addition, certain soils contain nitrates deposited during geologic times that can move to groundwater when water is percolated through them (Sullivan et al., 1979; White and Moore, 1972).

The widespread appearance of nitrates in groundwater is a consequence of a number of factors. Nitrate is very mobile in soils and is easily displaced from its point of origin by water additions. It is stable in soil except when biologically transformed by denitrification, which only occurs in very wet soil or inside soil aggregates at high moisture content (Broadbent and Clark, 1965). Although nitrate is taken up rapidly by plant roots, this removal mechanism only occurs very close to the soil surface. Thus, whenever there is a source of nitrogen and an excess of water applied to the soil, nitrates have the potential to reach groundwater. Since two of the major sources of nitrogen addition to soil, irrigated agriculture and septic tanks, are also sources of excess water, it is easy to see why these operations have been associated with nitrate pollution in the past.

Although deep groundwater tables require longer to reach than shallow ones, dissolved nitrates will eventually arrive in groundwater as long as water continues to enter the soil at a rate in excess of evaporative demand. Pratt et al. (1970) sampled underneath citrus groves throughout California and found nitrate plumes that had moved far below the root zone over a period of years but had not yet reached groundwater. In some locations they estimated that travel times in excess of 50 years would be required to reach groundwater. This explains why groundwater under land developed over former citrus groves has continued to experience high nitrate levels for years after the high additions of nitrate fertilizer to the surface ceased (Ayers and Branson, 1973).

One of the major goals of solute transport models has been to predict the movement of contaminants such as nitrate to groundwater. A serious limitation to achieving this goal has been the absence of a procedure for quantifying the nature of the dispersion process for large-scale emission of solutes from the soil surface.

To illustrate the effect that uncertainty about the dispersion process can have on groundwater predictions, Jury (1983) compared two models, the one-dimensional CDE with a constant dispersion coefficient (7.82) and the convective–lognormal transfer function (7.83) with a linearly increasing dispersion coefficient. As shown in Fig. 7.22, these two models virtually superimpose on each other when their parameters are optimized, but after calibration the CLT predicts much greater spreading than the CDE. Table 7.6 presents calculations by the two models of the time required to transport 1, 5, 10, 25, 50, 75, 95, and 99% of a pulse of nitrate from the surface to groundwater located at 10 m below the surface.

As the table illustrates, there is a substantial deviation between the predictions of the models for early and late arrival times, which reflects the way that they represent the dispersion process. However, the prediction of the mean arrival time is similar for both models, indicating that the piston flow model will provide rough estimates of the mean transport of a solute pulse or front. The leading edge or worst-case movement differs substantially, indicating that in the absence of further experimentation, transport models making this prediction of early arrival must be regarded as uncertain within the limits of these two models.

Fertilizer nitrogen is often applied through a knife applicator as a subsurface band beside crop rows. When fertilizer is banded, it is possible to make physical manipulations to the soil around and especially above the band in order to divert infiltrating water away from the band. By diverting water away from bands, nitrate leaching can be minimized. A nitrogen fertilizer injector has been designed to form a locally compacted soil layer and a surface ridge or dome [localized compaction and

TABLE 7.6   Predicted Travel Times (Days)
to Groundwater at $z = 10$ m of the First P
Percent of a Pulse of Nitrate[a]

P (%)	CDE	CLT
1	266	83
5	282	121
10	292	147
25	311	203
50	332	289
75	355	415
90	377	575
95	391	699
99	419	1007

[a] The models are calibrated at $z = 0.3$ m ($V = 3$ cm
day^{-1}; $D = 15$ cm^2 day^{-1}.)

doming (LCD)] over the injected fertilizer band (Ressler et al., 1997). The injector includes a knife with a triangular, horizontal base that smears the soil at the bottom of the knife slit to close any existing macropores. A cone disk guide wheel follows the knife, to close the knife slit and compress a soil layer over the fertilizer band. Another following disk completes the closure of the slit and mounds soil above the fertilizer band. These soil manipulations divert water from the fertilizer band and protect the fertilizer from leaching. In a lysimeter study (Ressler et al., 1998), less anion tracer applied by LCD appeared in drainage than did anion tracer applied by a conventional knife applicator. A field study in Iowa (Ressler et al., 1999) showed that when rainfall was above average, more nitrate remained in the top 0.8 m of soil when fertilizer was applied by LCD than when applied by conventional knife application. Corn yields in LCD plots were about 0.48 Mg ha^{-1} larger than corn yields in conventional knife plots. Greater understanding of chemical transport processes will lead to the development of improved management of soil chemicals.

Industrial Chemicals   Groundwater contamination from industrial chemicals such as the solvents trichloroethylene (TCE) and perchloroethylene (PCE) and petroleum hydrocarbons such as gasoline are increasingly found in groundwater as monitoring programs become more widespread. Soluble organic chemicals such as TCE and PCE behave similarly to pesticides in soil in that they undergo vapor- and liquid-phase movement and can adsorb to stationary soil solid materials. Thus, models such as those discussed in Section 7.4 can be used to estimate the environmental fate of these compounds. However, there is another class of compounds, loosely called *nonaqueous-phase liquids* (NAPLs), that are only marginally soluble in water and exist as a separate nonaqueous-phase liquid when added to soil (Dracos, 1987). NAPL transport differs from that of a soluble organic liquid in several ways. Because of their density and viscosity contrast with water, they are frequently prone to unstable flow. They also leave a portion of their mass trapped inside soil pores as isolated liquid globules as they move through the medium. The residual NAPL left in the soil might be as large as 20% of the wetted pore space in a saturated soil (Dracos, 1987) and will not leach out readily as water flows through the contaminated region. In the absence of artificial means for removing the compound, its only release from the stationary liquid phase thus formed is by slow dissolution into the surrounding mobile water phase. This slow release may take many years and contribute to contamination over a long period of time.

Pesticide Pollution of Groundwater   Agricultural pesticides form a special class of toxic organic compounds in that they are deliberately (rather than accidentally) added to soil and are accompanied by large and continued additions of water. Pesticide management to prevent groundwater contamination is problematic, because the compounds must be mobile enough to reach their target organism and persistent enough to eliminate it. However, persistence and mobility are not desirable properties for a toxicant to possess from an environmental perspective. Early compounds, such as DDT and dieldrin, had very low mobility in soil and therefore did not have the potential to reach groundwater. However, they were very persistent and thus had the potential to reach the food chain by exposure through the atmosphere or migration in surface waters. To avoid this exposure route, newer pesticides have been designed to have a far more rapid breakdown in soil. For this reason they must be applied over time and have reasonably high mobility in order to reach and control the target organism.

Modern legislation in the United States and worldwide has been directed toward regulating pesticides so as to prevent groundwater contamination. Implicit in this legislation is the idea that pesticides can be screened based on their environmental fate properties at the time of their development to make preliminary decisions about their pollution potential. Jury et al. (1987) produced a simplified version of their behavior assessment model (Jury et al., 1983) to assist in the decision-making process (see Problems 7.4 and 7.5). Figure 7.26 shows the results of this screening model applied to 50 pesticides in current or former use. The model calculations were constructed to screen compounds for a given set of environmental conditions so as to identify those compounds that when applied in a pulse of mass M_0, retained more than $0.0001 M_0$ of this surface mass when they migrated below the biologically active surface zone.

In Fig. 7.26, two sets of environmental conditions are constructed. The lower line, called the *low pollution potential condition*, represents agricultural settings that produce a low potential for pesticides to reach groundwater (i.e., high organic carbon, low annual drainage rate), while the higher line represents high potential conditions (low organic carbon, high annual drainage rate). The compounds are plotted as a function of their organic carbon partition coefficient and their effective biological half-life and thus appear as a single point on this curve. Compounds that appear to the right of the line formed by a given condition are deemed to possess a potential groundwater contamination threat according to this criterion. Thus, the combination of high biological half-life and low organic carbon partition coefficient identify the worst potential for migration.

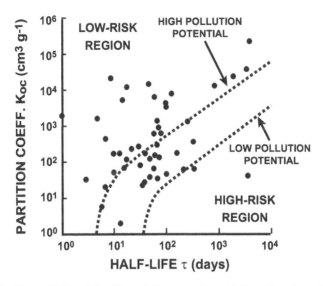

Figure 7.26 Plots of (K_{oc}, τ) for 50 pesticides, together with lines denoting relatively high and low groundwater pollution potential conditions. Compounds to the right of the line are considered to be a potential groundwater contamination risk. (After Jury et al., 1987a.)

PROBLEMS

7.1 A scientist is investigating a location where a hazardous waste spill occurred 25 years ago. In this accident, a large quantity of nitrates ($K_d = 0$) and an industrial solvent with $K_d = 0.5$ cm^3 g^{-1} were dumped on the soil surface as a massive pulse. Rainfall and evaporation records indicate that about 1000 cm of water has infiltrated the soil since that time. From soil coring it was determined that the average water content and bulk density of the soil profile are approximately $\theta = 0.25$ and $\rho_b = 1.5$ g cm^{-3}, respectively.

    (a) Calculate the approximate positions of the nitrate and solvent pulses today.

    (b) If the water table depth is 50 m, calculate the expected future arrival time of both pulses in groundwater, assuming that the climate does not change.

7.2 Calculate the chemical transport equations analogous to (7.24) and (7.25) assuming that the chemical obeys an equilibrium Freundlich adsorption isotherm (1.16). Calculate the ratio of the effective retardation factors of a chemical with $\beta = 0.75$ at concentrations of 1 and 1000 mg L^{-1} ($\theta = 0.25$, $\rho_b = 1.5$ g cm^{-3}). What does this say about the velocity of a waste spill of high concentration? (See Rao and Davidson, 1979.)

7.3 A large undisturbed soil core is brought to the laboratory and subjected to steady-state water flow. At $t = 0$ a narrow pulse of Cl$^-$ is added to the inlet end, and the measured outflow concentration–time record looks like the curve

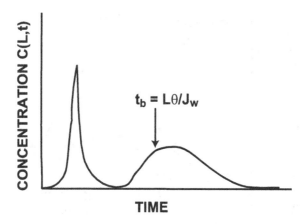

Figure 7.27 Breakthrough curve for the soil in Problem 7.3.

shown in Fig. 7.27. Give some plausible explanations for why the soil might show a breakthrough curve of the type in this figure.

7.4 According to the piston flow model, a chemical moving under steady-state water flux while adsorbing to equilibrium will break through at depth $z = L$ in a time

$$t_{bR} = \frac{L\theta R}{J_w} \qquad \text{where} \quad R = 1 + \frac{\rho_b K_d}{\theta}$$

If the chemical also decays by a first-order reaction, the fraction of a chemical application of mass M_0 that is still present in the soil at time t_{bR} is

$$\frac{M(t)}{M_0} = \exp(-\mu t_b) = \exp\left(-\frac{t_b \ln 2}{\tau_{1/2}}\right)$$

Using the properties of the six pesticides given in Table 7.3, calculate the mass fraction remaining at depth L under the conditions given in Table 7.4. Which pesticides have the greatest pollution potential? (See Rao et al., 1985.)

7.5 The residual mass fraction $\exp(-\mu R L\theta / J_w)$ in Problem 7.4 is required to be less than some small number β,

$$\exp\left(-\frac{\mu R L\theta}{J_w}\right) < \beta$$

Show that this relation may be written in the form

$$K_{oc} < a\tau_{1/2} - b$$

Indicate how this property may be used to select pesticide properties with low groundwater pollution potential. (See Jury et al., 1987a.)

7.6 The travel time required for a nonadsorbing solute to reach depth L of a crop root zone receiving water input at a steady input flux rate $Jw = i_0$ is

$$t(L) = \int_0^L \frac{\theta \, dz}{J_w(z)}$$

where $V = J_w/\theta$ is the solute velocity. For the special case of uniform water uptake (crop) and constant θ, calculate $t(L)$ as a function of i_0, ET, L, and θ. Using the definition of the leaching fraction f_L in (7.98), express the number of pore volumes

$$P(L) = \frac{i_0 t(L)}{L\theta}$$

of applied water required to move the solute to depth L as a function of f_L alone and plot the result. (See Jury et al., 1978).

7.7 According to the piston flow model discussed in Example 7.9, the solute concentration at the bottom of the root zone (7.99) is proportional to $1/f_L$.
   (a) What is the solute flux at $z = -L$ as a function of f_L?
   (b) What effect does chemical precipitation have on these estimates?
   (c) Discuss the recommended leaching fraction one would propose to (i) minimize the solute concentration below the root zone and (ii) minimize the solute flux below the root zone.

7.8 Derive (7.40).

7.9 Derive the appropriate form of the differential equations to describe the following process:
   (a) Chemical transport in a uniform medium with a constant volumetric air content a and water content θ
   (b) Zero water flow in the system
   (c) Diffusive transport of a nonreactive gas that is also soluble in water
   (d) The movement of chemical into and out of the dissolved phase is a rate-limited process proportional to the deviation from equilibrium as expressed in (7.39).

7.10 Derive the dimensionless form of the MIM equations (7.64)–(7.65) from (7.31)–(7.32) using the substitutions (7.60)–(7.61).

APPENDIX
Methods for Analyzing Spatial Variations of Soil Properties

Natural field soils are quite variable in both the vertical and horizontal directions. Moreover, variability is present at all distance scales. For example, soil porosity might vary significantly at large distance scales, where different soil series are encountered. However, it is also variable over very small distances, since it can have only two values: 0 inside soil minerals and 1 between soil minerals. Thus, porosity has a meaning that is relative to the volume scale over which it is averaged (Bear, 1972). When measurements of a property such as porosity are made on a scale that is small compared to the scale over which the quantity has a constant value, the measurements must be averaged to produce an estimate of the quantity. In this chapter we develop methods for making estimates from a sample set of measurements of properties that vary in space or in time.

A.1   VARIABILITY OF SOIL PROPERTIES

The most important characteristics of a property Z that assumes a range of values $Z_1, Z_2, \ldots, Z_N$ are its average value, or mean m, and its spread about the average value. The most common index of the spread is the variance s^2, which for N samples is calculated by (Hald, 1952)

$$s^2 = \frac{1}{N-1} \sum_{J=1}^{N} (Z_J - m)^2 \tag{A.1}$$

where

$$m = \frac{1}{N} \sum_{J=1}^{N} Z_J \tag{A.2}$$

is the sample mean.

TABLE A.1   Sample CVs Measured for Different Soil Water Properties in Unsaturated Fields

Parameter	Number of Studies	CV (%)
Porosity	4	7–11
Bulk density	8	3–26
% sand or clay	5	3–55
Water content at 0.1 bar	4	4–20
Water content at 15 bar	5	14–45
pH	4	2–15
Saturated hydraulic conductivity	12	48–320
Infiltration rate	5	23–97

Source: Jury (1985).

Since m is a constant, we may also express (A.1) as

$$s^2 = \frac{1}{N-1} \left(\sum_{J=1}^{N} Z_J^2 - Nm^2 \right) \tag{A.3}$$

Another common index used to express the extent of variability is the sample coefficient of variation (CV):

$$CV = \frac{s}{m} \tag{A.4}$$

where s is the standard deviation from the sample mean. The CV is often expressed as a percentage, by multiplying (A.5) by 100.

Table A.1 summarizes sample CVs of various soil physical properties measured in the surface zone of natural field sites. Other reviews of the spatial variability of soil properties are found in Jury et al. (1987b,c), Peck (1983), and Warrick and Nielsen (1980).

Several features are notable in this table. First, the transport properties, such as hydraulic conductivity as a rule are more variable than the retention properties, such as water content, even on the same field site. Second, the CV of the transport properties on some fields is extremely large, reaching values well in excess of 100% for quantities such as the saturated hydraulic conductivity. The implications of this extensive variation are explored in Example A.1.

Example A.1   The saturated hydraulic conductivity of a 100-m^2 field has a mean of 100 cm day^{-1} and a CV of 50%. Assuming that the distribution is normal, calculate the values of K_s that are ± 1 and 2 standard deviations from the mean.

SOLUTION: Since the CV is 50%, $s = 50$ cm day^{-1}. Thus, the K_s values 50, 150 are $\pm s$ and 0, 200 are $\pm 2s$ from the mean. If the population is really normal, approximately 2.3% of the values of K_s should be negative, which is not physically possible. Thus, populations of variables that have only positive values (called *positive definite*) and large CV cannot be normally distributed.

The large CV calculated for some distributions often arises from a skewed population with a relatively small number of very large values. This is illustrated further in the next example.

Example A.2   A soil survey team makes 20 measurements of infiltration rate on a uniform field, and a mean $m = 100$ cm day^{-1} and a CV of 20% are calculated from these numbers. They decide to take one more reading and quit for the day. Measurement 21 happens to be taken right over a wormhole, which produces a high infiltration rate value of 500 cm day^{-1}. Calculate the new sample mean and CV for the 21 values.

SOLUTION: The old mean was based on 20 values. Thus, the sum of all measurements is, from (A.2),

$$\sum_{J=1}^{20} Z_J = 20 \times 100 = 2000$$

The new mean, based on 21 values, is therefore

$$m = \frac{1}{21} \sum_{J=1}^{21} Z_J = \frac{1}{21}(2000 + 500) = 119 \text{ cm day}^{-1}$$

Thus, the mean has increased 19% because of the new value. The old variance was $s^2 = (m \times \text{CV})^2 = 400$ cm^2 day^{-1}. Using (A.3), we can calculate the sum of squares of the values of Z_J:

$$\frac{1}{19}\left(\sum_{J=1}^{20} Z_J^2 - 20 \times 100^2\right) = 400 \rightarrow \sum_{J=1}^{20} Z^2 = 207{,}600$$

Thus, the new sum of squares, including the final value, is

$$\sum_{J=1}^{21} Z^2 = 207{,}600 + 500^2 = 457{,}600$$

Therefore, using (A.3), we calculate the new variance as

$$s^2 = \frac{1}{20}\left(\sum_{J=1}^{21} Z_J^2 - 21 \times 119^2\right) = \frac{1}{20}(457{,}600 - 297{,}381) = 8011$$

Thus, the new CV = 89.5/119 = 0.75. Therefore, the CV has increased from 20% to 75% with the addition of one new large value to the sample set.

These examples point out that variability must be characterized in some detail to portray the features of the distribution as well as the average value of a set of measurements. The principal concepts of probability and statistics required to make such a characterization are given in what follows.

A.2   CONCEPTS OF PROBABILITY

Probability may be thought of as the fraction of a specific outcome from among all possible outcomes of an experiment (Himmelblau, 1970). A parameter whose values are characterized by a probability distribution is known as a *random variable*. Since the entire population of all possible outcomes of any event is never known to the observer, probability can only be estimated approximately by repeated observations. Certain events, such as a coin flip or a roll of a die, have only a finite number of possible outcomes, and the frequency of occurrence of a given outcome can be estimated relatively easily from a large number of trials. Other events, such as the saturated hydraulic conductivity of a large field, are not bounded a priori (except by being positive definite), and their frequency of occurrence must be inferred from the distribution of a sample set that is much smaller than the population of all possible experiments.

A.2.1   Random and Regional Variables

The application of probability to parameters whose values vary in space is different in several respects from an event that varies in time. Dice may be shaken repeatedly in separate trials, and each outcome is unknown until the trial is completed. If the throw of the dice is "random" each time (i.e., different force, angle of lift, etc.), we expect, based on long experience, that each face of the dice has a probability $\frac{1}{6}$ of appearing. In contrast, properties that vary in space, such as porosity, have a single fixed value within a given small volume of space. If the property can be measured at that location without disturbing the place where the measurement was made, its value is known with certainty at that location thereafter. Such parameters are often called *regional variables* (Journel and Huijbregts, 1978). To apply the concepts of probability to a regional variable, we must assume that it is a random function in space by treating measurements of the property at different locations as though they were repeated trials of the same event. Thus, for a variable to be random in space, it must have values at different locations that are characterized by a probability distribution. This distinction will become clearer after some definitions are made.

A parameter $Y(x, y, z)$ whose values vary at a given time as a function of position is called a regional variable. It may be treated as though it were a random variable Z whose value at a given point in space (x, y, z) is characterized by the probability distribution for the variable at that point. In other words, the value $Y(x, y, z)$ measured for Z at that point is regarded as a particular outcome from among a set of possibilities determined by chance. In another identical universe, perhaps an observer would have measured a different value at that point.

The problem with applying probability to spatially variable properties is that the laws of probability for a variable at a given point cannot be inferred from the single available measurement. The way around this dilemma is to assume that the value of the variable at every point in space is determined by the same probability law.[1] Then each measurement may be regarded as a separate sample of the random variable, and the frequency of occurrence of outcomes may be used to generate the frequency distribution function for the variable over the domain of measurement. The method by which frequency distributions are constructed is covered in the next section.

A.2.2 Frequency Distributions

The frequency distribution is the statement of the relative occurrence of a specific value from the possible outcomes of an experiment. When the number of outcomes generated is less than the total number of outcomes of the experiment, the frequency distribution generated is called a sample frequency distribution.

Cumulative Distribution Function   The cumulative distribution function (cdf) $P(Z)$ of a random variable Z is defined as

$$P(Z_0) = \text{probability that } Z \leq Z_0 \tag{A.5}$$

Thus, for a positive definite random variable that takes on only finite values, we can write the two limits

$$P(Z_0) = 0 \quad \text{if } Z < 0 \tag{A.6}$$

$$P(\infty) = 1 \tag{A.7}$$

Other values of $P(Z)$ are determined by experiment, as demonstrated in the next example.

Example A.3   Twenty values of infiltration rate $i = \{1, 3, 9, 6, 5, 3, 3, 7, 8, 2, 5, 11, 4, 2, 1, 7, 5, 13, 2, 6\}$ cm h^{-1} are measured with a double-ring infiltrometer at different locations in the field. Calculate and plot the sample estimate of $P(Z)$.

SOLUTION: By definition, each measurement counts 0.05 of the total sample probability generated from the 10 measurements. Therefore, we can construct the cdf values for values of i from 0 to 13 (Table A.2). A plot of $P(i)$ versus i is shown in Fig. A.1.

---

[1] This is called the *stationarity hypothesis* (Journel and Huijbregts, 1978).

TABLE A.2   Sample cdf Values Calculated from Infiltration Rate Measurements of Example A.3

i	$P(i)$	i	$P(i)$
0	0.00	7	0.80
1	0.10	8	0.85
2	0.25	9	0.90
3	0.40	10	0.90
4	0.45	11	0.95
5	0.60	12	0.95
6	0.70	13	1.00

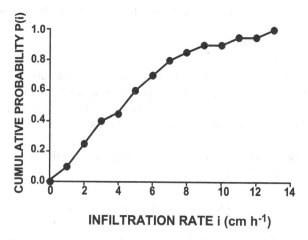

Figure A.1   Sample cdf constructed from the data in Table A.1.

Probability Density Function   The probability density function (pdf) $f(Z)$ for a continuous random variable Z is defined as

$$\lim_{\Delta Z \to 0} f(Z_0)\Delta Z = \text{probability that } Z_0 \leq Z \leq Z_0 + \Delta Z$$

$$= P(Z_0 + \Delta Z) - P(Z_0) \tag{A.8}$$

Thus, by definition,

$$f(Z) = \frac{dP(Z)}{dZ} \tag{A.9}$$

Therefore, we can construct a pdf from the same information used to calculate a cdf, as shown in the next example.

Example A.4   Calculate and plot $f(i)$ from the data given in Example A.3.

SOLUTION: If we assume that $P(i)$ is continuous and linearly interpolate between the values of $P(i)$ in Table A.1 using (A.8) with $\Delta Z = 1$, then $f(i)$ is the slope of

**TABLE A.3   Sample pdf Values Calculated from Infiltration
Rate Measurements of Example A.3**

i	$P(i)$	i	$P(i)$
0–1	0.10	7–8	0.05
1–2	0.15	8–9	0.05
2–3	0.15	9–10	0.00
3–4	0.05	10–11	0.05
4–5	0.15	11–12	0.00
5–6	0.10	12–13	0.05
6–7	0.10		

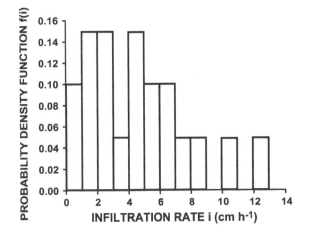

Figure A.2   Sample pdf constructed from the data in Table A.1.

the graph in Fig. A.1. Its values are given in Table A.3, and the corresponding graph
is given in Fig. A.2.

The probability density function $f(i)$ in Example A.4 has a crude structure because it has been constructed from very few observations.

Moments of the PDF   The pdf (A.8) represents the characteristics of the entire
population of values of the random variable. If Z takes on a continuous range of
values, $f(Z)$ will have a value at every Z in the population, and

$$\int f(Z)\,dZ = 1 \qquad (A.10)$$

where the integral limits span the entire range of values (i.e., from 0 to ∞ if Z is
positive definite, and $-\infty$ to ∞ if Z takes on all values). Equation (A.10), which
is equivalent to (A.7), simply states that the sum of the probability of occurrence of
each value of Z in the population is unity.

The average value of Z in the population is obtained as the sum of each value of Z multiplied by its probability of occurrence. Thus for continuous distributions, the sum is replaced by the integral

$$m = E[Z] = \int_0^\infty Zf(Z)\,dZ \tag{A.11}$$

where m is the population average or mean value of Z. It is also called the *expectation* or *ensemble average* $E[Z]$ of Z.

Similarly, the population average of Z^2 (called the *second moment* $E[Z^2]$ is

$$E[Z^2] = \int_0^\infty Z^2 f(Z)\,dZ \tag{A.12}$$

and

$$\mathrm{Var}[Z] = E[(Z - E[Z])^2] = \int_0^\infty (Z - E[Z])^2 f(Z)\,dZ \tag{A.13}$$

is the population variance of Z, or the average squared deviation from the mean.

Joint Distributions   Thus far we have considered only a single random variable Z. When there are two random variables Y and Z in an experiment, their joint probability of occurrence is defined by the *joint cdf*:

$$P(Y_0, Z_0) = \text{ probability that } Y \le Y_0 \text{ and } Z \le Z_0 \tag{A.14}$$

Similarly, the *joint pdf* is defined as

$$f(Y, Z) = \frac{\partial^2 P}{\partial Y\, \partial Z} \tag{A.15}$$

The moments of the joint distribution are calculated as follows. The pdf expressions for the individual random variables (called *marginal distributions*) are

$$f_Y(Y) = \int f(Y, Z)\,dZ \tag{A.16}$$

$$f_Z(Z) = \int f(Y, Z)\,dY \tag{A.17}$$

from which the moments of Y and Z can be calculated by (A.11) and (A.12).

The distribution of the product of Y and Z is frequently of importance. The expectation of YZ is defined as

$$E[YZ] = \int_0^\infty \int_0^\infty YZf(Y, Z)\,dY\,dZ \tag{A.18}$$

The deviation of this quantity from the product of the average values of Y and Z has a special name,

$$\text{Cov}[Y, Z] = E[YZ] - E[Y]E[Z]$$

$$= E[(Y - E[Y])(Z - E[Z])]$$

$$= \int_0^\infty \int_0^\infty (Y - E[Y])(Z - E[Z]) f(Y, Z) \, dY \, dZ \qquad (A.19)$$

is called the *covariance* $\text{Cov}[Y, Z]$ of Y and Z. The normalized covariance obtained by dividing (A.19) by $S_Y S_Z$, where $S_Y = \sqrt{\text{Var}(Y)}$, $S_Z = \sqrt{\text{Var}(Z)}$

$$\rho = E\left[\frac{Y - E[Y]}{S_Y} \frac{Z - E[Z]}{S_Z}\right] \qquad (A.20)$$

is called the *correlation coefficient* of Y and Z.

A special case of the joint distribution of two random variables occurs when the outcome of one does not affect the outcome of the other. In this case, the two variables are said to be *independent*, and

$$f(Y, Z) = f_Y(Y) f_Z(Z) \qquad (A.21)$$

It is possible for the correlation coefficient in (A.20) to be zero without (A.21) being true. In this case, Y and Z are said to be uncorrelated rather than independent.

Example A.5 Calculate the mean and variance of the sum $Y + Z$ of two random variables Y and Z.

SOLUTION: By definition, the mean is

$$E[Y + Z] = \int_0^\infty \int_0^\infty (Y + Z) f(Y, Z) \, dY \, dZ$$

$$= \int_0^\infty \int_0^\infty Y f(Y, Z) \, dY \, dZ + \int_0^\infty \int_0^\infty Z f(Y, Z) \, dY \, dZ$$

$$= \int_0^\infty Y f_Y \, dY + \int_0^\infty Z f_Z \, dZ = E[Y] + E[Z] \qquad (A.22)$$

By (A.13) and (A.22), we may write the variance as

$$\text{Var}[Y + Z] = E[(Y + Z)^2] - (E[Y + Z])^2$$

$$= E[Y^2 + 2YZ + Z^2] - (E[Y] + E[Z])^2$$

$$= E[Y^2] - E[Y]^2 + E[Z^2] - E[Z]^2$$

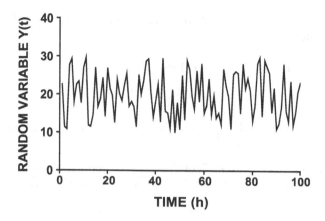

Figure A.3   Random function $Y(t)$ varying as a function of time.

$$+ 2(E[YZ] - E[Y]E[Z])$$

$$= \text{Var}[Y] + \text{Var}[Z] + 2\,\text{Cov}[Y, Z] \qquad (A.23)$$

For the special case when Y and Z are uncorrelated, $\text{Var}[Y + Z] = \text{Var}[Y] + \text{Var}[Z]$.

Random Functions of a Parameter   A random variable Z may be a function of a parameter x such as time or position. It is then denoted $Z(x)$, and the pdf of Z at x is denoted $f(Z; x)$ or $f(Z(x))$. In general, each value of the parameter has its own pdf, and $Z(x_1)$, $Z(x_2)$ must be described by the joint pdf $f(Z(x_1), Z(x_2))$. However, for reasons described in the preceding, it is impossible to measure the pdf if only one value $Z(x_i)$ is available at x_i to characterize the pdf $f(Z(x_i))$. To avoid this problem, it is generally assumed that the random variable is stationary, at least for its first two moments. A second-order stationary random variable $Z(x)$ has the following properties:

1. Constant mean:

$$E[Z(x)] = m = \text{constant for all x} \qquad (A.24)$$

2. Covariance of $Z(x_1)$ and $Z(x_2)$ is a function only of the separation[2] $h = x_2 - x_1$ between x_1 and x_2:

$$\text{Cov}[Z(x_1), Z(x_1 + h)] = \text{Cov}[h] \qquad (A.25)$$

The condition of stationary may be illustrated by a random variable $Y(t)$ fluctuating in time (Fig. A.3). At any point in time, the value of the variable is random but

[2] The separation may be a vector, in which case the covariance is anisotropic.

is fluctuating about the same mean value m. Moreover, the value of the variable at any instant of time t may be correlated with the values of the variable preceding it for some period of time rather than having a completely random value at each time.

These concepts are sufficient to develop the statistical theory of spatial variability used in the remainder of the chapter.

A.3   ANALYSIS OF FREQUENCY DISTRIBUTIONS

A.3.1   Model Probability Density Functions

Probability distributions are estimated from experimental observations, which are grouped as in Examples A.3 and A.4 to produce a tabular representation of the cdf $P(Z)$ and its derivative, the pdf $f(Z)$. In tabular or graphical form, these raw data provide a nonparametric representation of the probability distribution. An alternative approach that may be employed to generate distribution functions is to use model functions for $f(Z)$ with adjustable parameters to represent the structure of the sample measurements of the probability distribution optimally. Several commonly used model functions are described in what follows.

Normal Distribution   The pdf for the normal distribution is given by

$$f(Z) = \frac{1}{\sqrt{2\pi}\, s} \exp\left[-\frac{(Z-m)^2}{2s^2}\right] \tag{A.26}$$

where m and s are parameters. It may be shown by direct integration using (A.11) and (A.12) that

$$E[Z] = m \tag{A.27}$$

$$\mathrm{Var}(Z) = s^2 \tag{A.28}$$

A plot of $f(Z)$ for $m = 10$ and various values of s is shown in Fig. A.4.

The normal distribution is symmetric about its mean value m. Therefore, for large values of $CV = s/m$, part of the population assumes negative values, which makes the normal distribution a poor candidate to represent the distribution of positive definite variables with high CV (see Example A.1).

The cumulative normal distribution $P(Z)$ is defined as

$$P(Z) = \int_{-\infty}^{Z} \frac{1}{\sqrt{2\pi}\, s} \exp\left[-\frac{(x-m)^2}{2s^2}\right] dx \tag{A.29}$$

which may be written as

$$P(Z) = \frac{1}{\sqrt{2\pi}} \int_{-\infty}^{(Z-m)/s} \exp\left(-\frac{\xi^2}{2}\right) d\xi \tag{A.30}$$

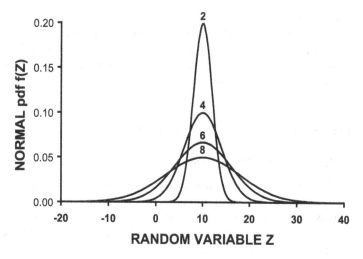

Figure A.4 Normal distribution pdf with $m = 10$ and various values of s (numbers above the curve peaks).

The function

$$N(u) = \frac{1}{\sqrt{2\pi}} \int_0^u \exp\left(-\frac{\xi^2}{2}\right) d\xi \qquad (A.31)$$

is called the *normal curve of error* and is tabulated in many statistics books. Since $N(-u) = -N(u)$ and $N(\infty) = 0.5$, $P(Z)$ in (A.30) may be written as

$$P(Z) = \frac{1}{2} + N\left(\frac{Z - m}{s}\right) \qquad (A.32)$$

in terms of the normal curve of error. The normalized variable $u = (Z - m)/s$ is often called a *u-variate*. It has zero mean and unit variance. The table of the normal curve of error can be used to calculate properties of normally distributed variables, as shown in the next example.

Example A.6   Use Table A.4 to calculate the values of Z that lie at the lowest and highest 1% of a normal distribution with a mean of 10 and variance of 36.

TABLE A.4   Selected Values from the Normal Curve of Error

u	$N(u)$
2.32	0.49
1.96	0.475
1.64	0.45
1.29	0.40

SOLUTION: The lower value of Z is the value Z_{min} that satisfies

$$P(Z_{min}) = 0.01 = 0.5 + N\left(\frac{Z_{min} - 10}{6}\right)$$

Therefore,

$$N\left(\frac{Z_{min} - 10}{6}\right) = -0.49 = -N\left(\frac{10 - Z_{min}}{6}\right)$$

From Table A.4, $N(u) = 0.49$ at $u = 2.32$. Thus,

$$\frac{Z_{min} - 10}{6} = 2.32 \rightarrow Z_{min} = 10 - 13.92 = -3.92$$

Similarly, the maximum value of Z is calculated from

$$P(Z_{max}) = 0.99 = 0.5 + N\left(\frac{Z_{max} - 10}{6}\right)$$

$$\frac{Z_{max} - 10}{6} = 2.32 \rightarrow Z_{max} = 10 + 13.92 = 23.92$$

Therefore, the values between 3.92 and 23.92 comprise 98% of the population of Z.

Example A.7   For the distribution in Example A.6, calculate the probability that Z is less than zero.

SOLUTION: By definition, $P(0)$ is the probability that $Z \leq 0$. Therefore, by (A.32),

$$P(0) = 0.01 = 0.5 + N\left(\frac{0 - 10}{6}\right) = 0.5 - N(1.67)$$

According to the normal value of error, $N(1.67) \sim 0.4525$. Therefore,

$$P(0) = 0.5 - 0.4525 = 0.0475$$

Thus, there is a 4.75% probability that Z is negative.

Lognormal Distribution   The pdf for the lognormal distribution is given by

$$f(Z) = \frac{1}{\sqrt{2\pi}\,\sigma Z} \exp\left[-\frac{(\ln Z - \mu)^2}{2\sigma^2}\right] \tag{A.33}$$

where μ and σ are parameters. It may be shown by direct integration using (A.11) and (A.12) that

$$E[Z] = \exp\left(\mu + \frac{\sigma^2}{2}\right) \tag{A.34}$$

$$\text{Var}[Z] = \exp(2\mu + \sigma^2)[\exp(\sigma^2) - 1] \tag{A.35}$$

(Aitcheson and Brown, 1976; Hald, 1952). Therefore, by (A.5),

$$\text{CV} = [\exp(\sigma^2) - 1]^{1/2} \tag{A.36}$$

A plot of $f(Z)$ for $E[Z] = 10$ and various values of σ is shown in Fig. A.5. In contrast to the normal distribution (Fig. A.4), the lognormal distribution is not symmetric and is defined only for positive Z.

The cumulative lognormal distribution $P(Z)$ is defined as

$$P(Z) = \int_0^Z \frac{1}{\sqrt{2\pi}\,\sigma x} \exp\left[-\frac{(\ln x - \mu)^2}{2\sigma^2}\right] dx \tag{A.37}$$

which may be written as

$$P(Z) = \frac{1}{\sqrt{2\pi}} \int_{-\infty}^{(\ln Z - \mu)/\sigma} \exp\left(-\frac{\xi^2}{2}\right) d\xi \tag{A.38}$$

Thus, using (A.31), we may write

$$P(Z) = \frac{1}{2} + N\left(\frac{\ln Z - \mu}{\sigma}\right) \tag{A.39}$$

in terms of the normal curve of error.

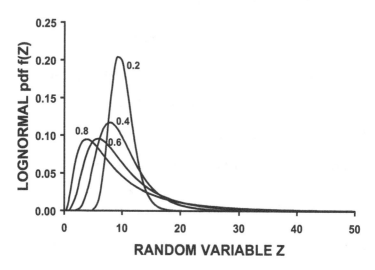

Figure A.5   Lognormal distribution pdf with $E[Z] = 10$ and various values of σ (numbers above the curve peaks).

Example A.8   Use $N(u)$ in Table A.4 to calculate the values of Z that bound the lowest and highest 1% of the population of a lognormal distribution with a mean of 10 and a variance of 36.

SOLUTION: By definition, CV = 0.6; thus, using (A.36),

$$\exp\left(\sigma^2\right) = 1 + \text{CV}^2 = 1.36 \rightarrow \sigma = 0.554$$

$$\exp\left(\mu + \frac{\sigma^2}{2}\right) = 10 \rightarrow \mu = 2.149$$

Using (A.31), we may calculate Z_{min} from

$$P(Z_{min}) = 0.01 = 0.5 + N\left(\frac{\ln Z_{min} - \mu}{\sigma}\right)$$

from which we determine that

$$N\left(\frac{\mu - \ln Z_{min}}{\sigma}\right) = 0.49 \rightarrow \frac{\mu - \ln Z_{min}}{\sigma} = 2.32$$

Therefore,

$$Z_{min} = \exp(\mu - 2.32\sigma) = 0.67$$

Similarly, $Z_{max} = \exp(\mu + 2.32\sigma) = 31$. Thus, the values between 0.67 and 31 form 98% of the population of Z. In contrast, the normal distribution with the same mean and variance (Example A.6) had a range of $(-3.92, 23.92)$ from the lowest to the highest 1% of the population.

Mode and Median   There are two other variable values besides the mean $m = \text{E}[Z] \equiv \bar{Z}$ that can be used to characterize a pdf. They are the mode $\hat{Z}$, or most probable value, and the median $\tilde{Z}$, or the middle value of the distribution. By definition, $f(Z)$ has a maximum at $\hat{Z}$. Therefore, the mode is an extremum of $f(Z)$, which means that its derivative must vanish there (Kaplan, 1984). Thus,

$$\frac{df}{dZ} = 0 \qquad \text{at } Z = \hat{Z} \tag{A.40}$$

Since $\tilde{Z}$ is the middle of the distribution, the median $\tilde{Z}$ is the value of Z that satisfies

$$P(Z) = 0.5 \qquad \text{at } Z = \tilde{Z} \tag{A.41}$$

Example A.9   Calculate $\hat{Z}$ and $\tilde{Z}$ for the normal distribution.

SOLUTION: After inserting (A.26) into (A.40), we obtain

$$\frac{df}{dZ} = -\frac{1}{s^2}(Z - m)\frac{\exp\left[-(Z - m^2)/2s^2\right]}{\sqrt{2\pi}\,s} = 0$$

This function is equal to zero at $z = m$. By (A.32),

$$P(Z) = 0.5 \qquad \text{when} \quad N\left(\frac{Z - m}{s}\right) = 0$$

which occurs when $z = m$. Thus, for the normal distribution, $\bar{Z} = \tilde{Z} = \hat{Z} = m$.

Example A.10   Calculate $\hat{Z}$ and $\tilde{Z}$ for the lognormal distribution.

SOLUTION: After inserting (A.33) into (A.40), we obtain

$$\frac{df}{dZ} = -\frac{f}{Z} - \frac{f}{\sigma^2 Z}(\ln Z - \mu) = -\frac{f}{\sigma^2 Z}(\ln Z - \mu + \sigma^2)$$

This function is equal to zero at

$$\hat{Z} = \exp(\mu - \sigma^2) \tag{A.42}$$

By (A.39),

$$P(Z) = 0.5 \quad \text{when } N\left(\frac{\ln Z - \mu}{\sigma}\right) = 0$$

Therefore,

$$\tilde{Z} = \exp(\mu) \tag{A.43}$$

Thus, for the normal distribution, $\hat{Z} < \tilde{Z} < \bar{Z}$. In Example A.8, $\hat{Z} = 6.31$, $\tilde{Z} = 8.58$, and $\bar{Z} = 10$.

Most of the soil properties that have been measured in the field have been found to be either normally or lognormally distributed. The transport properties with large CV are usually better described as lognormal (Jury, 1985). In the next section we discuss several methods of testing whether a sample set fits a distribution.

A.3.2   Methods for Determining a Frequency Distribution

Selection of a model frequency distribution to represent the population using a sample set of finite numbers can be a difficult task, particularly if there is no information prior to the sampling about the expected form of the distribution. For the common soil properties of interest that are routinely measured in the field, we may exclude a priori any unusual distributions (e.g., bimodal) and concentrate on determining whether the distribution of the property of interest is better described as normal or lognormal. There are several established methods for doing this.

Fractile Diagram   The *fractile diagram* is a graphical method that can be used to provide visual information about the distribution of the property (Hald, 1952). It is based on (A.32) for the normal distribution and (A.39) for the lognormal distribution.

Imagine that a sample set of N values $\{Z_1, \ldots, Z_N\}$ of the quantity of interest has been measured. If they are arranged in order so that $Z_1 < Z_2 < \cdots < Z_N$, the sample estimate of cumulative probability P is approximately equal to

$$P(Z_J) = \Pr[Z < Z_J] = \frac{J - 0.5}{N} \qquad J = 1, \ldots, N \qquad (A.44)$$

If Z is normally distributed, the theoretical value of $P(Z_J)$ is given by (A.32), or

$$P(Z_J) = 0.5 + N\left(\frac{Z_J - m}{s}\right) \qquad (A.45)$$

where m and s are calculated from the sample set using (A.1) and (A.2). If Z is normally distributed, a plot of (A.44) versus (A.45) should yield a straight line with zero intercept and unit slope.

Similarly, if Z is lognormally distributed, the theoretical value of $P(Z_J)$ is given by (A.39), or

$$P(Z_J) = 0.5 + N\left(\frac{\ln Z_J - \mu}{\sigma}\right) \qquad (A.46)$$

where μ and σ are calculated from the log-transformed sample set using (A.1) and (A.2). Fractile diagram analysis can be performed simply by using probability paper, which is scaled to represent $P(Z)$ on the ordinate when the unit variance, zero-mean deviate $(Z - m)/s$, or $(\ln Z - \mu)/\sigma$ is plotted. Alternatively, (A.45) or (A.46) can be calculated using a table or calculation[3] of $N(u)$ values and plotted versus (A.44) on regular paper. This is illustrated in the next example.

Example A.11   In the field study of Nielsen et al. (1973), the following 20 steady-state infiltration rates (in cm day^{-1}) were measured on ponded 6.3×6.3 m^2 plots over 150 ha: 5.12, 19.05, 0.54, 3.05, 11.73, 5.21, 11.98, 12.73, 1.96, 6.35, 13.26, 9.24, 40.23, 45.72, 1.16, 38.90, 12.23, 32.60, 14.20, 7.56. Calculate m, s, μ, and σ for the sample set and evaluate whether the distribution is better described as normal or lognormal using the fractile diagram.

SOLUTION: The sample mean and variance of these 20 values are $m = 14.64$ cm day^{-1} and $s^2 = 188.51$ cm^2 day^{-2}, while the corresponding values for the log-transformed data are $\mu = 2.18$ and $\sigma^2 = 1.39$. Figure A.6 shows the fractile diagrams of the sample probability (A.44) versus the theoretical probability (A.45) and (A.46) for the two assumed distributions. By inspection, the distribution appears to be somewhat better represented as lognormal than normal.

---

[3] There are functional approximations of $N(u)$ over its useful range that allow its values to be calculated directly (see Abramowitz and Stegun, 1970).

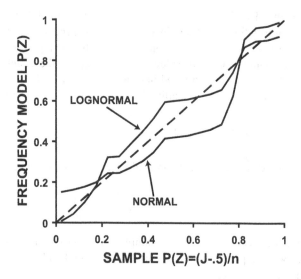

Figure A.6 Fractile diagrams of the infiltration data in Example A.11 tested against normal and lognormal distributions.

Kolmogorov–Smirnov Statistic   The Kolmogorov–Smirnov (KS) statistic is a test to determine whether a sample set of N numbers belongs to a theoretical distribution. The criterion used for evaluating the sample set is the absolute maximum deviation,

$$D = \max\{|S_J - P_J|\} \qquad J = 1, \ldots, N \qquad (A.47)$$

where S_J is the sample cumulative probability (A.44) and P_J is the theoretical cumulative probability of the distribution being tested [Equations (A.45) and (A.46)]. Tables of KS values of D that reject the null hypothesis (that the sample set belongs to the distribution) have been tabulated for various values of N. Table A.5 gives D values for $N = 20$ at various levels of precision.

Example A.12   Use the KS statistic to evaluate whether the 20 infiltration values in Example A.11 are normally or lognormally distributed.

SOLUTION: The maximum deviations for the normal and lognormal models are 0.213 and 0.146, respectively. Thus, the KS test accepts both distributions at $P \leq 0.2$. The KS test would reject the normal hypothesis at $P = 0.3$.

TABLE A.5 *D* **Values of KS Statistic That Reject the Null Hypothesis at Various Levels of Precision** *P* **for** *N* **= 20 Samples**

P	0.20	0.15	0.10	0.05	0.01
D	0.231	0.246	0.264	0.294	0.352

A.3.3   Confidence Limits for the Mean

If a set of N samples are drawn from a population that is normally distributed with mean m and variance s^2, the average $\bar{X}$ of the N samples is a normally distributed random variable with the following properties (Hald, 1952):

$$E[\bar{X}] = m \tag{A.48}$$

$$\mathrm{Var}[\bar{X}] = \frac{s^2}{N} \tag{A.49}$$

Confidence limits for the mean value m of the population may be estimated from the sample measurements using the t-distribution if the population variance s^2 is not known a priori. In this case, the probability is $P_2 - P_1$ that the population mean m is bounded by

$$\bar{X} - \frac{t_{P_1}\bar{s}}{\sqrt{N}} < m < \bar{X} + \frac{t_{P_2}\bar{s}}{\sqrt{N}} \tag{A.50}$$

where $\bar{s}$ is the sample variance (A.1) and t_{Pi} is the value of the t-variate at probability Pi.

Nonnormal Populations   If the population of values from which the sample is drawn is not normally distributed, the distribution of the mean will approach a normal distribution only in the limit of large sample size. This is called the *central limit theorem* (Himmelblau, 1970). For small sample size, there is no exact procedure for generating confidence limits for the mean of a set of samples. In practice, (A.48) and (A.49) usually are assumed to hold for arbitrary distributions.

Example A.13   Calculate the 95% confidence limits for the mean value of the 20 infiltration rates given in Example A.11.

SOLUTION: The $t_{0.025}$-statistic for $N = 20$ samples is 2.086. Since the data in Example A.11 have $\bar{m} = 14.64$ cm day^{-1} and $\bar{s} = 10.89$ cm day^{-1}, according to (A.50) the probability is $P_2 - P_1 = 0.95$ that

$$\bar{m} - \frac{2.086\bar{s}}{\sqrt{20}} = 9.56 < m < \bar{X} + \frac{2.086\bar{s}}{\sqrt{20}} = 19.71$$

Thus, there is a 95% probability that the true mean is between 9.56 and 19.71.

A.4   ANALYSIS OF SPATIAL STRUCTURE

An important aspect of sampling, which is not taken into account by such traditional statistical methods as analysis of variance, is that the samples may be correlated. Intuitively, it seems reasonable that measurements of a parameter at different places in a field are more likely to be similar to each other if they are taken close together than if they are taken far apart. Knowledge of the spatial correlation structure, or the distance over which properties are correlated with each other, is essential to the design of optimal sampling grids and interpolation methods (McBratney and Webster, 1981; Warrick et al., 1986). A function that is used to characterize the correlation structure is the variogram.

A.4.1 Variogram

The *variogram* $\gamma(h)$ is a measure of the variance of a given parameter with respect to samples that are a fixed distance apart in space. Thus, for a given separation distance h,

$$\gamma(h) = \frac{1}{2N(h)} \sum_{J=1}^{N} [Z(x_J) - Z(x_J + h)]^2 \qquad (A.51)$$

where $Z(x_J)$ is the value of the random variable Z at $x = x_J$, $Z(x_J + h)$ the value of Z at a distance h from x_J, and $N(h)$ the number of pairs of points that are a distance h apart. If the random variable Z is second-order stationary [i.e., it obeys (A.24) and (A.25)], all random variables $Z(x)$ at each x have the same mean value, and the covariance $\text{Cov}[Z(x_1), Z(x_2)]$ depends only on the separation distance $h = x_2 - x_1$ between the points. In this case $\gamma(h)$ becomes a statistical measure of the degree to which random variables $Z(x)$ at different x become alike as the distance between them decreases.

Figure A.7 shows an ideal variogram function and labels its significant features. The function increases uniformly from a small value at $h = 0$ (called the *nugget*

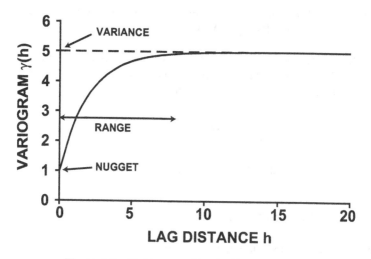

Figure A.7 Variogram and its significant features.

variance) to an asymptotic value (called the *sill*) at large h. The distance over which $\gamma(h)$ is changing, called the *range*, may be identified with the distance over which values of the random variable $Z(x)$ are correlated with each other.

The nugget variance (the variance at zero lag) may be a source of confusion, since it implies that a property measured at the same point may have some variance with itself. The nugget variance is an extrapolated value, since in practice there is some minimum separation distance h in the sample set of measurements used to estimate $\gamma(h)$. The sill should be equal to the population variance if the random variable $Z(x)$ is second-order stationary, since (A.51) is an estimate of the variance s^2 if its sample members are uncorrelated. Warrick et al. (1986) discuss different variogram shapes and their interpretation.

Estimation of $\gamma(h)$ The estimation of $\gamma(h)$ is straightforward using (A.51), although in practice many pairs of points $N(h)$ at each lag distance h are needed to obtain an accurate representation of the spatial structure (Journel and Huijbregts, 1978; Warrick et al., 1986). Russo and Jury (1987a,b) studied the variogram estimation problem theoretically by computer simulation. They created a "field" of points with a known statistical structure and tried to deduce this structure from various sampling schemes. Their most important conclusions were as follows:

- Two-dimensional sampling grids yield more accurate information about $\gamma(h)$ than do transects (one-dimensional lines of samples) with the same number of points.
- The minimum grid sample spacing must be less than half of the range to detect any spatial structure.
- The sample schemes that had less than 72 pairs of points per lag commonly estimated the range of the true spatial structure with an error of 25% or greater. In practice, this means that a square grid with at least seven points on a side is required for accurate analysis.
- If the field is not stationary but possesses a drift in addition to random spatial structure, the spatial structure is more difficult to detect.

A.4.2   Autocorrelation Function

The variogram $\gamma(h)$ is related to other indices of spatial structure. If the field is stationary and a population variance s^2 can be defined, the autocorrelation function $\rho(h)$ is defined as (Warrick and Nielsen, 1980)

$$\rho(h) = 1 - \frac{\gamma(h)}{s^2} \qquad (A.52)$$

Figure A.8 shows the autocorrelation function corresponding to the idealized variogram $\gamma(h)$ in Fig. A.7. It has a maximum value at $h = 0$ and decreases to zero at a separation distance comparable to the range of the variogram.

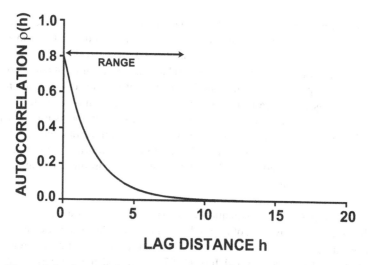

Figure A.8   Autocorrelation function calculated from Fig. A.7 with (A.52).

A quantitative measure of the range of correlation of a random spatial structure may be calculated from the autocorrelation function (A.52). This measure, called the *integral scale* I^*, is calculated by (Bakr et al., 1978; Lumley and Panofsky, 1964; Russo and Bresler, 1981)

$$I^* = \int_0^\infty \rho(h)\, dh \qquad (A.53)$$

if the spatial structure was calculated from a one-dimensional row of data points, and

$$I^* = \left[\int_0^\infty h\rho(h)\, dh \right]^{1/2} \qquad (A.54)$$

if the data points are in two dimensions. Jury (1985) and Warrick et al. (1986) present tables of measured values of I^* from various field studies. In general, I^* is not an intrinsic property of the field but depends on the scale over which it is measured.

Model Variogram Functions   A number of variogram model functions are used to represent the structure of a set of data. The most common ones, taken from Journel and Huijbregts (1978) and Warrick et al. (1986), are as follows:

$$\text{Linear: } \gamma(h) = \begin{cases} C_0 + \dfrac{h(C_1 - C_0)}{a} & 0 \leq h \leq a \\ C_1 & h \geq a \end{cases} \qquad (A.55)$$

$$\text{Spherical: } \gamma(h) = \begin{cases} C_0 + (C_1 - C_0)\left[\dfrac{3}{2}\left(\dfrac{h}{a}\right) - \dfrac{1}{2}\left(\dfrac{h}{a}\right)^3\right] & 0 \le h \le a \\ C_1 & h \ge a \end{cases} \tag{A.56}$$

$$\text{Gaussian: } \gamma(h) = C_0 + h(C_1 - C_0)[1 - \exp\left(-\frac{h^2}{\lambda^2}\right) \tag{A.57}$$

$$\text{Exponential: } \gamma(h) = C_0 + h(C_1 - C_0)[1 - \exp\left(-\frac{h}{\lambda}\right) \tag{A.58}$$

Figure A.9 shows plots of the four functions, which have been given the same integral scale I^*.

A.4.3   Kriging

The random variable Z may be estimated at a particular point x_0 in space by interpolating values of Z around the point that are within the range of correlation. Thus, a linear estimator $Z^*(x_0)$ may be defined as (Journel and Huijbregts, 1978)

$$Z^*(x_0) = \sum_{J=1}^{N} \lambda_J Z_J \tag{A.59}$$

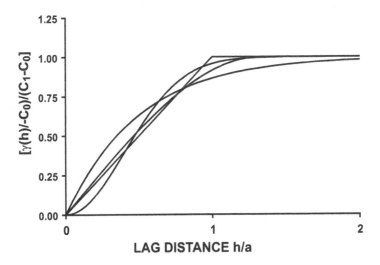

Figure A.9   Model variogram functions (A.55)–(A.58), optimized to have the same integral scale I^* as defined in (A.53).

where $Z_J = Z(x_J)$, $J = 1, \ldots, N$, are the N measured points around x_0 that are within the zone of correlation. Kriging derives values of λ_J that satisfy two criteria (assuming that Z is second-order stationary):

1. Z^* has the same mean value as each of the Z_J.
2. Z^* has the minimum possible variance about the true value $Z(x_0)$.

Derivation of the correct values of λ that satisfy these two criteria is straightforward but involves techniques beyond the scope of this book. Interested readers may consult Journel and Huijbregts (1978). Numerous computer software programs are available to calculate λ_J from a set of data.

When the assumptions of the theory are well met, (A.59) becomes an excellent method for constructing interpolated values of the variable between the measured points, from which contour lines may be drawn. In addition, the theory allows a calculation of the estimation variance, which may be used to determine where values of the function are the most uncertain. This information could be very useful if a second sampling of the same field is planned.

A.4.4   Drift

When the differences in values of a variable $Z(x)$ at different points are systematic rather than random, the field of points is said to be *nonstationary*. A nonstationary field cannot be analyzed by the methods described above because the covariance, variogram, and autocorrelation function depend on the locations of the points as well as on the space between them. In such a case, a spatial structure analysis may still be performed if the *deterministic* or *drift component* of the field is first removed. Formally, one may write

$$Z(x) = m(x) + \epsilon(x) \qquad (A.60)$$

where $m(x)$ is a deterministic drift function and $\epsilon(x)$ is a zero-mean, second-order stationary random function. Removal of the drift function is problematical. Only values of $Z(x)$ are measured, and it is not known a priori what fraction of $Z(x)$ is drift and what fraction is random. Russo and Jury (1987b) discuss various techniques for drift identification and removal. Usually, simple drift functions are postulated and the residual function $\epsilon(x)$ is analyzed to determine whether it has the attributes of a second-order stationary function. The most common method of identification of stationarity is the shape of the variogram, which should approach an asymptotic value at large lags if the process is stationary (Journel and Huijbregts, 1978).

PROBLEMS

A.1   Samples of infiltration rate i and percentage of clay (PC) are taken at random locations throughout a very large field. The data are summarized in Table

TABLE A.6   Infiltration Rate Measurements and Percentage of Clay at 25 Locations in a Large Field Site

Sample No.	i (cm day^{-1})	% Clay	Sample No.	i (cm day^{-1})	% Clay
1	2.0	3.5	14	6.0	1.3
2	5.0	2.0	15	7.5	0.9
3	25.0	1.1	16	2.0	2.6
4	3.0	1.3	17	8.0	1.0
5	15.0	0.8	18	3.0	1.5
6	4.0	2.2	19	2.0	2.0
7	5.0	0.9	20	5.0	1.3
8	7.0	1.2	21	15.0	0.8
9	6.0	1.0	22	7.0	1.0
10	5.0	3.0	23	3.0	2.5
11	9.0	2.8	24	3.5	1.6
12	0.5	5.3	25	2.0	1.5
13	4.5	1.0			

A.6. Construct cumulative distribution functions for the two variables using an interval of 2 cm day^{-1} for i and 0.5 for PC. Calculate the sample mean and variance of the data and the correlation coefficient. What does the correlation coefficient tell you about the nature of the relationship between i and PC?

A.2 Repeat Problem A.1 using the log-transformed data for the infiltration rote and an interval of 0.5 for $\ln i$. Plot i and $\ln i$ on probability paper and decide whether the distribution of values is better described as normal or lognormal.

A.3 The saturated hydraulic conductivity K_s was measured at 20 sites in the field study of Nielsen et al. (1973) and was found to be lognormally distributed with sample estimates of $E[\ln K_s] = \mu = 2.58$ and $Var[\ln K_s] = \sigma^2 = 1.01$.

(a) Calculate the mean, mode, and median of the K_s distribution.

(b) Calculate the cumulative probability $P(M) = \Pr(K_s \leq M)$ of obtaining a sample that is less than the mean $M = E[K_s]$ Hint: $P(M) = 0.5 + N[(\ln M - \mu)/\sigma]$.

A.4 An agricultural field has a lognormal steady-state infiltration rate with $E[\ln i] = \mu = 1.5$, $Var[\ln i] = \sigma = 1.5$.

(a) If water is ponded continuously over the entire field, calculate the steady-state mean infiltration rate.

(b) Calculate the fraction of the field that has an infiltration rate greater than the mean M. This is equal to $1 - P(M)$.

(c) Calculate the fraction of the flow F that is entering through the part of the field that has an infiltration rate greater than the mean M. The flow fraction through this part of the field is

$$F = \frac{\int_M^\infty i f(i)\, di}{\int_0^\infty i f(i)\, di} = \frac{1}{M} \int_M^\infty \frac{1}{\sqrt{2\pi}\,\sigma} \exp\left[\frac{(\ln i - \mu)^2}{2\sigma^2}\right] di$$

(d) If a chemical tracer was added with the ponded water, what would the spatial distribution of the front edge of the plume look like?

A.5 A set of spatial measurements of clay content are made at 100 locations spaced at regular 1-m intervals along a line of 100 m in the field. Because of a nonuniform deposition process, the mean clay content increases linearly along the transect. Thus, an approximate random model of the percentage of clay (PC) is

$$PC(x) = ax + b + \xi(x)$$

where a and b are constants and $\xi(x)$ is a zero-mean random variable. Show that the contribution to the variogram (A.51) from the linear drift term is a quadratic function of the spacing between points. Discuss how this variogram shape might be used to detect the presence of drift in a data set.

A.6 Theoretical analyses (e.g., Russo and Jury, 1987a,b) have shown that the variogram function (A.51) must be constructed from about 100 or more pairs of points per lag interval. Assuming that the data are taken along a square grid, calculate the lag distances and the number of pairs of points per lag on a grid of 36 points in a 6×6 pattern of separation L. What would the result have been if the 36 points were laid out at equal spaces L along a line?

A.7 A chemical tracer pulse is added to the surface of a soil and is leached downward under continuous irrigation at a rate $i_0 = 1$ cm day^{-1}. The CDE and CLT models are fit to the pulse data obtained by solution samplers at the 1-m depth with parameters $\mu = 3.20$, $\sigma = 0.5$, $V = 3.6$, and $D = 15.26$, which makes the two models virtually superimpose on each other. If the models are regarded as travel-time pdfs, it can be shown that the predicted probability of arriving at depth z in a time less than t_0 is given by

$$P(t \leq t_0) \approx 0.5 + N \left(\frac{V t_0 - z}{\sqrt{2D t_0}} \right) \qquad \text{CDE model}$$

$$P(t \leq t_0) = 0.5 + N \left[\frac{\ln(t_0 L/z) - \mu}{\sigma} \right] \qquad \text{CLT model}$$

Using Table A.5, calculate the amount of time t_0 predicted by each model for the first $P(t_0) = 0.01, 0.05, 0.1$ of the pulse to reach $z = 500$ cm. Since these models fit the data equally well, discuss the problem of predicting arrival times to groundwater from surface measurements.

Solutions to Problems

Chapter 1

1.1 The time in seconds required to move a distance $X = 10$ cm at a velocity V is simply

$$t = \frac{X}{V} = \frac{18\eta X}{(\rho_s - \rho_\ell)D^2 g} = \frac{0.0011}{D^2}$$

for D in centimeters, where (1.5) has been used for the velocity and the water properties have been taken from Table 2.2. The values requested are listed in Table 1.

TABLE 1 Results for Problem 1.1

D (cm)	t (s)	t (hr)
0.2	0.03	—
0.005	43.2	0.01
0.0002	27,010	7.5
0.0001	108,043	30.00

1.2 The 50 g of soil contains 5 g of clay and 45 g of sand and silt. The 1000 mL of water has a mass of 1000 g. If we ignore the volume change, the initial mass is 1050 g in 1000 cm^3, or a density of 1.05 g cm^{-3}. After the sand and silt settle out, the new density falls to 1.005 g cm^{-3}.

The volume change may be taken into account if the particle density is known. If we assume that the sand and silt have a density of 2.7 g cm^{-3} and the clay, 2.65 g cm^{-3}, the volume of the 50 g of soil is

$$\frac{5}{2.65} + \frac{45}{2.7} = 18.6 \text{ cm}^3$$

so that the initial volume of the 1050 g of soil and water system is 1018.6 cm^3, and the initial density is $1050/1018.6 = 1.031$ g cm^{-3}. The volume of the clay mass is only $5/2.65 = 1.89$ cm^3, so that the final density is $1005/1001.89 = 1.003$ g cm^{-3}.

1.3 Both the surface area and the mass of the disk are nearly proportional to the square of the radius. Since the specific surface is the ratio of the surface area to the mass, it does not depend on R.

1.4 If we define an effective K_d as the slope of the isotherm, then for the Freundlich model (1.16) the K_d is equal to

$$K_d \equiv \frac{dC_a}{dC_\ell} = \beta K_f C_\ell^{\beta-1}$$

which depends on the dissolved concentration. If $\beta = 0.8$, the ratio of K_d at the low concentration to that at the high concentration is

$$\frac{K_d(1)}{K_d(1000)} = \frac{\beta K_f(1)^{\beta-1}}{\beta K_f(1000)^{\beta-1}} = 1000^{.2} \approx 4.0$$

Thus, low concentrations are sorbed more strongly than high concentrations.

1.5 The bulk density of the system may be calculated by regarding the aggregates as "particles" of density $\rho_p = 1.3$ g cm^{-3}. Thus, the bulk density is

$$\rho_b = \rho_p(1 - \phi_i) = 1.3 \times 0.6 = 0.78 \text{ g cm}^{-3}$$

where $\phi_i = 0.4$ is the interaggregate porosity. The aggregate porosity ϕ_a is calculated from the normal relationship between particle density and bulk density:

$$\phi_a = 1 - \frac{\rho_b}{\rho_m} = 1 - \frac{1.3}{2.65} = 0.51$$

The overall porosity is calculated from the overall bulk density and the true particle density as

$$\phi = 1 - \frac{\rho_b}{\rho_m} = 1 - \frac{0.78}{2.65} = 0.706$$

This can also be calculated as the sum of the interaggregate porosity ϕ_i and the porosity of the aggregates multiplied by the aggregate volume fraction $(1-\phi_i)$. Thus,

$$\phi = \phi_i + \phi_a(1 - \phi_i) = 0.4 + 0.6 \times 0.51 = 0.706$$

If the aggregates are saturated, their wet bulk density ρ_{bw} is elevated by filling the 0.51 volume fraction of pore space with water of density 1.0 g cm^{-3}. Thus, $\rho_{bw} = 1.3 + 0.51 = 1.81$ g cm^{-3}. The wet bulk density of the soil is then

$$\rho_b = \rho_{bw}(1 - \phi_i) = 1.81 \times 0.6 = 1.086 \text{ g cm}^{-3}$$

In practice, the aggregates would always have water in them.

Chapter 2

2.1 Let the initial velocities be denoted by v and the final ones by υ. Since the second sphere is initially at rest ($v_2 = 0$) then by conservation of momentum

$$m_1 v_1 = m_1 \upsilon_1 + m_2 \upsilon_2$$

where υ_i is the final velocity of the ith ball after collision. The corresponding energy conservation equation is

$$\tfrac{1}{2} m_1 v_1^2 = \tfrac{1}{2} m_1 \upsilon_1^2 + \tfrac{1}{2} m_2 \upsilon_2^2$$

which may be factored and simplified into

$$m_1 (v_1 - \upsilon_1)(v_1 + \upsilon_1) = m_2 \upsilon_2^2$$

Now we rearrange the momentum equation to the form

$$m_1 (v_1 - \upsilon_1) = m_2 \upsilon_2$$

and substitute it into the factored energy equation, producing after cancellation:

$$v_1 + \upsilon_1 = \upsilon_2$$

This equation can be reinserted into the rearranged momentum equation and the result solved for the final velocity υ_1 of the first ball:

$$\upsilon_1 = v_1 \frac{m_1 - m_2}{m_1 + m_2}$$

This result may then be plugged into the equation above it to solve for the final velocity υ_2 of the second ball:

$$\upsilon_2 = v_1 \left(1 + \frac{m_1 - m_2}{m_1 + m_2} \right) = \frac{2 v_1 m_1}{m_1 + m_2}$$

The special case $m_2 = m_1$ produces the simplification

$$\upsilon_1 = 0$$

$$\upsilon_2 = v_1$$

This case offers an illustration of the scattering of a neutron by a hydrogen nucleus; the velocity is slowed down from its initial state and passed on to the receiving body. In the more general case where the second ball has an initial velocity, the two balls of equal mass simply trade velocities. The special case $m_2 \gg m_1$ produces the simplification

$$v_1 = v_1 \frac{m_1 - m_2}{m_1 + m_2} \approx -v_1$$

$$v_2 = \frac{2v_1 m_1}{m_2} \approx 0$$

In this case, the second ball acts like an immovable wall and the first ball bounces back with its original velocity in the opposite direction, just like a neutron colliding with a heavy nucleus.

2.2 From the information given we deduce the following: mass of water $m_w = 2$ g; mass of dry soil $m_s = 13$ g; volume of soil $V_s = 10$ cm^3; particle density $\rho_m = 2.65$ g cm^{-3}. Then, by definition:

(a) Bulk density $\rho_b = m_s / V_s = 1.3$ g cm^{-3}

(b) Volumetric water content $\theta_v = m_w / \rho_w / V_s = 0.2$

(c) Gravimetric water content $\theta_g = m_w / m_s = 0.154$

(d) Porosity $\phi = 1 - \rho_b / \rho_m = 0.51$

(e) Volumetric air content $a = \phi - \theta_v = 0.31$

2.3 Intuitively, we deduce that the system will eventually reach an equilibrium state where the air pressure is above atmospheric, and the height of rise is less than it would be in free air. The solution proceeds as follows:

- We can imagine the water column (without its interface) as a rigid body pushing down on the water below it with a liquid pressure P_{atm} (causing an upward force on itself) and pushing up on the air–water interface above it with a smaller pressure P_{liq} (causing a downward force on itself). The net upward pressure-induced force is counteracted by the force of gravity:

$$F_{up} = P_{water} A = P_{atm} A$$

$$F_{down} = F_{grav} + F_{liq} = \rho_w g A H + P_{liq} A$$

- We calculate P_{liq} in two stages:

— The final air pressure P_{air} compressed in the tube is given by Boyle's law:

$$P_{air} V_{air} = P_{air} A (L - H) = \text{const} = P_{atm} A L \longrightarrow P_{air} = P_{atm} \frac{L}{L - H}$$

— The liquid pressure is given by the capillarity equation:

$$P_{liq} = P_{air} - \frac{2\sigma}{R} = P_{atm} \frac{L}{L - H} - \frac{2\sigma}{R}$$

- The force balance $F_{up} = F_{down}$ produces the result

$$\rho_w g A H = \left(P_{atm} - P_{liq} \right) A = \left(P_{atm} - \frac{L}{L - H} P_{atm} + \frac{2\sigma}{R} \right) A$$

which may be rearranged to produce

$$H = \frac{2\sigma}{\rho_w g R} - \frac{P_{atm}}{\rho_w g} \frac{H}{L - H}$$

- We may now define two special "lengths" for our problem:
  — Capillary length H_c:

$$H_c \equiv \frac{2\sigma}{\rho_w g R} = \text{height of rise when } P = P_{atm}$$

— Atmospheric pressure head H_{atm}:

$$H_{atm} \equiv \frac{P_{atm}}{\rho_w g} = \text{atmospheric } P \text{ in head units}$$

- With these definitions we may rewrite the force balance equation as

$$H = H_c - \frac{H_{atm} H}{L - H}$$

which produces the quadratic equation

$$H^2 - (H_c + H_{atm} + L) H + H_c L = 0$$

This equation has two roots, one of which is discarded on physical grounds because it is greater than the maximum[1] possible height of rise L. The physical root is

$$H = \tfrac{1}{2}(H_c + H_{atm} + L) - \tfrac{1}{2} \left[(H_c + H_{atm} + L)^2 - 4H_c L \right]^{1/2}$$

The variables in the problem have the following values:

$$H_c = 148 \text{cm}$$

$$H_{atm} = 1033 \text{cm}$$

$$L = 200 \text{cm}$$

This produces a value of $H \approx 22$ cm.

Notice how easy this is as an energy balance. At the bottom we define $P_0 = P_{atm}$, $z_0 = 0$, which makes $\psi_t \equiv 0$ everywhere. At the air–water interface:

- $\psi_s = \psi_p = 0$, by definition
- $\psi_z = \rho_w g H$
- $\psi_a = P_{air} - P_{atm} = P_{atm} H / (L - H) \equiv \rho_w g H_{atm} H / (L - H)$
- $\psi_m = -2\sigma / R \equiv -\rho_w g H_c$

---

[1] Note that $P_{air} \longrightarrow \infty$ as $H \longrightarrow L$.

- $\psi_z + \psi_a + \psi_m = 0$ at the interface, which is the same as the final equation derived by the force balance

2.4 Let the soil be point A, where $z = 0$. Then the total potential head is $h_t = h = -100$ cm. This must equal the total potential head at point B (the top of the tree), where the potential head components are gravitational and osmotic. Thus,

$$h_t = -1000 = z_b + s_b = 10{,}000 + s_b \text{ cm} \rightarrow s_b = -11{,}000 \text{ cm}$$

or about 11 atm of osmotic pressure.

2.5 Begin by defining $z = 0$ at the surface, and $P_0 = P_{atm}$. We focus on two points of interest. Point A is at the mercury–water interface in the tube, and point B is in the soil next to the tensiometer cup. Then at point A:

- $z = 20$ cm, by definition
- $a = 0$ (no air in tube)
- $s = 0$ (no membranes in system)
- $h = 0$ (no soil)
- $p = -10\rho_m g / \rho_w g = -136$ cm (water lifts mercury column)
- $h_t = z + a + s + h + p = -116$ cm

At point B:

- $z = -15$ cm, by definition
- $a = 0$ (by assumption; it could be otherwise)
- $s = 0$ (no membranes in system)
- $h =?$
- $p =?$
- $h_t = -116$ cm $= -15 + 0 + 0 + h_B + p_B \longrightarrow h + p = -101 \longrightarrow h = -101; p = 0$

Note that this solution is a direct application of (2.20) with $L = 15$ cm, $H = 10$ cm, and $X = 10$ cm. Thus,

$$h = H + L - 12.6X = -101 \text{ cm}$$

2.6 This is a reproduction of Example 2.10 with the final position known but not the solute potential. Its analysis leads to (2.18), with $s = -\pi/\rho_w g$, since $\pi = -\psi_s$ is the osmotic pressure. Thus,

$$s = H - \frac{2\sigma}{\rho_w g R} = 100 - 298 = -198 \text{ cm}$$

2.7 Begin by defining $z = 0$ at the surface, and $P_0 = P_{atm}$. The two points of interest are the alloy-water interface in the tube (point A) and the soil next to the tensiometer cup (point B). Let x be the unknown height of rise of the alloy above the alloy reservoir height. Then at point A:

- $z = 10 + x$ cm, by definition
- $a = 0$ (no air in tube)
- $s = 0$ (no membranes in system)
- $h = 0$ (no soil)
- $p = -\rho_X gx / \rho_w g = -8x$ cm (water lifts alloy column)
- $h_t = z + a + s + h + p = 10 - 7x$ cm

At point B:

- $z = -50$ cm, by definition
- $a = 0$ (by assumption; it could be otherwise)
- $s = 0$ (no membranes in system)
- $h = -150$ (given)
- $p = 0$ (since soil is unsaturated)
- $h_t = -50 + 0 + 0 - 150 + 0 = -200 = 10 - 7x \longrightarrow x = 30$ cm, a total of 40 cm above the soil surface

2.8 The background for the second scan is the empty column, which has a scan density of N_0. Thus, we may write

$$N_1 = N_0 \exp(-\rho_b v_m L)$$

since there is no water in the column at this time. Therefore,

$$\log \frac{N_1}{N_0} = -\rho_b v_m L \rightarrow v_m = -\frac{1}{\rho_b L} \log \frac{N_1}{N_0}$$

2.9 First, calibration experiments are run so that each beam's mass absorption coefficient for water and soil is known. Then, imagine that a single location is scanned at two different times by each beam, so that the following four equations are written:

$$N_{A1} = N_{A0} \exp[-\rho_b(t_1) v_{Am} L - \rho_w v_{Aw} \theta(t_1) L]$$

$$N_{A2} = N_{A0} \exp[-\rho_b(t_2) v_{Am} L - \rho_w v_{Aw} \theta(t_2) L]$$

$$N_{B1} = N_{B0} \exp[-\rho_b(t_1) v_{Bm} L - \rho_w v_{Bw} \theta(t_1) L]$$

$$N_{B2} = N_{B0} \exp[-\rho_b(t_2) v_{Bm} L - \rho_w v_{Bw} \theta(t_2) L]$$

If we divide the second and fourth equations by the first and third, take the logarithm of both sides and rewrite, we obtain

$$\log \frac{N_{A1}}{N_{A2}} = v_{Am} L \, \Delta\rho_b + \rho_w v_{Aw} L \, \Delta\theta$$

$$\log \frac{N_{B1}}{N_{B2}} = v_{Bm} L \, \Delta\rho_b + \rho_w v_{Bw} L \, \Delta\theta$$

where $\Delta\rho_b = \rho_b(t_2) - \rho_b(t_1)$, and so on. These are two algebraic equations in two unknowns $(\Delta\rho_b, \Delta\theta)$, which may be solved by standard methods.

2.10 The area of a sphere is $4\pi R^2$ and the volume is $4\pi R^3/3$. Thus, the mass is $\rho(4\pi R^3)/3$ and the surface area per mass is $3/\rho R$. The circular disk has two circular surfaces and an annular side, of total area $2\pi R^2 + 2\pi R\tau = 2\pi R(R+\tau)$. Its volume is $\pi R^2\tau$, so its mass is $\rho\pi R^2\tau$ and its surface area per mass is $2(R + \tau)/\rho R\tau \approx 2/\rho\tau$. The surface per mass values from Example 2.12 are 8,000,000 and 500 cm^2 g^{-1} for clay and sand. The particle density is $\rho = 2.7$ g cm^{-3}. Thus, the sand radius is $R = 3/\rho S = 0.0022$ cm. The clay particle thickness $\tau = 2/\rho S = 9.26 \times 10^{-8}$ cm.

Chapter 3

3.1 This problem involves an application of the finite-difference form of the Buckingham–Darcy flux law:

$$K\left(\frac{h_1 + h_2}{2}\right) = K(\bar{h}) \approx \frac{-E\,\Delta z}{\Delta H}$$

where $E = 150/100 = 1.5$ cm day^{-1}. The results of the calculations are given in Table 2 and in Fig. 1.

TABLE 2   Results for Problem 3.1

z cm	h cm	H cm	ΔH cm	Δz cm	$K(\bar{h})$ cm day^{-1}	$\bar{h}$ cm
120	−750	−630				
			−430	20	0.07	−525
100	−300	−200				
			−95	30	0.47	−237.5
70	−175	−105				
			−75	30	0.60	−122.5
40	−70	−30				
			−20	20	1.50	−50
20	−30	−10				
			−9	15	2.50	−18
5	−6	−1				
			−1	5	7.50	−3
0	0	0				

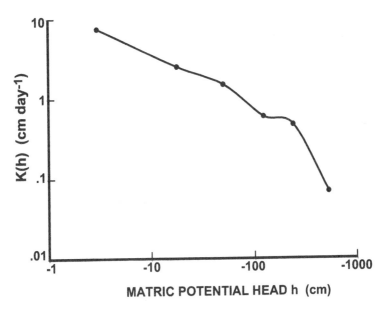

Figure 1 Log-log plot of $K(h)$ from Problem 3.1.

3.2 This problem is solved by creating an effective homogeneous column with the same hydraulic resistance as the layered column.

(a) The first step is to calculate the effective hydraulic conductivity using (3.26):

$$\frac{20}{K_{\text{eff}}} = \frac{10}{5} + \frac{10}{10} = 3 \longrightarrow K_{\text{eff}} = \frac{20}{3} \text{ cm day}^{-1}$$

Then Darcy's law is written across the effective column, which has a total hydraulic head of 30 cm at the top and 0 at the bottom. Thus,

$$J_w = -K_{\text{eff}} \frac{H_2 - H_1}{z_2 - z_1} = -\frac{20}{3} \frac{30}{20} = -10 \text{ cm day}^{-1}$$

(b) The water pressure is calculated by writing Darcy's law across the bottom half of the column:

$$J_w = -10 = -K_2 \frac{z_A + p}{z_A} = -10 \frac{10 + p}{10} \longrightarrow p = 0$$

(c) Since the pressure is exactly zero at the interface when the height of water above the column is 10 cm, as it decreases, p will become negative, eventually desaturating the column if the air entry suction is exceeded.

3.3 Since the pressure is given at the interface, the hydraulic conductivity of each layer can be calculated directly by Darcy's law. If we set $z = 0$ at the bottom (point 1), then $H_1 = 0$. At the interface (point 2), $H_2 = 50 + 50 = 100$ cm. At the top of the column (point 3), $H_3 = 100 + 20 = 120$ cm. Thus, applying Darcy's law across the sand layer, we obtain

$$J_w = -20 = -K_{sand} \frac{H_3 - H_2}{z_3 - z_2} = -K_{sand} \frac{20}{50} \longrightarrow K_{sand} = 50 \text{ cm h}^{-1}$$

Similarly, for the clay layer we obtain

$$J_w = -20 = -K_{clay} \frac{H_2 - H_1}{z_2 - z_1} = -K_{clay} \frac{100}{50} \longrightarrow K_{clay} = 10 \text{ cm h}^{-1}$$

Finally, the effective conductivity is obtained across the entire column as

$$J_w = -20 = -K_{eff} \frac{H_3 - H_1}{z_3 - z_1} = -K_{eff} \frac{120}{100} \longrightarrow K_{eff} = \frac{100}{6} \text{ cm h}^{-1}$$

Note that we can also obtain K_{eff} from

$$\frac{100}{K_{eff}} = \frac{50}{K_{sand}} + \frac{50}{K_{clay}}$$

3.4   The cross-sectional area $A = 100$ cm^2, volume flow rate $Q = 1000$ cm^3 h^{-1}, column height $L = 50$ cm, and ponding height $d = 10$ cm.

(a) By definition, the flux $J_w = Q/A = -10$ cm h^{-1}.

(b) By Darcy's law, $K_s = -J_w L/(L + d) = 8.33$ cm.

3.5   (a) By Poiseuille's law, the volume flow through the 0.05-cm radius hole is

$$Q = \frac{\rho_w g R^4 (L + d)}{8 L \nu} = 1039 \text{ cm}^3 \text{ h}^{-1}$$

using the water properties in Table 2.1.

(b) We can assume that the volume flow through the soil part of the column remains at 1000, since it has lost negligible area. Therefore, the flux through the system is

$$J_{eff} = -\frac{Q_1 + Q_2}{A_1 + A_2} = -20.39 \text{ cm h}^{-1}$$

(c) By Darcy's law,

$$K_{eff} = -J_{eff} \frac{L}{d + L} = 16.99 \text{ cm h}^{-1}$$

Thus, the conductivity of the medium more than doubles by the addition of the single hole.

3.6 Let the thickness of the lens be x and the total column height L.

(a) Without the lens, we use Darcy's law:

$$J_w = -K_S \frac{d+L}{L} = -220 \text{ cm day}^{-1}$$

(b) If the actual flux is only -15 cm day^{-1}, we first calculate K_{eff} for the layered system by Darcy's law:

$$J_w = -15 = -K_{\text{eff}} \frac{d+L}{L} \longrightarrow K_{\text{eff}} = \frac{15 \times 100}{110} = \frac{150}{11} \text{ cm day}^{-1}$$

Now we imagine a three-layer column, consisting of sand with clay in the middle somewhere. Letting the top layer thickness be z, the clay layer x, and the bottom $100 - x - z$, we use the layered soil formula (3.26)

$$\frac{100}{K_{\text{eff}}} = \frac{z}{K_S} + \frac{100 - x - z}{K_S} + \frac{x}{K_C} = \frac{100 - x}{K_S} + \frac{x}{K_C} \longrightarrow x = 0.6837 \text{ cm}$$

(c) There are two possibilities:

(i) When the clay is under the sand, we write Darcy's law across the clay layer:

$$J_w = -15 = -K_C \frac{p+x}{x} \longrightarrow p = 101.86 \text{ cm}$$

(ii) When the clay is on top of the sand, we write Darcy's law across the clay layer:

$$J_w = -15 = -K_C \frac{110 - (100 - x + p)}{x} \longrightarrow p = -91.87 \text{ cm}$$

(d) In the second case the system will definitely be unsaturated underneath the clay layer.

3.7 Using the theoretical curve, we calculate

(a) $h(\theta) = -147.0, -68.7, -40.0, -22.5, 0.$

(b) The model porous medium has four sizes of capillary tube. The radius of a given tube is (3.33)

$$R_J = -\frac{2\sigma}{\rho_w g h_J} \approx -\frac{0.147}{h_J}$$

and the corresponding number of tubes per area that fill a subregion $\Delta\theta = 0.1$ is, from (3.34),

$$n_J = \frac{\Delta\theta}{\pi R_J^2} = \frac{\rho_w^2 g^2 \Delta\theta h_J^2}{4\pi \sigma^2} \approx 1.474 h_J^2$$

(c) The total volume flow is given by (3.31)

$$Q_T = \frac{\pi \rho_w g A}{8\nu} \frac{\Delta H}{L} \sum_{J=1}^{4} n_J R_J^4 = \Omega \sum_{J=1}^{4} R_J^2$$

where

$$\Omega = \frac{\Delta \theta \rho_w g A}{8\nu} \frac{\Delta H}{L}$$

Thus, the flow fraction in a given interval K is

$$f_K = \frac{R_K^2}{\sum_{J=1}^{4} R_J^2}$$

Using these formulas, the results are calculated (Table 3).

TABLE 3   Results for Problem 3.7

$\Delta \theta$	R_J (cm)	n_J	f_J
0.1–0.2	0.0010	31,857	0.016
0.2–0.3	0.0021	6,958	0.072
0.3–0.4	0.0037	2,359	0.223
0.4–0.5	0.0065	746	0.689

3.8 The problem proceeds analogously to the vertical case covered in the text:

(a) This is converted to an integral in the following way:

$$J_w = E = -K(h)\frac{dh}{dx} \longrightarrow E\,dx = -K(h)dh \longrightarrow$$

$$E\int_{x_1}^{x_2} dx = -\int_{h_1}^{h_2} K(h)\,dh$$

(b) Substituting in the model form for $K(h)$ and putting in the appropriate limits, we obtain

$$E_{\max} = -\frac{K_S}{L}\int_0^{-\infty} \frac{dh}{1+(h/a)^2} = -\frac{K_S a}{L}\int_0^{\infty} \frac{dy}{1+y^2} = -\frac{K_S a\pi}{2L}$$

where the substitution $y = h/a$ ($a < 0$) has been made and the definite integral

$$\int_0^{\infty} \frac{dy}{1+y^2} = \frac{\pi}{2}$$

has been substituted.

(c) For $N = 2$, the vertical formula is equal to

$$E_{\text{max}} = \frac{K_s a^2 \pi^2}{4L^2}$$

The ratio of the vertical to the horizontal is, therefore,

$$\frac{E_{\text{ver}}}{E_{\text{hor}}} = \frac{a\pi}{2L}$$

This clearly approaches zero for large L, as is physically reasonable. The apparent contradiction that it approaches ∞ as L becomes small is an artifact of the approximation made ($E/K_s \ll 1$) in deriving the vertical case. If the exact solution is worked out, the true vertical evaporation rate is equal to

$$E_{\text{max}} = \frac{K_s}{2}\left(\sqrt{1 + \frac{a^2\pi^2}{4L^2}} - 1\right)$$

Using this value, the ratio always is less than unity.

3.9 The steady-state water conservation equation (3.80) becomes in this case

$$\frac{dJ_w}{dz} = -a(z+L) \longrightarrow \int_{-\iota_0}^{J_w} dJ_w = -a\int_0^z (z+L)\,dz$$

$$= J_w + \iota_0 = -a\left(\frac{z^2}{2} + Lz\right) \longrightarrow J_w = -\iota_0 - a\left(\frac{z^2}{2} + Lz\right)$$

By definition,

$$\text{ET} = \int_{-L}^0 r_w(z)\,dz = a\int_{-L}^0 (z+L)\,dz = a\frac{L^2}{2} \longrightarrow a = \frac{2\text{ET}}{L^2}$$

3.10 If we apply a mass balance to the column and assume gravity flow within, we obtain

$$L\frac{d\theta}{dt} = -K(\theta) = -K_S \exp[\beta(\theta - \theta_s)]$$

We transform this to an integral form in the usual way, applying the boundary condition $\theta = \theta_s$ at $t = 0$:

$$\int_0^t \frac{dt}{L} = -\int_{\theta_s}^{\theta} \frac{d\theta}{K_S \exp[\beta(\theta - \theta_s)]}$$

$$= \frac{t}{L} = -\int_0^{\theta - \theta_s} \frac{dy}{K_S \exp(\beta y)}$$

$$= -\frac{1}{\beta K_S}\{1 - \exp[-\beta(\theta - \theta_s)]\}$$

(a) Inverting this to solve for θ, we obtain

$$\theta = \theta_s - \frac{1}{\beta}\log\left(1 + \frac{\beta K_s t}{L}\right) = 0.4 - 0.05\log(1 + 40t)$$

(b) The water content continues to decrease with time to a final value of $-\infty$.

(c) $0.4 - 0.05\log(1 + 40t) = 0 \rightarrow t = [\exp(8) - 1]/40 = 74$ days

(d) The problem lies in the model conductivity, which is nonzero when $\theta = 0$. This means that the column loses water when there is no water! A different conductivity function, such as (3.84), which approaches 0 as $\theta \longrightarrow 0$ would behave physically. In practice, the exponential model is only valid over a small range of water content.

3.11 (a) Let us first set up the problem symbolically, using d as the initial ponding height, L as the column height, and t as the elapsed time. Then by the permeameter formula (3.21), the exact answer is

$$K_s = \frac{L}{t}\log\frac{d + L}{L} = 18.23$$

(b) The approximate form of Darcy's law is

$$J_w \approx -\frac{d}{t} = -K_s\frac{d/2 + L}{L} \longrightarrow K_s = \frac{dL}{(d/2 + L)t} = 18.18$$

(c) These are remarkably close, for the simple reason that the hydraulic head gradient doesn't change very much (from 120 to 100) during the experiment, so that using the constant value of 110 is a good approximation.

3.12 In this case, the two formulas give 8.05 and 6.67 using the exact and approximate formulas. There is significantly more error this time because the hydraulic head gradient changes from 25 to 5 during the experiment, so that the constant value of 15 is not a good approximation.

3.13 For this problem, the flux equation may be written as

$$J_w = -K(z)\frac{dH}{dz}$$

We can rearrange this and integrate from $z = 0$, where $H = H_1$, to $z = L$, where $H = H_2 = H_1 + \Delta H$:

$$\int_{H_1}^{H_2} dH = \Delta H = -J_w\int_0^L \frac{dz}{K(z)}$$

We may rewrite this as

$$J_w = -K_{\text{eff}}\frac{\Delta H}{L}$$

where

$$\frac{L}{K_{\text{eff}}} = \int_0^L \frac{dz}{K(z)}dz$$

which is a generalization of (3.26). Now we consider the special case where

$$L = \sum_{J=1}^N L_J \qquad K(z) = K_J \quad \text{when} \quad \sum_{K=1}^{J-1} L_K < z < \sum_{K=1}^{J} L_K$$

so that the effective conductivity formula becomes

$$\frac{L}{K_{\text{eff}}} = \sum_{J=1}^N \frac{L_J}{K_J}$$

which is the same as (3.26).

Chapter 4

4.1 The model we use is that of a traveling front that is saturated (i.e., constant K_s, θ_s) and has a constant matric potential h_f at the top where $z = L(t)$. Assume for simplicity that the water content above the rising front is negligible. Thus, Darcy's law is

$$J_w = \theta_s\frac{dL}{dt} = -K_s\frac{L+h_f}{L}$$

We may rewrite this as

$$\int_0^L \frac{LdL}{L+h_f} = -\frac{K_s t}{\theta_s}$$

This may be easily integrated, with the result that

$$L - h_f \ln\left(1 + \frac{L}{h_f}\right) = -\frac{K_s t}{\theta_s}$$

We denote $L_{\max} = -h_f > 0$, and change to dimensionless variables $X = L/L_{\max}, T = K_s t/\theta_s L_{\max}$, which produces

$$X + \ln(1 - X) = -T$$

This curve is plotted in Fig. 2. It shows that the upward rise height slowly approaches a maximum $L_{\max}$ at large time. In a real soil, the shape of the wetting front changes over time and is not as sharp as the Green–Ampt model assumption. Upward flow in a real soil is illustrated in Fig. 3.17.

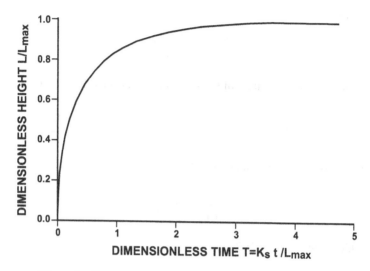

Figure 2   Green–Ampt model solution derived in Problem 4.1.

4.2   For this model, the gravity flow equation is

$$L\frac{d\theta}{dt} = -K(\theta) = -K_s\left(\frac{\theta}{\theta_s}\right)^N \Rightarrow \int_{\theta_s}^{\theta}\left(\frac{\theta_s}{\theta}\right)^N d\theta = -\int_0^t \frac{K_s\,dt}{L}$$

If we let $S = \theta/\theta_s$, $\tau = K_s t/L\theta_s$, this becomes

$$\int_1^S \frac{dS}{S^N} = -\int_0^\tau d\tau = -\frac{1}{(N-1)}\left(\frac{1}{S^{N-1}} - 1\right) = -\tau$$

$$\frac{1}{S^{N-1}} = 1 + (N-1)\tau \Rightarrow S = \left[\frac{1}{1+(N-1)\tau}\right]^{1/(N-1)}$$

Therefore, the water content and flux are

$$\theta = \theta_s\left[\frac{1}{(1+(N-1)K_s t/L)}\right]^{1/(N-1)}$$

$$J_w = K_s\left(\frac{\theta}{\theta_s}\right)^N = K_s\left[\frac{1}{(1+(N-1)K_s t/L)}\right]^{N/(N-1)}$$

Both θ and $K(\theta)$ approach 0 as $t \longrightarrow \infty$. This is in contrast to the result calculated in Problem 3.10, because our present conductivity function is zero when $\theta = 0$.

4.3   The exercise is merely to use the finite-difference definitions of the hydraulic properties to calculate the functions from the data.

$$K(\bar{h}) = -E\frac{\Delta z}{\Delta H}$$

$$C(\bar{h}) = \frac{\Delta \theta}{\Delta h}$$

$$D(\bar{h}) = \frac{K}{C}$$

See Figure 3.

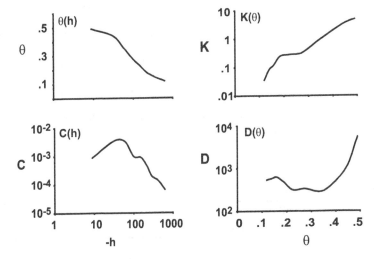

Figure 3   Functions calculated in Problem 4.3.

4.4   This problem requires converting the tabular information into a consistent set of units. Let us convert everything into centimeters (Table 4). The weight change ΔW is given in kilograms, which is converted into centimeters by dividing by the cross-sectional area A (4 m^2) and by the water density in units of 1000 kg m^{-3}. This produces the water storage change ΔS in meters, which is multiplied by 100. Thus, the conversion factor is a division by 40. The drainage rate d is in m^3, which is converted to centimeters by dividing by the cross-sectional area A and multiplying by 100. Thus, the conversion factor is a multiplication by 25. The rainfall change r is just the change in cumulative rainfall, which is already given in cm. After these conversions, the evapotranspiration rate ET is given by

$$\text{ET} = \frac{r - \Delta S - d}{\Delta t}$$

Table 5 gives the results.

TABLE 4   Solution to Problem 4.3

z (cm)	h (cm)	H (cm)	θ	$\overline{h}$ (cm)	$\overline{\theta}$	K (cm day^{-1})	C (cm^{-1})	D (cm^2 day^{-1})
180	−780	−600	0.11					
				−630	0.12	0.036	6.67×10^{-5}	536
160	−480	−320	0.13					
				−410	0.14	0.083	1.43×10^{-4}	583
140	−340	−200	0.15					
				−290	0.16	0.125	2.00×10^{-4}	625
120	−240	−120	0.17					
				−210	0.185	0.250	5.00×10^{-4}	500
100	−180	−80	0.20					
				−152.5	0.225	0.286	9.09×10^{-4}	315
80	−125	−45	0.25					
				−100	0.275	0.330	1.00×10^{-3}	330
60	−75	−15	0.30					
				−60	0.35	1.000	3.33×10^{-3}	300
40	−45	−5	0.40					
				−31.5	0.44	3.333	3.47×10^{-3}	958
20	−22	−2	0.48					
				−9	0.49	5.000	9.09×10^{-4}	5500
0	0	0	0.50					

TABLE 5   Results for Problem 4.4

t (days)	R (cm)	W (kg)	D (m^3)	r (cm)	d (cm)	ΔS (cm)	ET (cm day^{-1})
0	0	0	0				
				2	1	0	0.5
2	2	0	0.04				
				1	0	0	0.5
4	3	0	0.04				
				2	0	1	0.5
6	5	40	0.04				
				2	1	0	0.5
8	7	40	0.08				
				0	0	−0.75	0.375
10	7	10	0.08				
				1	0	0.5	0.025
12	8	30	0.08				
				1	0	0.5	0.25
14	9	50	0.08				
				5	3	1.25	0.375
16	14	100	0.20				
				3	3	−0.5	0.25
18	17	80	0.32				
				1	1	−0.5	0.25
20	18	60	0.36				

There is 18 cm of water input, 9 cm of drainage, 1.5 cm of storage increase, and 7.5 cm of ET. The average f_L based on the drainage is thus 0.5.

4.5 The first part is just a plug into the equation

$$V_F = \frac{K(\theta_o) - K(\theta_i)}{\theta_o - \theta_i}$$

using $\theta_0 \equiv \theta_s$ for initial water contents of $\theta_i = 0.0, 0.1, 0.2, 0.3, 0.4,$ and 0.49. The amount of water that has entered the soil at the different initial water contents is trickier, and assuming that this model is valid, we determine it as follows:

- If the front advances a distance $z = V_F t$, the profile now holds an additional amount of water $W_0 = z(\theta_s - \theta_i)$.
- During that time an amount of water $W_i = K(\theta_i)t$ drained out of the profile below the distance z.
- Therefore, the amount infiltrated is

$$W_{net} = W_0 + W_i = z(\theta_s - \theta_i) + K(\theta_i)t = V_f t(\theta_s - \theta_i) + K(\theta_i)t \equiv K_s t$$

Thus, in all cases the same amount of water, 50 cm, infiltrated, regardless of the initial profile. This is because the capillary effect was turned off in this calculation.

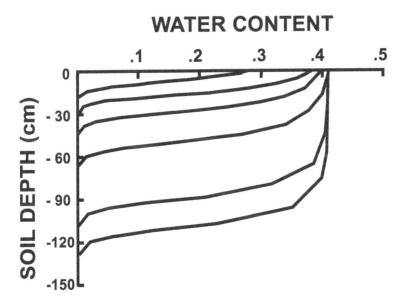

Figure 4 Water content profiles during constant flux infiltration from Problem 4.6.

4.6 This problem can be approached qualitatively by thinking through what happens during the constant-potential case. The flux into the soil is initially much greater than the final value, because the capillary attraction adds on to the effect of gravity. On the other hand, in constant flux infiltration at a rate i, the water intake is constant, so that initially the water content in the surface must be less than in the constant-potential case. Gradually, as the capillary effect decreases, the water content must increase, because after a long period of time it will be equal to the constant value $\theta = K^{-1}[i]$. Figure 4 is a reasonable representation of what happens.

4.7 Let the constant flux rate be P. Then the equation for t_{min} is given by the solution to

$$P = i_0 + (i_f - i_0)\exp(-\beta t_{min}) \longrightarrow t_{min} = \frac{1}{\beta}\log\frac{P - i_f}{i_0 - i_f}$$

The corresponding equation for t_{max} is given by the solution to

$$Pt_{max} = i_0 t_{max} + \frac{i_f - i_0}{\beta}\left[1 - \exp(-\beta t_{max})\right]$$

This cannot be analytically inverted to solve for t_{max}. A quick way to calculate t_{max} numerically is by the Newton–Raphson root-finding method of evaluating the function $F(x) = 0$. To use this procedure, an initial value x_0 is estimated. Then new values are calculated from the iterative equation,

$$x_{N+1} = x_N - \frac{F(x_N)}{F'(x_N)}$$

and this continues until the change from one iteration to the next is negligible. In this case, the function F for calculating t_{max} is

$$F(t_{max}) = 0 = i_0 t_{max} + \frac{i_f - i_0}{\beta}[1 - \exp(-\beta t_{max})] - Pt_{max}$$

The results of the calculation for the values furnished in the problem are as follows:

P	t_{max}	t_{min}
2.0	2.2	9.0
4.0	1.1	2.80
6.0	0.6	1.30

Chapter 5

5.1 The problem is a straightforward application of (5.54).

$$d = \frac{z_2 - z_1}{\log(\Delta T_2/\Delta T)} = 224 \text{ cm}$$

$$K_T = \frac{\pi d^2}{\tau} = 435 \text{ cm}^2 \text{ day}^{-1}$$

$$T_0 = \frac{T_{max} + T_{min}}{2} = 10°C$$

$$A = \frac{T_{max} - T_{min}}{2 \exp(z/d)} = \frac{T_{max} - T_0}{\exp(z/d)} = 15.0°C$$

$$T_{max} = T_0 + A \exp(-250 \text{ day}^{-1}) = 15.51°C$$

$$T_{min} = T_0 - A \exp(-250 \text{ day}^{-1}) = 4.49°C$$

5.2 The important principle in this problem is that it can be worked out by the same procedure as that used for water transport in saturated horizontal layered soil, because the heat flow equation has the same form as Darcy's law. Therefore, the first thing we do is calculate the thermal conductivity λ_{equiv} of a homogeneous column that offers the same thermal resistance as the layered one, by using the layered formula

$$\frac{50}{\lambda_Q} + \frac{50}{\lambda_{silt}} = \frac{100}{\lambda_{equiv}} \longrightarrow \lambda_{equiv} = \frac{1}{75} = 0.0133 \text{ cal cm}^{-1} \text{ s}^{-1} \text{ °C}^{-1}$$

The heat flux is then calculated from Fourier's law:

$$J_H = -\lambda_{equiv} \frac{T_2 - T_1}{2L} = -\frac{2}{750} = -0.00267 \text{ cal cm}^{-2} \text{ s}^{-1}$$

The temperature at the midpoint is determined by writing Fourier's law across half of the column:

$$J_H = -\lambda_Q \frac{T - T_1}{L} = -\frac{2}{750} \longrightarrow T = \frac{20}{3} °C$$

5.3 At a depth z, the maximum and minimum temperatures are given by

$$T_{max}(z) = T_A + A \exp(z/d)$$

$$T_{min}(z) = T_A - A \exp(z/d)$$

For this problem, $d = \sqrt{2K_T/\omega} = 215$ cm. Therefore, the maximum and minimum temperatures are 29.7 and 20.3°C. The time delay is $\Delta t = z/2\pi d = 67$ days.

5.4 The average value of K_T between 0.1 and 0.4 is about 5.6, as shown in Figure 5. The maximum and minimum vary about ±0.6 about the average, or about 10%.

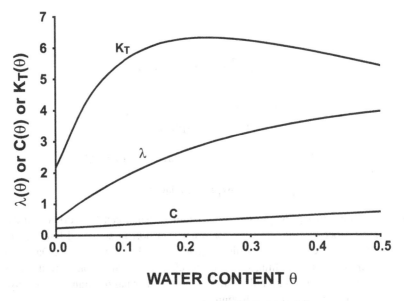

Figure 5   Functions calculated in Problem 5.4.

5.5   The effective conductivity $\lambda_e = 259$ cal cm^{-1} day^{-1} °C^{-1}. The rest of the quantities (Table 6) are obtained by insertion into the equation given for the flux. The ratio is

$$r = \frac{H_v J_v}{\lambda_e}$$

Since $H_v J_v$ is 25.3 at 20°C, λ^* is 233.7 cal cm^{-1} day^{-1} °C^{-1}.

TABLE 6   Results for Problem 5.5

T	λ_e	$H_v J_v$	r
20	259	25.3	0.097
30	275	40.8	0.148
40	300	66.0	0.220
50	342	108.0	0.320
60	411	177.0	0.430

5.6   (a) The effective conductivity has the form

$$\lambda_e(T) = a + b \exp(cT) = 234 + 8.775 \exp(0.05T)$$

Therefore, the steady-state heat flux equation is

$$J_H = \text{const} = -\lambda_e(T)\frac{dT}{dz}$$

To calculate the heat flux, this equation is integrated over the limits given:

$$J_H \int_0^L dz = - \int_{T_0}^{T_L} \left[a + b \exp(cT) \right] dT = J_H L$$

$$= -a(T_L - T_0) + \frac{b}{c}[\exp(cT_0) - \exp(cT_L)]$$

which produces $J_H = -347 \ \text{cal cm}^{-2} \ \text{day}^{-1}$.

(b) Given this value, the temperature at a given point z is obtained by integrating again from a fixed point to an arbitrary z:

$$J_H \int_0^z dz = - \int_0^T [a + b \exp(cT)] dT = J_H z$$

$$= -a(T - T_0) + \frac{b}{c}[\exp(cT_0) - \exp(cT)]$$

This provides z as a function of T.

5.7 The period of the ripples is 1 h, which has a damping depth of $d \approx 2$ cm. The damping depth of the main wave is about 10 cm, so that at $z = 15$ cm, it will be attenuated by $e^{-1.5} = 0.22$. The wave at 15 cm will have no ripples remaining in it and will be 22% as large and time-shifted by about 0.25 day.

5.8 This problem can accumulate roundoff error, so it is a good idea to use fractions. To calculate the heat flux, we first determine the equivalent thermal conductivity of the layered system:

$$\frac{40}{\lambda_C} + \frac{5}{\lambda_S} = \frac{45}{\lambda_{\text{equiv}}} \longrightarrow \lambda_{\text{equiv}} = \frac{45}{50,040}$$

Then the heat flux is given by Fourier's law:

$$J_H = -\lambda_{\text{equiv}} \frac{20}{45} = -\frac{20}{50,040}$$

The temperatures at the internal points are determined by writing Fourier's law across the copper regions alone:

$$J_H = - \lambda_C \frac{T_A - 5}{20} = -\lambda_C \frac{25 - T_B}{20} \longrightarrow T_A = 5 + \frac{40}{5004} \approx 5.008$$

$$T_B \approx 24.992$$

5.9 Because of superposition, we can treat each term in the infinite series as a separate input condition of sinusoidal form, with an angular frequency of

$$\omega_N = \frac{(4N - 2)\pi}{\tau}$$

and a damping depth

$$d_N = \sqrt{\frac{2K_T}{\omega_N}}$$

Therefore, the overall solution to the problem is a series of terms of the form given in (5.52):

$$\frac{T(z)}{T_0} = \frac{1}{2} + \frac{2}{\pi} \sum_{N=1}^{\infty} \frac{1}{2N-1} \sin\left(\omega_N t + \frac{z}{d_N}\right) \exp\left(\frac{z}{d_N}\right)$$

This is plotted in Fig. 6.

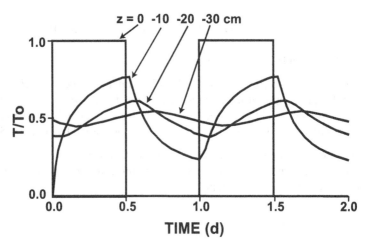

Figure 6   Solution to Problem 5.9.

Chapter 6

6.1 (a) The easiest way to solve the equation is to let $F = r^2 J_g$, so that

$$\frac{1}{r^2}\frac{dF}{dr} + \Omega_g = 0 \qquad F(0) = 0$$

Then we can turn the equation directly into an integral as follows:

$$\int_0^F dF = F = r^2 J_g = -\Omega_g \int_0^r r^2 \, dr = -\Omega_g \frac{r^3}{3}$$

$$\longrightarrow J_g(r) = -\Omega_g \frac{r}{3} = -D_g^s \frac{dC}{dr}$$

(b) Next, we convert the equation in C into an integral as follows:

$$\int_C^{C_0} dC = C_0 - C = \frac{\Omega_g}{3D_g^s} \int_r^R r\,dr = \frac{\Omega_g}{6D_g^s}\left(R^2 - r^2\right)$$

$$\longrightarrow C(r) = C_0 - \frac{\Omega_g}{6D_g^s}\left(R^2 - r^2\right)$$

6.2   The critical radius is by definition the point at which $C(R_c) = 0$. Therefore,

$$0 = C_0 - \frac{\Omega_g}{6D_g^s}\left(R^2 - R_c^2\right) \longrightarrow R_c = \sqrt{R^2 - \frac{6C_0 D_g^s}{\Omega_g}}$$

Plugging in (6.9), we obtain

$$R_c = \sqrt{R^2 - \frac{6C_0 D_g^a 0.66a}{\Omega_g}}$$

6.3   If $\alpha(-h)^N \gg 1$ and $\theta_r = 0$, (3.82) becomes

$$\theta = \frac{\theta_s}{\alpha^M (-h)^{NM}}$$

Taking the log of both sides, we have

$$\ln\theta = \ln\theta_s - M\ln\alpha - NM\ln(-h)$$

Therefore,

$$b = \frac{d\,\ln(-h)}{d\,\ln\theta} = -\frac{1}{NM}$$

6.4   Initially, they would look the same until the gas reached the barrier. Then the profile would begin to approach a steady state, which it would reach in finite time.

6.5   The ratio is

$$r = \frac{J_{PD}}{J_g} = \frac{-\xi_g(\phi)\,D_v^a\,d\rho_v^*/dT\,\omega\,dT/dz}{-\xi_g(a)\,D_v^a\,d\rho_v^*/dT\,dT/dz} = \omega\left(\frac{\phi}{a}\right)^{10/3} \approx \frac{2.5}{a^{10/3}}$$

if $\phi = 0.5$. The table gives the ratio at the air contents indicated:

a	r
0.5	2.5
0.4	5.3
0.3	13.7

6.6 The conservation of total solute mass leads to

$$\frac{\partial C_T}{\partial t} + \frac{\partial J_s}{\partial z} + r_s = 0$$

which reduces in this case to (assuming constant properties and $r_s = 0$)

$$a\frac{\partial C_g}{\partial t} + \theta\frac{\partial C_\ell}{\partial t} = D_g^s\frac{\partial^2 C_g}{\partial z^2} + D_\ell^s\frac{\partial^2 C_\ell}{\partial z^2}$$

Letting $C_g = K_H C_\ell$, we obtain

$$\left(a + \frac{\theta}{K_H}\right)\frac{\partial C_g}{\partial t} = \left(D_g^s + \frac{D_\ell^s}{K_H}\right)\frac{\partial^2 C_g}{\partial z^2}$$

from which the relation given in the problem follows after division of both sides by $a + (\theta/K_H)$.

Chapter 7

7.1 **(a)** The average water flux is $J_w = 1000/25 = 40$ cm yr^{-1}, and the corresponding pore water velocity is $V = J_w/\theta = 160$ cm yr^{-1}. The retardation factor of the solvent is $R = 1 + \rho_b K_d/\theta = 4$, so its velocity is $V/R = 40$ cm yr^{-1}. Therefore, the nitrate has traveled a distance $z = Vt = 4000$ cm, and the pesticide, 1000 cm.

(b) The total time to reach $L = 5000$ cm is $t = L/V = 31.25$ yr for the nitrate and 125 yr for the pesticide, which is 6.25 and 100 yr in the future, respectively.

7.2 The Freundlich isotherm is

$$C_s = K_f C_l^\beta$$

The solute mass balance equation is (assuming no gas phase)

$$\rho_b\frac{C_s}{\partial t} + \theta\frac{C_l}{\partial t} = D_l^s\frac{\partial^2 C_l}{\partial z^2} - J_w\frac{\partial C_l}{\partial z} = (\theta + \rho_b\beta K_f C_l^{\beta-1})\frac{C_l}{\partial t}$$

Dividing through by θ, we obtain

$$R_{\text{eff}}\frac{\partial C_l}{\partial t} = D\frac{\partial^2 C_l}{\partial z^2} - V\frac{\partial C_l}{\partial z}$$

where $R_{\text{eff}} = 1 + \rho_b\beta K_f C_l^{\beta-1}/\theta$. The ratio of the two R values is

$$\frac{R_2}{R_1} = \frac{1 + \rho_b\beta K_f C_2^{\beta-1}/\theta}{1 + \rho_b\beta K_f C_2^{\beta-1}/\theta} \approx \left(\frac{C_2}{C_1}\right)^{\beta-1}$$

where the approximation is valid whenever $\rho_b \beta K_f C_l^{\beta-1}/\theta \gg 1$. This ratio has the value 5.25, showing that the speed of the chemical should be much greater when it is added at high concentration. This is what Rao and Davidson observed.

7.3 This curve indicates that some of the solute is moving through a preferential flow channel, like a crack, that causes it to be transported without interacting with the rest of the medium. The matrix is permeable, so that chemical moves through it also.

7.4 Assume the root zone thickness $L = 50$ cm. Using the tables and formulas in the text, we calculate the values given in Table 7. From the table we see that DBCP and bromacil are the most likely to pollute groundwater.

TABLE 7   **Results for Problem 7.4**

Name	R	$t_{1/2}$	t_b	M/M_0
Atrazine	10.6	64	265	0.057
Bromacil	5.2	350	130	0.774
DBCP	8.8	225	220	0.509
DDT	14,401	3,837	360,000	0
Lindane	79.0	266	1,975	0.006
Phorate	40.6	82	1,015	0.0002

7.5 Plugging in the definitions for μ and R, and taking the natural logarithm, we obtain

$$\frac{\ln(2)L(\theta + \rho_b f_{oc} K_{oc})}{J_w \tau_{1/2}} < \ln(1/\beta)$$

This is a linear inequality in K_{oc} and $\tau_{1/2}$. Therefore, if pesticides are plotted as points on a $K_{oc} - \tau_{1/2}$ graph, their position relative to the line formed by the equation above will indicate whether they satisfy the inequality, or equivalently, whether they will leach more than the specified fraction to groundwater.

7.6 For simplicity, assume that the downward direction is positive. For uniform water uptake,

$$J_w = i_0 - \frac{\text{ET} \cdot z}{L}$$

Plugging this into the integral, we have

$$t(L) = \int_0^L \frac{\theta \, dz}{i_0 - \text{ET} \cdot z/L} = \frac{\theta L}{\text{ET}} \int_{i_0 - \text{ET}}^{i_0} \frac{dy}{y} = \frac{\theta L}{\text{ET}} \ln \frac{1}{f_L}$$

In terms of pores volumes, this is

$$P(L) = \frac{i_0}{ET} \ln \frac{1}{f_L} = \frac{1}{1 - f_L} \ln \frac{1}{f_L}$$

As shown in Jury et al. (1978), this result is also valid for an arbitrary water uptake function.

7.7 The piston flow model assumptions produce the following:

(a) J_s is constant and does not depend on f_L.

(b) At high concentration, precipitation will increase. Therefore, the solute flux probably will decrease as f_L decreases.

(c) The concentration is minimized by dilution, or as large a f_L value as possible. Conversely, the flux is minimized by a small f_L value.

7.8 We begin with the dimensional form of Henry's law:

$$P_v = k_H X$$

and the ideal gas law:

$$P_v V = \frac{mRT}{M} \rightarrow P_v = \frac{C_v RT}{M}$$

The mole fraction X is equal to

$$X = \frac{\text{moles of solute}}{\text{moles of solute} + \text{moles of water}} = \frac{m_s/M_s}{m_s/M_s + m_w/M_w}$$

If we let $m_w = \rho_w V_w$ and $C_l = m_s/V_w$ and multiply the denominator by M_s/V_w, we obtain

$$X = \frac{C_l}{C_l + \rho_w M_s/M_w} \approx \frac{C_l M_w}{\rho_w M_s}$$

assuming that $C_l \ll \rho_w M_s/M_w$. Thus,

$$C_g \frac{RT}{M_s} = k_H \frac{M_w}{\rho_w M_s} C_l \rightarrow C_g = \frac{k_H M_w}{\rho_w RT} C_l = K_H C_l$$

7.9 We begin with the general equation (7.15):

$$\frac{\partial}{\partial t}(\rho_b C_a + \theta C_l + a C_g) = \frac{\partial}{\partial z}\left(D_g^s \frac{\partial C_g}{\partial z}\right) + \frac{\partial}{\partial z}\left(D_e \frac{\partial C_l}{\partial z}\right)$$

$$- \frac{\partial}{\partial z}(J_w C_l) - (\rho_b r_a + \theta r_l + a r_g)$$

(a) The only change in the general equation is that the coefficients are all constant and can be removed from the derivatives. Thus,

$$\rho_b \frac{\partial C_a}{\partial t} + \theta \frac{\partial C_l}{\partial t} + a \frac{\partial C_g}{\partial t} = D_g^s \frac{\partial^2 C_g}{\partial z^2} + D_e \frac{\partial^2 C_l}{\partial z^2} - J_w \frac{\partial C_l}{\partial z} - (\rho_b r_a + \theta r_l + a r_g)$$

(b) The J_w term drops out.

$$\rho_b \frac{\partial C_a}{\partial t} + \theta \frac{\partial C_l}{\partial t} + a \frac{\partial C_g}{\partial t} = D_g^s \frac{\partial^2 C_g}{\partial z^2} + D_e \frac{\partial^2 C_l}{\partial z^2} - (\rho_b r_a + \theta r_l + a r_g)$$

(c) The reaction terms drop out and the sorption term drops out.

$$\theta \frac{\partial C_l}{\partial t} + a \frac{\partial C_g}{\partial t} = D_g^s \frac{\partial^2 C_g}{\partial z^2} + D_e \frac{\partial^2 C_l}{\partial z^2}$$

(d) A second equation relating the two phases is added.

$$\theta \frac{\partial C_l}{\partial t} = \alpha (K_H C_l - C_g)$$

7.10 We begin by making the substitutions (7.60) and (7.61) to the MIM equations (7.31) and (7.32).

$$\frac{J_w \theta_m}{L\theta} \frac{\partial C_m}{\partial T} + \frac{J_w \theta_{im}}{L\theta} \frac{\partial C_{im}}{\partial T} = \frac{D_e}{L^2} \frac{\partial^2 C_m}{\partial Y^2} - \frac{J_w}{L} \frac{\partial C_m}{\partial Y}$$

$$\frac{J_w \theta_{im}}{L\theta} \frac{\partial C_{im}}{\partial T} = \alpha (C_m - C_{im})$$

We next divide these two equations by J_w/L:

$$\frac{\theta_m}{\theta} \frac{\partial C_m}{\partial T} + \frac{\theta_{im}}{\theta} \frac{\partial C_{im}}{\partial T} = \frac{D_e}{J_w L} \frac{\partial^2 C_m}{\partial Y^2} - \frac{\partial C_m}{\partial Y}$$

$$\frac{\theta_{im}}{\theta} \frac{\partial C_{im}}{\partial T} = \frac{\alpha L}{J_w} (C_m - C_{im})$$

and apply the definitions given in (7.66).

Appendix

A.1 Straightforward application of the formulas yields the results given in Table 8. The cdf values are shown in Fig. 7. The negative correlation implies that higher infiltration is associated with smaller amounts of clay and the converse.

TABLE 8   Results for Problem A.1

Quantity	Value
Mean i	6.20
S.D. i	5.31
CV i	0.86
Mean % C	1.76
S.D. % C	1.06
CV % C	0.60
Correlation	−0.445

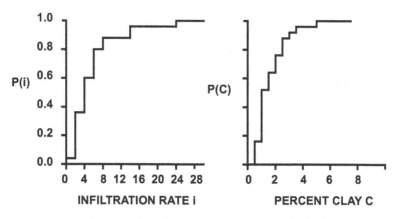

Figure 7 Cumulative distribution functions from the data in Problem A.1.

A.2 The log transform produces a mean, SD, and CV of 1.53, 0.80, and 0.53 for the infiltration rate. The fractile diagram shown in Fig. 8 compares the sample probability $(J - 0.5)/N$ with the theoretical probability

$$P_T(i) = 0.5 + N \left(\frac{i - E[i]}{SD[i]} \right)$$

for the normal distribution and

$$P_T(i) = 0.5 + N \left(\frac{\ln(i) - E[\ln(i)]}{SD[\ln(i)]} \right)$$

for the lognormal distribution. It heavily favors the lognormal distribution.

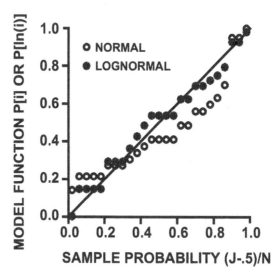

Figure 8 Fractile distributions for the infiltration data in Problem A.1.

A.3   (a) The formulas are given in the text:

Name	Formula	Value
Mean	$\exp(\mu + \sigma^2/2)$	21.86
Median	$\exp(\mu)$	13.20
Mode	$\exp(\mu - \sigma^2)$	4.80

(b) The probability that $K_s < E[K_s]$ is

$$P\left[K_s < \exp\left(\mu + \frac{\sigma^2}{2}\right)\right] = 0.5 + N\left(\frac{\mu + \sigma^2/2 - \mu}{\sigma}\right)$$

$$= 0.5 + N\left(\frac{\sigma}{2}\right) = 0.693$$

A.4   The problem requires an extension of the concepts learned in the text.

(a) Ponding over the entire field provides the mean infiltration rate

$$\exp\left(\mu + \frac{\sigma^2}{2}\right) = 9.5$$

(b) This is given by

$$P\left[i > \exp\left(\mu + \frac{\sigma^2}{2}\right)\right] = 0.5 - N\left(\frac{\mu + \sigma^2/2 - \mu}{\sigma}\right) = 0.5 - N\left(\frac{\sigma}{2}\right) = 0.27$$

(c) If we change variables in the equation to $y = (\ln i - \mu)/\sigma$, we obtain

$$f = \frac{1}{\sqrt{2\pi}\, M} \int_{\sigma/2}^{\infty} \exp\left(-\frac{y^2}{2} + \mu + \sigma y\right) dy$$

$$= \frac{1}{\sqrt{2\pi}} \int_{\sigma/2}^{\infty} \exp\left[-\frac{(y-\sigma)^2}{2}\right] dy$$

$$= \frac{1}{\sqrt{2\pi}} \int_{-\sigma/2}^{\infty} \exp\left[-\frac{(z)^2}{2}\right] dz = 0.5 + N\left(\frac{\sigma}{2}\right) = 0.73$$

(d) Of course we cannot say exactly what the distribution would be, but we do know that the position of a particular location on the front would be proportional to the local infiltration rate. Thus, the distribution of front positions would be lognormal, with log variance $\sigma^2 = 1.5$. This would be quite variable. For example, ratio of the depths of the 5% of the field having the shallowest and deepest penetration would be about $e^{3.28\sigma} = 56$.

A.5   Assume that the clay content is a pure linear drift $PC(x) = a + bx$. Construct a variogram out of samples taken at intervals of Δx. The variogram function for a separation $h = K\,\Delta x$ is

$$\gamma(K\,\Delta x) = \frac{1}{2N}\sum_{J=1}^{N}[\text{PC}(J\Delta x) - \text{PC}((J+K)\,\Delta x)]^2 = \frac{b^2}{2N}$$

$$= \sum_{J=1}^{N}(K\,\Delta x)^2 = \frac{(Kb\,\Delta x)^2}{2}$$

If there is an apparent quadratic increase in the variogram function, one could postulate a linear drift and remove it. An easy way to test this is to calculate the variogram of the difference between adjacent values. If the drift is linear, the new variogram will be stationary.

A.6   There are formulas for calculating the number of pairs of a square grid, but it is simple enough to count them directly in a small grid. For the 6×6 array, the pairs are as listed in Table 9. In contrast, a 36-unit-long linear array has $36 - J$ pairs of separation J.

TABLE 9   Results for Problem A.6

Spacing	Number	Spacing	Number
L	60	$L\sqrt{17}$	40
$L\sqrt{2}$	50	$L\sqrt{18}$	18
$2L$	48	$L\sqrt{20}$	32
$L\sqrt{5}$	80	$5L$	36
$L\sqrt{8}$	32	$L\sqrt{26}$	20
$3L$	36	$L\sqrt{29}$	16
$L\sqrt{10}$	60	$L\sqrt{34}$	2
$L\sqrt{13}$	48	$L\sqrt{41}$	8
$4L$	24	$L\sqrt{50}$	2

A.7   The probability is used to find the appropriate u value from the table. For example, for $P = 0.01$,

$$0.01 = 0.5 + N(\xi) \longrightarrow \xi = 2.32$$

For the CDE, this involves solving the equation

$$\frac{Vt_0 - z}{\sqrt{2Dt_0}} = u \longrightarrow t_0 = \left(\frac{\sqrt{2Du^2 + 4zV} - u\sqrt{2D}}{/2V}\right)^2$$

whereas for the CLT the equation for t_0 is

$$t_0 = \frac{z}{L}\exp(\mu + u\sigma)$$

These formulas produce the times listed in Table 10. Obviously, if the CLT process extends far below the surface, then the risk of groundwater contamination is greatly underestimated by the CDE.

TABLE 10   Results for Problem A.7

Probability	t_0 CDE	t_0 CLT
0.01	112	39
0.05	120	54
0.10	124	65
0.50	140	123

REFERENCES

Abramowitz, M. and I. A. Stegun. 1970. *Handbook of Mathematical Functions*. Dover, New York.

Addiscot, T. M. 1977. A simple computer model for leaching in structured soils. *J. Soil Sci.* 28:554–563.

Aitcheson, J. and J. A. C. Brown. 1976. *The Lognormal Distribution*. Cambridge University Press, New York.

Alexander, M. E. 1965. Nitrification. In: J. W. Bartholomew and F. E. Clark (Eds.), *Soil Nitrogen*. Monograph 10. American Society of Agronomy, Madison, WI.

Al-Jabri, S. A. 2001. Field estimation of hydraulic and chemical transport properties. Ph.D. dissertation. Iowa State University, Ames, IA (ISU 2001 A45).

Al-Jabri, S. A., R. Horton, and D. B. Jaynes. 2002a. A point source method for rapid estimation of soil hydraulic and chemical transport properties. *Soil Sci. Soc. Am. J.* 66:12–18.

Al-Jabri, S. A., R. Horton, D. B. Jaynes, and A. Gaur. 2002b. Field determination of soil hydraulic and chemical transport properties. *Soil Sci.* 167:353–368.

Allison, L. E. 1956. Soil and plant responses to VAMA and HPAN soil conditioners in the presence of high sodium. *Soil Sci. Soc. Am. Proc.* 20:147–151.

Allmaras, R. R., W. C. Burrows, and W. E. Larson. 1964. Early growth of corn as affected by soil temperature. *Soil Sci. Soc. Am. Proc.* 28:271–275.

Anderson, M. P. 1979. Using models to simulate the movement of contaminants through groundwater flow systems. *CRC Crit. Rev. Environ. Control* 9:97–156.

Andraski, B. J. and B. R. Scanlon. 2002. Thermocouple psychometry. pp. 609–642. In: J. Dane and G. Topp (Eds.), *Methods of Soil Analysis*, Part 4, *Physical Methods*. Soil Science Society of America, Madison, WI.

Ankeny, M. D., T. C. Kaspar, and R. Horton. 1988. Design for an automated tension infiltrometer. *Soil Sci. Soc. Am. J.* 52:893–896.

Ankeny, M. D., M. Ahmed, T. C. Kaspar, and R. Horton. 1991. Simple field method for determining unsaturated hydraulic conductivity. *Soil Sci. Soc. Am. J.* 55:467–469.

Arfken, G. 2000. *Mathematical Methods for Physicists*, 5th ed. Academic Press, San Diego, CA. 1112 pp.

Arndt, W. 1965a. The nature of the mechanical impedance to seedlings by soil surface seals. *Aust. J. Soil Res.* 3:45–54.

Arndt, W. 1965b. The impedance of soil seals and the forces of emerging seedlings. *Aust. J. Soil Res.* 3:56–58.

Atterberg, A. 1912. Die mechanische Bodenanalyse und die Klassifizierung der Mineraloboden Schwedens. *Int. Mitt. Bodenk.* 2:312–342.

Auzet, A. V., R. Angulo-Jaramillo, H. Damiron, M. Pluyms, B. Ambroise, and M. Vauclin. 1999. Tension disc infiltrometry to determine the near-saturated hydraulic conductivity of organic and coarse-textured soils on granites. pp. 571–578. In: M. Th. van Genuchten et al. (Eds.), *Characterization and Measurement of the Hydraulic Properties of Unsaturated Porous Media*. University of California–Riverside, Riverside, CA.

Ayers, R. S. and R. L. Branson. 1973. Nitrates in the upper Santa Ana River basin in relation to groundwater pollution. *Calif. Agric. Exp. Stn. Bull.* 861:1–60.

Babcock, K. L. and R. Overstreet. 1957. The extra-thermodynamics of soil moisture. *Soil Sci.* 83:455–464.

Bachmann, J., R. Horton, R. R. van der Ploeg, and S. Woche. 2000. Modified sessile drop method for assessing initial soil–water contact angle of sandy soil. *Soil Sci. Soc. Am. J.* 64:564–567.

Bachmann, J., R. Horton, S. A. Grant, and R. R. van der Ploeg. 2002. Temperature dependence of water retention curves for wettable and water repellent soils. *Soil Sci. Soc. Am. J.* 66:44–52.

Baker, R. S. and D. Hillel. 1990. Laboratory tests of a theory of fingering during infiltration into layered soils. *Soil Sci. Soc. Am. J.* 54:20–30.

Baker, J. and R. Norman. 2002. Evaporation from natural surfaces. pp. 1047–1071. In: J. Dane and G. Topp (Eds.), *Methods of Soil Analysis*, Part 4, *Physical Methods*. Soil Science Society of America, Madison, WI.

Bakr, A. A., L. W. Gelhar, A. L. Gutjahr, and H. R. Macmillan. 1978. Stochastic analysis of spatial variability of subsurface flow: I. Comparison of one and three dimensional flows. *Water Resour. Res.* 14:263–272.

Ball, B. C. and P. Schjonning. 2002. Air permeability. pp. 1141–1158. In: G. C. Topp and J. H. Dane (Eds.), *Methods of Soil Analysis*, Part 4, *Physical Methods*. Soil Science Society of America, Madison, WI.

Barry, D. A., J.-Y. Parlange, R. Haverkamp, and P. J. Ross. 1995. Infiltration under ponded conditions: 4. An explicit predictive infiltration formula. *Soil Sci. Soc. Am. J.* 160:8–17.

Bass, D. H., N. A. Hastings, and R. A. Brown. 2000. Performance of air sparging systems: a review of case studies. *J. Hazard. Mater.* 72:101–119.

Batchelor, G. K. 2000. *An Introduction to Fluid Dynamics*. Cambridge University Press, Cambridge. 615 pp.

Baver, L. D. and G. M. Homer. 1933. Water content of soil colloids as related to their chemical composition. *Soil Sci.* 30:329–353.

Baver, L. D. and H. F. Rhoades. 1932. Aggregate analysis as an aid to the study of soil structure relationships. *Agric. J.* 24:920–930.

Bear, J. 1972. *Dynamics of Fluids in Porous Media*. Elsevier, New York.

Beavers, A. H. and C. E. Marshall. 1951. The cataphoresis of clay minerals and factors affecting their separation. *Soil Sci. Soc. Am. Proc.* 15:142–145.

Benjamin, J. G., A. D. Blaylock, H. J. Brown, and R. M. Cruse. 1990. Ridge tillage effects on simulated water and heat transport. *Soil Tillage Res.* 18:167–180.

Benner, M. L., R. H. Mohtar, and L. S. Lee. 2002. Factors affecting air sparging remediation systems using field data and numerical simulations. *J. Hazard. Mater. B* 95:305–329.

Bertrand, A. R. and H. Kohnke. 1957. Subsoil conditions and their effects on oxygen supply and the growth of corn roots. *Soil Sci. Soc. Am. Proc.* 21:135–139.

Beven, K. and P. Germann. 1982. Macropores and water flow in soils. *Water Resour. Res.* 18:1311–1325.

Biggar, J. W. and D. R. Nielsen. 1967. Miscible displacement and leaching phenomena. *Agronomy* 11:254–274.

Bird, R. B., W. Stewart, and E. Lightfoot. 2001. *Transport Phenomena*, 2nd ed. Wiley, New York. 895 pp.

Bittelli, M., G. S. Campbell, and M. Flury. 1999. Characterization of particle–size distribution in soils with a fragmentation model. *Soil Sci. Soc. Am. J.* 63:782–788.

Black, T. A., W. R. Gardner, and G. W. Thurtell. 1969. The prediction of evaporation, drainage, and soil water storage for a bare soil. *Soil Sci. Soc. Am. Proc.* 33:655–660.

Blake, G. R. and K. H. Hartge, 1986. Bulk density. pp. 363–375. In: A. Klute (Ed.), *Methods of Soil Analysis*, Part 1. Monograph 9. American Society of Agronomy, Madison, WI.

Bohn, H., B. L. McNeal, and G. O'Connor. 1979. *Soil Chemistry*. Wiley, New York.

Bolt, G. H. 1955. Analysis of the validity of Gouy–Chapman theory of the electron double layer. *J. Colloid. Sci.* 10:206–218.

Bolt, G. H. 1976. Soil physics terminology. *Bull. Int. Soc. Soil Sci.* 49:26–36.

Bolt, G. H. 1982. *Soil Chemistry*, Part B, *Physico-Chemical Models*. Elsevier, Amsterdam.

Bolt, G. and M. G. M. Bruggenwert. 1976. *Soil Chemistry*, Part A, *Basic Elements*. Elsevier, Amsterdam.

Borchardt, G. A. 1988. Montmorillonite and other smectite minerals. In: J. B. Dixon and S. B. Weed (Eds.), *Minerals in Soil Environments*, 2nd ed. Soil Science Society of America, Madison, WI.

Boulier, J. F., J. Y. Parlange, M. Vauclin, D. A. Lockington, and R. Haverkamp. 1987. Upper and lower bounds of the ponding time for near-constant surface flux. *Soil Sci. Soc. Am. J.* 51:1424–1428.

Bouwer, H. 1966. Rapid field measurement of air entry value and hydraulic conductivity of soil as significant parameters in flow systems analysis. *Water Resour. Res.* 2:729–738.

Bouyoucos, G. J. 1913. An investigation of soil temperature and some of the most important factors influencing it. *Mich. Agric. Exp. Stn. Tech. Bull.* 17.

Bowers, S. A. and R. J. Hanks. 1962. Specific heat capacity of soils and minerals as determined with a radiation calorimeter. *Soil Sci.* 94:392–396.

Bradley, W. F. 1940. The structural scheme of attapulgite. *Am. Mineral.* 25:405–410.

Braida, W. J. and S. K. Ong. 2000. Modeling of air sparging of VOC-contaminated soil columns. *J. Contam. Hydrol.* 41:385–402.

Braida, W. J. and S. K. Ong. 2001. Air sparging effectiveness: laboratory characterization of air-channel mass transfer zone for VOC volatilization. *J. Hazard. Mater. B* 87:241–258.

Brewer, R. 1964. *Fabric and Mineral Analysis of Soils*. Wiley, New York.

Brindley, G. W. 1951. The kaolin minerals. Chap. 11. In: G. W. Brindley and G. Brown (Eds.), *X-Ray Identification and Crystal Structures of Clay Minerals*. Mineralogical Society of London, London.

Brinsfield, R., M. Yaramanoglu, and F. Wheaton. 1984. Ground level solar radiation prediction model including cloud cover effects. *Solar Energy* 33:493–499.

Bristow, K. L. 2002. Thermal conductivity. pp. 1209–1226. In: G. C. Topp and J. H. Dane (Eds.) *Methods of Soil Analysis*, Part 4, *Physical Methods*. Soil Science Society of America, Madison, WI.

Bristow, K. L., G. S. Campbell, R. I. Papendick, and L. F. Elliott. 1986. Simulation of heat and moisture transfer through a surface residue–soil system. *Agric. For. Meteorol.* 36:193–214.

Bristow, K. L., G. J. Kluitenberg, and R. Horton. 1994. Measurement of soil thermal properties with a dual-probe heat-pulse technique. *Soil Sci. Soc. Am. J.* 58:1288–1294.

Bristow, K., J. Bilskie, G. Kluitenberg, and R. Horton. 1995. Comparison of techniques for extracting soil thermal properties from dual-probe heat-pulse data. *Soil Sci.* 160:1–7.

Broadbent, F. E. and F. E. Clark. 1965. Denitrification. In: J. W. Bartholomew and F. E. Clark (Eds.), *Soil Nitrogen*. Monograph 10. American Society of Agronomy, Madison, WI.

Brooks, R. H. and A. T. Corey. 1964. *Hydraulic Properties of Porous Media*. Hydrology Paper 3. Colorado State University, Fort Collins, CO.

Brooks, R. H. and A. T. Corey. 1966. Properties of porous media affecting fluid flow. *J. Irr. Dr. Div. Am. Soc. Civ. Eng.* 92:61–88.

Bruce, R. R. and A. Klute. 1956. The measurement of soil water diffusivity. *Soil Sci. Soc. Am. Proc.* 20:458–462.

Bruce, R. R. and R. J. Luxmoore. 1986. Water retention: field methods. pp. 663–684. In: A. Klute (Ed.), *Methods of Soil Analysis*, Part 1. Monograph 9. American Society of Agronomy, Madison, WI.

Brunauer, S., P. H. Emmett, and E. Teller. 1938. Adsorption of gases in multimolecular layers. *J. Am. Chem. Soc.* 60:309–319.

Buckingham, E. 1904. *Contributions to Our Knowledge of the Aeration of Soils*. Bulletin 25. U.S. Department of Agriculture Bureau of Soils, Washington, DC.

Buckingham, E. 1907. *Studies on the Movement of Soil Moisture*. Bulletin 38. U.S. Department of Agriculture Bureau of Soils, Washington, DC.

Burchill, S., M. H. B. Hayes, and D. J. Greenland. 1978. Adsorption. In: D. J. Greenland and M. H. B. Hayes (Eds.), *The Chemistry of Soil Processes*. Wiley, Chichester, West Sussex, England.

Butters, G. L., W. A. Jury, and F. Ernst. 1989. Field scale transport of bromide in an unsaturated soil: I. Experimental methodology and results. *Water Resour. Res.* 25:1575–1581.

Cahill, A. T. and M. B. Parlange. 1998. On water vapor transport in field soils. *Water Resour. Res.* 34:731–739.

Call, F. 1957. The mechanism of sorption of ethylene dibromide on moist soils. *J. Sci. Food Agric.* 8:630–639.

Campbell, G. S. 1974. A simple method for determining unsaturated conductivity from moisture retention data. *Soil Sci.* 117:311–314.

Campbell, G. S. 1985. *Soil Physics with BASIC: Transport Models for Soil–Plant Systems*. Elsevier, New York.

Campbell, G. S. and G. W. Gee. 1986. Water potential: miscellaneous methods. pp. 619–632. In: A. Klute (Ed.), *Methods of Soil Analysis*, Part 1. Monograph 9. American Society of Agronomy, Madison, WI.

Carslaw, H. S. and J. C. Jaeger. 1959. *Conduction of Heat in Solids*. Oxford University Press, London.

Cary, J. W. and D. D. Evans. 1975. Soil crusts. *Ariz. Agric. Exp. Stn. Tech. Bull.* 214.

Casey, F. and N. Derby. 2002. Improved design for an automated tension infiltrometer. *Soil Sci. Soc. Am. J.* 66:64–67.

Casey, F. X., S. L. Logsdon, R. Horton, and D. B. Jaynes. 1997. Immobile water content and mass exchange coefficient of a field soil. *Soil Sci. Soc. Am. J.* 61:1030–1036.

Casey, F. X., S. L. Logsdon, R. Horton, and D. B. Jaynes. 1998. Measurement of field soil hydraulic and solute transport parameters as a function of water pressure head. *Soil Sci. Soc. Am. J.* 62:1172–1178.

Cass, A., G. S. Campbell, and T. L. Jones. 1984. Enhancement of thermal water vapor diffusion in soil. *Soil Sci. Soc. Am. J.* 48:25–32.

Chang, J. H. 1961. *Climate and Agriculture*. Aldine, Chicago.

Chang, J. H. 1968. Microclimate of sugar cane. *Hawaii. Plant. Rec.* 56:195–225.

Chang, J. H., R. B. Campbell, H. W. Brodie, and L. D. Baver. 1965. Evapotranspiration research at the HSP A experiment station. *Proc. 12th Congr., International Society for Sugarcane Technology*, pp. 10–24.

Chang, W. L., J. W. Biggar, and D. R. Nielsen. 1994. Fractal description of wetting front instability in layered soils. *Water Resour. Res.* 30:125–132.

Chapman, H. D. 1965. Cation exchange capacity. pp. 891–901. In: C. A. Black et al. (Eds.), *Methods of Soil Analysis*. Part 2. Monograph 9. American Society of Agronomy, Madison, WI.

Chepil, W. S. 1953. Field structure of cultivated soils with special reference to erodability by wind. *Soil Sci. Soc. Am. Proc.* 17:185–190.

Chepil, W. S. 1962. A compact rotary sieve and the importance of dry sieving on physical soil analysis. *Soil Soc. Am. Proc.* 26:4–6.

Childs, E. C. and N. Collis-George. 1950. The permeability of porous materials. *Proc. R. Soc. London. Ser. A* 201:392–405.

Chong, S. K., R. E. Green, and L. R. Ahuja. 1982. Infiltration prediction based on estimation of Green–Ampt wetting front pressure head from measurement of soil water redistribution. *Soil Sci. Soc. Am. J.* 46:235–238.

Chudnovskii, A. F. 1966. Plants and light: I. Radiant energy. pp. 1–51. In: *Fundamentals of Agrophysics*. Israel Program for Scientific Translations, Jerusalem.

Chung S. O. and R. Horton. 1987. Soil heat and water flow with a partial surface mulch. *Water Resour. Res.* 23: 2175–2186.

Chuoke, R. L., P. van Meurs, and C. van der Poel. 1959. The instability of slow, immiscible, viscous liquid–liquid displacements in permeable media. *Trans. Am. Inst. Min. Eng.* 216:188–194.

Clark, F. E. and W. D. Kemper. 1967. Microbial activity in relation to soil water and soil aeration. pp. 447–480. In: H. R. Haise and T. W. Edminster (Eds.), *Irrigation of Agricultural Lands*. Monograph 11. American Society of Agronomy, Madison, WI.

Clausnitzer, V., J. W. Hopmans, and J. L. Starr. 1998. Parameter uncertainty analysis of common infiltration models. *Soil Sci. Soc. Am. J.* 62:1477–1487.

Clothier, B. E. and D. R. Scotter. 2002. Unsaturated water transmission parameters obtained from infiltration. pp. 879–896. In: J. Dane and G. Topp (Eds.), *Methods of Soil Analysis*, Part 4, *Physical Methods*. Soil Science Society of America, Madison, WI.

Clothier, B. E., D. R. Scotter, and A. E. Green. 1983. Diffusivity and one-dimensional absorption experiments. *Soil Sci. Soc. Am. J.* 47:641–644.

Clothier, B. E., M. B. Kirkham, and J. E. McLean. 1992. In situ measurements of the effective transport volume for solute moving through soil. *Soil Sci. Soc. Am. J.* 56:733–736.

Coats, K. H. and B. D. Smith. 1956. Dead end pore volume and dispersion in porous media. *Soc. Pet. Eng. J.* 4:73–84.

Constantz, J., S. W. Tyler, and E. Kwicklis. 2003. Temperature-profile methods for estimating percolation rates in arid environments. *Vadose Zone J.* 2:12–24.

Currie, J. A. 1962. The importance of diffusion in providing the right conditions for plant growth. *J. Sci. Food Agric.* 13:380–385.

Currie, J. A. 1970. Movement of gases in soil respiration. *Soc. Chem. Ind. Monogr.* 37:152–171.

Curtis, A. A. and K. K. Watson. 1984. Hysteresis-affected water movement in scale-heterogeneous profiles. *Water Resour. Res.* 20:719–726.

Dalton, F. N. and M. Th. Van Genuchten. 1986. The time domain reflectometry method for measuring soil water content and salinity. *Geoderma* 38:237–250.

Darcy, H. 1856. *Les Fontaines publiques de la ville de Dijon.* Dalmont, Paris.

Davidson, J. M., D. R. Nielsen, and J. W. Biggar. 1963. The measurement and description of water flow through Columbia silt loam and Hesperia sandy loam. *Hilgardia* 34:601–617.

Day, P. R. 1942. The moisture potential of soils. *Soil Sci.* 54:391–400.

De Vries, D. A. 1958. Simultaneous transfer of heat and moisture in porous media. *Trans. Am. Geophys. Union* 39:909–916.

De Vries, D. A. 1963. Thermal properties of soils. In: W. R. Van Wijk (Ed.), *Physics of Plant Environment.* North-Holland, Amsterdam.

De Vries, D. A. and A. J. Peck. 1968. On the cylindrical probe method of measuring thermal conductivity with special reference to soils. *Aust. J. Phys.* 11:255–271.

Diment, G. A. and K. K. Watson. 1985. Stability analysis of water movement in unsaturated porous materials: 3. Experimental studies. *Water Resour. Res.* 21:979–984.

Dirksen, C. 1975. Determination of soil water diffusivity by sorptivity measurements. *Soil Sci. Soc. Am. Proc.* 39:22–27.

Dirksen, C. 1979. Flux-controlled sorptivity measurements to determine soil hydraulic property functions. *Soil Sci. Soc. Am. J.* 43:827–834.

Dixon, J. B. and S. B. Weed. 1988. *Minerals in Soil Environments,* 2nd ed. Soil Science Society of America, Madison, WI.

Dobsen, M. C., F. T. Ulaby, M. T. Hallikainen, and M. A. El-Rayes. 1986. Microwave dielectric behavior of wet soil: 2. Dielectric mixing models. *IEEE Trans. Geosci. Remote Sens.* GE-23:35–46.

Doorenbos, J. and W. Pruitt, 1976. *Crop Water Requirements.* Irrigation and Drainage Paper 24. FAO, Rome.

Douglas, L. A. 1988. Vermiculite. pp. 259–292. In: J. B. Dixon and S. B. Weed (Eds.), *Minerals in Soil Environments,* 2nd ed. Soil Science Society of America, Madison, WI.

Dracos, T. 1987. Immiscible transport of hydrocarbons infiltrating in unconfined aquifers. In: J. H. Vandermeulen and S. E. Hrudney (Eds.), *Oil in Freshwater: Chemistry, Biology, Countermeasure Technology.* Pergamon Press, New York.

Drees, L. R., L. P. Wilding, N. E. Smeck, and A. L. Senaki. 1988. Silica in soils: quartz and disordered silica polymorphs. In: J. B. Dixon and S. B. Weed (Eds.), *Minerals in Soil Environments,* 2nd ed. Soil Science Society of America, Madison, WI.

Dullien, F. A. L. 1979. *Porous Media: Fluid Transport and Pore Structure.* Academic Press, San Diego, CA. 379 pp.

Dyson, J. S. and R. E. White. 1986. The effect of irrigation rate on solute transport in soil during steady water flow. *J. Hydrol.* 107:19–29.

Edelman, C. H. and J. C. L. Favagee. 1940. On the crystal structure of montmorillonite and halloysite. *Z. Krist.* 102:417–431.

Elliot, E. T. 1986. Aggregate structure and carbon, nitrogen, and phosphorus in native and cultivated soils. *Soil Sci. Soc. Am. J.* 50:627–633.

Evans, D. D., and S. W. Buol. 1968. Micromorphological study of soil crusts. *Soil Sci. Soc. Am. Proc.* 32:19–22.

Fanning, D. S., V. R. Keramidas, and M. A. El-Desoky. 1988. Micas. In: J. B. Dixon and S. B. Weed (Eds.), *Minerals in Soil Environments*, 2nd ed. Soil Science Society of America, Madison, WI.

Farrell, D. A., E. L. Greacen, and C. G. Gurr. 1966. Vapor transfer in soil due to air turbulence. *Soil Sci.* 102:305–313.

Farrell, R., E. de Jong, and J. Elliott. 2002. Gas sampling and analysis. pp. 1076–1140. In: J. Dane and G. Topp (Eds.), *Methods of Soil Analysis*, Part 4, *Physical Methods*. Soil Science Society of America, Madison, WI.

Fatt, I. and H. Dykstra. 1951. Relative permeability studies. *Pet. Trans. AIME* 192:249–255.

Faybishenko, B. A. 1999. Comparison of laboratory and field methods for determining the quasi-saturated hydraulic conductivity of soils. pp. 279–293. In: M. Th. van Genuchten et al. (Eds.), *Characterization and Measurement of the Hydraulic Properties of Unsaturated Porous Media*. University of California–Riverside, Riverside, CA.

Fieldes, M. 1962. The nature of the active fraction of soils. pp. 62–78. In: *Trans. Jt. Mtg. Comm. IV–V*. International Society of Soil Science, New Zealand.

Fieldes, M. and R. K. Schofield. 1960. Mechanism of ion adsorption by inorganic soil colloids. *N.Z. J. Sci.* 3:563–569.

Fields, K., J. Gibbs, W. Condit, A. Leeson, and G. Wickramanayake. 2002. *Air-Sparging: A Project Manager's Guide*. Battelle Press, Columbus, OH. 170 pp.

Flegg, P. B. 1953. The effect of aggregation on diffusion of gases and vapors through soils. *J. Sci. Food Agric.* 4:104–108.

Flint, A. and L. Flint. 2002. Particle density. pp. 229–240. In: J. Dane and G. Topp (Eds.), *Methods of Soil Analysis*, Part 4, *Physical Methods*. Soil Science Society of America, Madison, WI.

Flury, M., H. Fluhler, W. Jury, and J. Leuenberger. 1994. Susceptibility of soils to preferential flow of water: a field study. *Water Resour. Res.* 30:1945–1954.

Fredlund, D. G. and S. K. Vanapalli. 2002. Shear strength of unsaturated soils. pp. 329–362. In: J. H. Dane and G. C. Topp (Eds.), *Methods of Soil Analysis*, Part 4, *Physical Methods*. Soil Science Society of America, Madison, WI.

Freeze, R. A. 1969. The mechanism of natural ground water recharge and discharge. *Water Resour. Res.* 5:153–171.

Fried, J. J. 1975. *Groundwater Pollution*. Elsevier, New York.

Fripiat, J. J. 1964. Surface chemistry and soil science. pp. 3–13. In: *Experimental Pedology, Proc. 11th Easter School Agric. Sci.*, University of Nottingham.

Fuchs, M. and A. Hadas. 1972. The heat flux density in a non-homogeneous bare loessial soil. *Boundary-Layer Meteorol.* 3:191–200.

Gapon, Y. N. 1933. On the theory of exchange adsorption in soils. *J. Gen. Chem. USSR* (Engl. transl.) 3:144–160.

Gardner, W. R. 1958. Some steady state solutions of the unsaturated moisture flow equation with application to evaporation from a water table. *Soil Sci.* 85:228–232.

Gardner, W. H. 1986. Water content. pp. 493–544. In: A. Klute (Ed.), *Methods of Soil Analysis*, Part 1. Monograph 9. American Society of Agronomy, Madison, WI.

Gardner, W. R. and M. Fireman. 1958. Laboratory studies of evaporation from soil columns in the presence of a water table. *Soil Sci.* 85:244–249.

Gardner, W. R. and D. Kirkham. 1952. Determination of soil moisture by neutron scattering. *Soil Sci.* 73:392–401.

Gardner, W. R., D. I. Hillel, and Y. Benyamini. 1970. Post irrigation movement of soil water: I. Redistribution. *Water Resour. Res.* 6:851–861.

Gast, R. 1977. Surface and colloid chemistry. In: J. B. Dixon and S. B. Weed (Eds.), *Minerals and Soil Environments*. Soil Science Society of America, Madison, WI.

Gee, G. W. and D. Or. 2002. Particle-size analysis. pp. 255–293. In: J. Dane and G. Topp (Eds.), *Methods of Soil Analysis*, Part 4, *Physical Methods*. Soil Science Society of America, Madison, WI.

Geiger, R. 1965. *The Climate near the Ground.* Harvard University Press, Cambridge, MA.

Geiger, S. L. and D. S. Durnford. 2000. Infiltration in homogeneous sands and a mechanistic model of unstable flow. *Soil Sci. Soc. Am. J.* 64:460–469.

Gelhar, L. W. and C. Axness. 1983. Three dimensional stochastic analysis of macrodispersion in aquifers. *Water Resour. Res.* 19:161–180.

Gelhar, L. W., A. Mantaglou, C. Welty, and K. R. Rehfeldt. 1985. *A Review of Field Scale Physical Solute Transport Processes in Unsaturated and Saturated Porous Media.* EPRI Topical Report EA-4190. Electric Power Research Institute, Palo Alto, CA.

Glass, R. J., J. Y. Parlange, and T. S. Steenhuis. 1989. Wetting front instability: 1. Theoretical discussion and dimensional analysis. *Water Resour. Res.* 25:1187–1194.

Goates, J. R. and S. J. Bennett. 1957. Thermodynamic properties of water adsorbed on soil minerals: kaolinite. *Soil Sci.* 83:325–330.

Gomez, J. A., J. V. Giraldez, and E. Fereres. 2001. Analysis of infiltration and runoff in an olive orchard under no-till. *Soil Sci. Soc. Am. J.* 65:291–299.

Green, R. E. 1974. Pesticide–clay–water interactions. pp. 3–38. In: W. D. Guenzi (Ed.), *Pesticides in Soil and Water*. Soil Science Society of America, Madison, WI.

Green, W. H. and G. A. Ampt. 1911. Studies in soil physics: I. The flow of air and water through soils. *J. Agric. Sci.* 4:1–24.

Green, R. E. and J. C. Corey. 1971. Calculation of hydraulic conductivity: a further evaluation of predictive methods. *Soil Sci. Soc. Am. Proc.* 35:3–8.

Green, D. W., H. Dabiri, C. F. Weinaug, and R. Prill. 1970. Numerical modeling of unsaturated ground water flow and comparison of the model to field experiment. *Water Resour. Res.* 6:862–864.

Greenwood, D. J. 1971. Soil aeration and plant growth. *Rep. Prog. Soil Chem.* 55:423–431.

Grim, R. E. 1962. *Applied Clay Mineralogy.* McGraw-Hill, New York.

Grossman, R. B. and T. G. Reinsch. 2002. Bulk density and linear extensibility. pp. 201–228. In: J. Dane and G. Topp (Eds.), *Methods of Soil Analysis*, Part 4, *Physical Methods*. Soil Science Society of America, Madison, WI.

Gupta, S. C., J. K. Radke, and W. E. Larson. 1981. Predicting temperature of bare and residue covered soils with and without a corn crop. *Soil Sci. Soc. Am. J.* 45:405–410.

Gurr, C. G., T. J. Marshall, and J. T. Hutton. 1952. Movement of water in soil due to temperature gradients. *Soil Sci.* 74:333–345.

Hald, A. 1952. *Statistical Theory with Engineering Applications*. Wiley, New York.

Halliday, D., R. Resnick, and J. Walker. 2001. *Fundamentals of Physics*, 6th ed. Wiley, New York, 384 pp.

Ham, J. M. and G. J. Kluitenberg. 1994. Modeling the effect of mulch optical properties and mulch–soil contact resistance on soil heating under plastic mulch culture. *Agric. For. Meteorol.* 71:403–424.

Hamaker, J. W. 1972. Decomposition: quantitative aspects. pp. 253–340. In: C. A. I. Goring and M. Hamaker (Eds.), *Organic Chemicals in the Soil Environment*. Marcel Dekker, New York.

Hamaker, J. W. and J. M. Thompson. 1972. Adsorption. In: C. A. I. Goring and M. Hamaker (Eds.), *Organic Chemicals in the Soil Environment*. Marcel Dekker, New York.

Hanks, R. J. 1960. Soil crusting and seedling emergence. *Trans. 7th Int. Congr. Soil Sci.*, 1:340–346.

Hares, M. A. and M. D. Novak. 1992. Simulation of surface energy balance and soil temperature under strip tillage: I. Model description. *Soil Sci. Soc. Am. J.* 56:22–28.

Haverkamp, R., P. J. Ross, K. Smettem, and J. Y. Parlange. 1994. Three-dimensional analysis of infiltration from the disc infiltrometer: 2. Physically based infiltration equation. *Water Resour. Res.* 30:2931–2936.

Hendricks, S. G. 1942. Lattice structure of clay minerals and some properties of clays. *J. Geol.* 50:276–290.

Hendrickx, J. M. H., L. W. Dekker, and O. H. Boersma. 1993. Unstable wetting fronts in water repellent field soils. *J. Environ. Qual.* 22:109–118.

Hill, D. E. and J.-Y. Parlange. 1972. Wetting front instability in layered soils. *Soil Sci. Soc. Am. Proc.* 36:697–702.

Hillel, D. 1960. Crust formation in loessial soils. *Trans. 7th Int. Congr. Soil Sci.* 1:330–339.

Himmelblau, D. M. 1970. *Process Analysis by Statistical Methods*. Sterling Swift, Manchecka, TX.

Hofmann, V., K. Endell, and D. Wilm. 1933. Kristallstruktur und Quellung von Montmorillinit. *Z. Krist.* 86A:304–348.

Holmes, J. W. and J. S. Colville. 1970. Forest hydrology in a Karstic region of Southern Australia. *J. Hydrol.* 10:59–74.

Hopmans, J. W., J. Simunek, N. Romano, and W. Durner. 2002. Inverse methods. pp. 963–1003. In: J. Dane and G. Topp (Eds.), *Methods of Soil Analysis*, Part 4, *Physical Methods*. Soil Science Society of America, Madison, WI.

Horton, R. E. 1933. The role of infiltration in the hydrologic cycle. *Trans. Am. Geophys. Union, 14th Annu. Meet.*, pp. 446–460.

Horton, R. E. 1939. Analysis of runoff-plot experiments with varying infiltration capacity. *Trans. Am. Geophys. Union, 20th Annu. Meet.*, Part IV, pp. 693–694.

Horton, R. E. 1940. An approach towards a physical meaning of infiltration capacity. *Soil Sci. Soc. Am. Proc.* 5:399–417.

Horton, R. 1989. Canopy shading effects on soil heat and water-flow. *Soil Sci. Soc. Am. J.* 53:669–679.

Horton, R. 2002. Soil thermal diffusivity. Chap. 5-4. In: J. Dane and C. Topp (Eds.), *Methods of Soil Analysis*, Part 4, *Physical Methods*. Soil Science Society of America, Madison, WI.

Horton, R. and P. J. Wierenga. 1983. Estimating the soil heat flux from observations of soil temperature near the surface. *Soil Sci. Soc. Am. J.* 47:14–20.

Horton, R., P. J. Wierenga, and D. R. Nielsen. 1983. Evaluation of methods for determining the apparent thermal-diffusivity of soil near the surface. *Soil Sci. Soc. Am. J.* 47:25–32.

Horton, R., K. Bristow, and G. Kluitenberg. 1996. Crop residue effects on surface radiation and energy balance: review. *Theor. Appl. Climatol.* 54:27–37.

Houghton, H. G. 1954. On the annual heat balance of the northern hemisphere. *J. Meteorol.* 11:1–9.

Hsu, P. H. 1988. Aluminum hydroxides and oxyhydroxides. In: J. B. Dixon and S. B. Weed (Eds.), *Minerals in Soil Environments*, 2nd ed. Soil Science Society of America, Madison, WI.

Hunt, A. G. and R. P. Ewing. 2003. On the vanishing of solute diffusion in porous media at a threshold moisture content. *Soil Sci. Soc. Am. J.* 67:1701–1702.

Hutchinson, G. L. and G. P. Livingston. 2002. Soil–atmosphere gas exchange. pp. 1159–1182. In: G. C. Topp and J. H. Dane (Eds.), *Methods of Soil Analysis*, Part 4, *Physical Methods*. Soil Science Society of America, Madison, WI.

Idso, S. B., R. J. Reginato, R. D. Jackson, B. A. Kimball, and F. S. Nakayama. 1974. The three stages of drying of a field soil. *Soil Sci. Soc. Am. Proc.* 38:831–837.

Idso, S. B., R. D. Jackson, R. J. Reginato, B. A. Kimball, and F. S. Nakayama. 1975. The dependence of bare soil albedo on soil water content. *J. Appl. Meteorol.* 14:109–113.

IPCC. 2001. *Climate Change 2001: The Scientific Basis*. Report of the Intergovernmental Panel on Climate Change. Cambridge University Press, Cambridge. 944 pp.

Jacinthe, P. A., R. Lal, and J. M. Kimble. 2002. Carbon budget and seasonal carbon dioxide emission from a central Ohio Luvisol as influenced by wheat residue amendment. *Soil Tillage Res.* 67:147–157.

Jackson, M. L. 1964. Chemical composition of soils. pp. 71–141. In: F. E. Bear (Ed.), *Chemistry of Soil*. Reinhold, New York.

Jackson, R. D. 1972. On the calculation of hydraulic conductivity. *Soil Sci. Soc. Am. Proc.* 36:380–383.

Jackson, R. D. and S. A. Taylor. 1965. Heat transfer. pp. 349–360. In: C. Black (Ed.), *Methods of Soil Analysis*, Vol. 1. American Society of Agronomy, Madison, WI.

Jaynes, D. B. and R. Horton. 1998. Field parameterization of the mobile/immobile domain model. pp. 297–310. In: H. M. Selim and L. Ma (Ed.), *Physical Nonequilibrium in Soils: Modeling and Application*. Ann Arbor Press, Chelsea, MI.

Jaynes, D. B., S. L. Logsdon, and R. Horton. 1995. Field method for measuring mobile/immobile water content and solute transfer rate. *Soil Sci. Soc. Am. J.* 59:352–356.

Jenny, H. and R. F. Reitemeier. 1935. Ionic exchange in relation to the stability of colloidal systems. *J. Phys. Chem.* 39:593–604.

Jones, M. J. and K. K. Watson. 1987. Effect of soil water hysteresis on solute movement during intermittent leaching. *Water Resour. Res.* 23:1251–1256.

Journel, A. and Ch. J. Huijbregts. 1978. *Mining Geostatistics*. Academic Press, London.

Jury, W. A. 1973. Simultaneous movement of heat and water through a medium sand. Ph.D. dissertation. University of Wisconsin, Madison, WI.

Jury, W. A. 1982. Simulation of solute transport using a transfer function mode. *Water Resour. Res.* 18:363–368.

Jury, W. A. 1983. Chemical transport modeling: current approaches and unresolved problems. In: D. W. Nelson, D. E. Elrich, and K. K. Tanji (Eds.), *Chemical Mobility and Reactivity in Soil Systems.* Special Publication 42. American Society of Agronomy, Madison, WI.

Jury, W. A. 1985. *Spatial Variability of Soil Physical Parameters in Solute Migration: A Critical Literature Review.* EPRI Topical Report EA 4228. Electric Power Research Institute, Palo Alto, CA.

Jury, W. A. and J. Letey. 1979. Water vapor movement in soil: reconciliation of theory and experiment. *Soil Sci. Soc. Am. J.* 43:823–827.

Jury, W. A. and E. E. Miller. 1974. Measurement of the transport coefficients for coupled flow of heat and moisture in a medium sand. *Soil Sci. Soc. Am. Proc.* 38:551–557.

Jury, W. A. and K. Roth. 1990. *Transfer Functions and Solute Transport through Soil: Theory and Applications.* Berkhäuser, Basel.

Jury, W. A. and G. Sposito. 1985. Field calibration and validation of solute transport models for the unsaturated zone. *Soil Sci. Soc. Am. J.* 49:1331–1341.

Jury, W. A., W. R. Gardner, C. B. Tanner, and P. Saffigna. 1976. Model for predicting movement of nitrate and water through a loamy sand. *Soil Sci.* 122:36–43.

Jury, W. A., H. Fluhler, H. Frenkel, D. Devitt, and L. H. Stolzy. 1978. Use of saline water and minimal leaching for crop production. *Hilgardia* 46:169–193.

Jury, W. A., J. Letey, and T. Collins. 1982a. Analysis of chamber methods used to measure N_2O production in the field. *Soil Sci. Soc. Am. J.* 46:250–256.

Jury, W. A., L. H. Stolzy, and P. Shouse. 1982b. A field test of the transfer function model for predicting solute movement. *Water Resour. Res.* 18:368–374.

Jury, W. A., W. F. Spencer, and W. J. Farmer. 1983. Behavior assessment model for trace organics in soil: I. Description of model. *J. Environ. Qual.* 12:558–564.

Jury, W. A., W. J. Farmer, and W. F. Spencer. 1984a. Behavior assessment mode for trace organics in soil: II. Chemical classification and parameter sensitivity. *J. Environ. Qual.* 13:567–572.

Jury, W. A., W. F. Spencer, and W. J. Farmer. 1984b. Behavior assessment model for trace organics in soil: III. Application of screening model. *J. Environ. Qual.* 13:573–579.

Jury, W. A., W. F. Spencer, and W. J. Farmer. 1984c. Behavior assessment model for trace organics in soil: IV. Review of experimental evidence. *J. Environ. Qual.* 13:580–585.

Jury, W. A., G. Sposito, and R. E. White. 1986. A transfer function model of solute movement through soil. I. Fundamental concepts. *Water Resour. Res.* 22:243–247.

Jury, W. A., D. D. Focht, and W. J. Farmer. 1987a. Evaluation of pesticide ground water pollution potential from standard indices of soil-chemical adsorption and biodegradation. *J. Environ. Qual.* 16:422–428.

Jury, W. A., D. Russo, G. Sposito, and H. Elabd. 1987b. The spatial variability of water and solute transport properties in unsaturated soil: I. Analysis of property variation and spatial structure. *Hilgardia* 55:1–32.

Jury, W. A., D. Russo, and G. Sposito. 1987c. The spatial variability of water and solute transport properties in unsaturated soil: II. Analysis of scaling theory. *Hilgardia* 55:33–57.

Jury, W. A., J. S. Dyson, and G. L. Butters. 1990. A transfer function model of field scale solute transport under transient water flow. *Soil Sci. Soc. Am. J.* 54:327–332.

Jury, W. A., Z. Wang, and A. Tuli. 2003. A conceptual model of unstable flow in unsaturated soil during redistribution. *Vadose Zone J.* 2:61–67.

Kaplan, W. 1984. *Advanced Calculus.* Addison-Wesley, Reading, MA.

Kelley, W. P., W. H. Dore, and S. M. Brown. 1931. The nature of base-exchange material for bentonite, soils, and zeolites as revealed by chemical investigations and x-ray analysis. *Soil Sci.* 41:259–274.

Kemper, W. D. 1961a. Movement of water as affected by free energy and pressure gradients: I. Application of classic equations for viscous and diffusive movements to the liquid phase in finely porous media. *Soil Sci. Soc. Am. Proc.* 25:255–260.

Kemper, W. D. 1961b. Movement of water as affected by free energy and pressure gradients: II. Experimental analysis of porous systems in which free energy and pressure gradients act in opposite directions. *Soil Sci. Soc. Am. Proc.* 25:261–265.

Kemper, W. D. 1965. Aggregate stability. pp. 511–519. In: C. A. Black et al. (Eds.), *Methods of Soil Analysis*, Part 1. Monograph 9. American Society of Agronomy, Madison, WI.

Kemper, W. D. and R. C. Rosenau. 1986. Aggregate stability and size distribution. pp. 425–442. In: A. Klute (Ed.), *Methods of Soil Analysis*, Part 1. American Society of Agronomy, Madison, WI.

Kersten, M. S. 1949. Thermal properties of soils. *Univ. Minn. Inst. Technol. Bull.* 28.

Khan, A. U. H. 1988. A laboratory test of the dispersion scale effect. Ph.D. dissertation. University of California–Riverside.

Khan, A. U. H. and W. A. Jury. 1990. A laboratory test of the dispersion scale effect. *J. Contam. Hydrol.* 5:119–132.

Kluitenberg, G., J. Ham, and K. Bristow. 1993. Error analysis of the heat pulse method for measuring the volumetric heat capacity of soil. *Soil Sci. Soc. Am. J.* 57:1444–1451.

Kluitenberg, G. 2002. Heat capacity and specific heat. pp. 1201–1208. In: J. Dane and G. Topp (Eds.), *Methods of Soil Analysis*, Part 4, *Physical Methods*. Soil Science Society of America, Madison, WI.

Klute, A. and C. Dirksen. 1986. Hydraulic conductivity and diffusivity: laboratory methods. In: A. Klute (Ed.), *Methods of Soil Analysis*, Part 1. Monograph 9. American Society of Agronomy, Madison, WI.

Knight, J. 1992. The sensitivity of time domain reflectometry measurements to lateral variations in soil water content. *Water Resour. Res.* 28:2345–2352.

Kristensen, K. J. and E. R. Lemon. 1964. Soil aeration and plant–root relations: III. Physical aspects of oxygen diffusion in the liquid phase of the soil. *Agron. J.* 56:295–301.

Kung, K. J. S. 1990. Preferential flow in a sandy vadose zone: 1. Field observations. *Geoderma* 46:51–58.

Kutilek, M. 1980. Constant rainfall infiltration. *J. Hydrol.* 45:289–303.

Lang, C. 1878. Über Wärmekapazität der Bodenkonstituieren. *Fortschr. Geb. Agr. Phys.* 1:109–147.

Langmuir, I. 1918. The adsorption of gases on plane surfaces of glass, mica, and platinum. *J. Am. Chem. Soc.* 40:1361–1403.

Lee, J., R. Horton, and D. B. Jaynes. 2000a. A time domain reflectometry method to measure immobile water content and mass exchange coefficient. *Soil Sci. Soc. Am. J.* 64:1911–1917.

Lee, J., R. Horton, and D. B. Jaynes. 2002. The feasibility of shallow time domain reflectometry probes to describe solute transport through undisturbed soil cores. *Soil Sci. Soc. Am. J.* 66:53–57.

Lemon, E. R. 1962. Soil aeration and plant root relations: I. *Theory Agric. J.* 54:167–170.

Lemon, E. R. and A. E. Erickson. 1952. The measurement of oxygen diffusion in soil with a platinum electrode. *Soil Sci. Soc. Am. Proc.* 16:160–163.

Lemos, P. and J. F. Lutz. 1957. Soil crusting and some factors affecting it. *Soil Sci. Soc. Am. Proc.* 21:485–491.

Letey, J. and W. Farmer. 1973. Mass transfer. In: *Pesticides in Soil and Water*. American Society of Agronomy, Madison, WI.

Letey, J. and L. H. Stolzy. 1964. Measurement of oxygen diffusion rates with the platinum electrode: I. Theory and equipment. *Hilgardia* 35:545–554.

Libardi, P. L., K. Reichardt, D. R. Nielsen, and J. W. Biggar. 1980. Some simple field methods for estimating soil hydraulic conductivity. *Soil Sci. Soc. Am. J.* 44:3–6.

Lide, D. R. (Ed.-in-Chief). 2002. *Handbook of Chemistry and Physics*, 83rd ed. CRC Press, Boca Raton, FL. 2664 pp.

Logsdon, S. D. 2002. Determination of preferential flow model parameters. *Soil Sci. Soc. Am. J.* 66:1095–1103.

Low, P. F. 1959. The viscosity of water in clay systems. In: *Proc. 8th National Clay Conf.* Pergamon Press, New York.

Low, P. F. 1961. Physical chemistry of the clay–water interaction. *Adv. Agron.* 13:269–327.

Lowery, B. and J. Morrison. 2002. Soil penetrometers and penetrability. pp. 363–388. In: J. Dane and G. Topp (Eds.), *Methods of Soil Analysis*, Part 4, *Physical Methods*. Soil Science Society of America, Madison, WI.

Lumley, J. L. and A. Panofsky. 1964. *The Structure of Atmospheric Turbulence*. Wiley, New York.

Lyles, L., J. Dickerson, and L. Disrud. 1970. Modified rotary sieve for improved accuracy. *Soil Sci.* 109:207–210.

Marshall, C. E. 1935a. Mineralogical methods for the study of silts and clays. *Z. Kristallogr. A* 90:8–34.

Marshall, C. E. 1935b. Layer lattices and base exchange clays. *Z. Kristallogr. A* 91:433–449.

Marshall, C. E. 1964. *The Physical Chemistry and Mineralogy of Soils*. Wiley, New York.

Marshall, C. E. and G. Garcia. 1959. Exchange equilibria in a carboxylic resin and in attapulgite clay. *J. Phys. Chem.* 63:1663–1668.

Marshall, T. J. 1958. A relation between permeability and size distribution of pores. *J. Soil Sci.* 9:1–8.

Marshall, T. J. 1959. The diffusion of gas through porous media. *J. Soil Sci.* 10:79–82.

Marshall, T. J. and J. P. Quirk. 1950. Stability of structural aggregates of dry soil. *Aust. J. Agric. Res.* 1:266–275.

Matthias, A. D., A. M. Blackmer, and J. M. Bremner. 1980. A simple chamber technique for field measurement of emissions of nitrous oxide from soils. *J. Environ. Qual.* 9:251–256.

McBratney, A. B. and R. Webster. 1981. The design of optimal sampling schemes for local estimation and mapping of regionalized variables. *Comput. Geosci.* 7:335–365.

McBride, M. B. 1994. *Environmental Chemistry of Soils*. Oxford University Press, New York. 406 pp.

McIntyre, D. S. 1958. Permeability measurements of soil crusts formed by raindrop impact. *Soil Sci.* 85:185–189.

McIntyre, D. S. 1971. The platinum microelectrode method for soil aeration measurements. *Adv. Agron.* 22:235–283.

Millington, R. J. and J. P. Quirk. 1959. Permeability of porous media. *Nature* 183:387–388.

Millington, R. J. and J. P. Quirk. 1961. Permeability of porous solids. *Trans. Faraday Soc.* 57:1200–1207.

Mohanty, B. P., R. Horton, and M. D. Ankeny. 1996. Infiltration and macroporosity under row crop agricultural field in a glacial till soil. *Soil Sci.* 161:205–213.

Moldrup, P., T. G. Poulsen, P. Schjonning, T. Olesen, and T. Yamaguchi. 1998. Gas permeability in undisturbed soils: measurements and predictive models. *Soil Sci.* 163:180–189.

Moldrup, P., T. Olesen, P. Schjonning, T. Yamaguchi, and D. E. Rolston. 2000. Predicting the gas diffusion coefficient in undisturbed soil from soil water characteristics. *Soil Sci. Soc. Am. J.* 64:94–100.

Moldrup, P., T. Olesen, T. Komatsu, P. Schjonning, and D. E. Rolston. 2001. Tortuosity, diffusivity, and permeability in the soil liquid and gas phases. *Soil Sci. Soc. Am. J.* 65:613–623.

Monteith, J. L., G. Szeicz, and K. Yabuki. 1964. Crop photosynthesis and the flux of carbon dioxide below the canopy. *J. Appl. Ecol.* 6:321–337.

Morton, C. T. and W. F. Buchele. 1960. Emergence energy of plant seedlings. *Agric. Eng.* 41:428–431.

Mualem, Y. 1976a. *A Catalogue of the Properties of Unsaturated Soils.* Research Report 442. Technion, Israel Institute of Technology, Haifa, Israel.

Mualem, Y. 1976b. A new model for predicting the hydraulic conductivity of unsaturated porous media. *Water Resour. Res.* 12:513–522.

Mualem, Y. 1992. Modeling the hydraulic conductivity of unsaturated porous media. pp. 15–36. In: M. Th. van Genuchten et al. (Eds.), *Indirect Methods for Estimating the Hydraulic Properties of Unsaturated Soils.* University of California–Riverside, Riverside, CA.

Murray, F. W. 1967. On the computation of saturation vapor pressure. *J. Appl. Meteorol.* 6:203–205.

Nassar, I. N. and R. Horton. 1989. Water transport in unsaturated nonisothermal, salty soil. *Soil Sci. Soc. Am. J.* 53:1323–1337.

Nassar, I. N. and R. Horton. 1997. Heat, water, and solute transfer in unsaturated porous media: I. Theory development and transport coefficient evaluation. *Transp. Porous Media* 27:17–38.

Nassar, I. N. and R. Horton. 1999a. Salinity and compaction effects on soil water evaporation and water and solute distributions. *Soil Sci. Soc. Am. J.* 63:752–758.

Nassar, I. N. and R. Horton. 1999b. Transport and fate of volatile organic chemicals in unsaturated, nonisothermal salty porous media: I. Theoretical development. *J. Hazard. Mater.* 69:151–167.

Nassar, I. N., L. Ukrainczyk, and R. Horton. 1999. Transport and fate of volatile organic chemicals in unsaturated, nonisothermal salty porous media: II. Experimental and numerical studies for benzene. *J. Hazard. Mater.* 69:169–185.

Nazeroff, W. W., S. R. Lewis, S. M. Doyle, B. A. Moed, and A. E. Nero. 1987. Experiments on pollutant transport from soil into residential basements by pressure-driven airflow. *Environ. Sci. Technol.* 21:459–466.

Nero, A. E. and W. W. Nazeroff. 1985. Characterizing the source of radon indoors. pp. 23–29. In: *Radiation Protection Dosimetry*, Vol. 7, No. 1–4, Nuclear Technology Publishing, Ashford, Kent, England.

Nielsen, D. R., J. W. Biggar, and K. T. Em. 1973. Spatial variability of field-measured soil–water properties. *Hilgardia* 42:215–259.

Nimmo, J. and K. Perkins. 2002. Aggregate stability and size distribution. pp. 317–328. In: J. Dane and G. Topp (Eds.), *Methods of Soil Analysis*, Part 4, *Physical Methods*. Soil Science Society of America, Madison, WI.

NOAA (National Oceanic and Atmospheric Administration). 1986. *Climatological Data—California: 1986*, Vol. 110, *Annual Summary*. NOAA, Ashville, NC.

Novotny, V. and G. Chesters. 1981. *Handbook of Nonpoint Pollution*. Van Nostrand Reinhold, New York.

Ochsner, T. E., R. Horton, and T. Ren. 2001. Simultaneous water content, air-filled porosity, and bulk density measurements with thermal-time domain reflectometry. *Soil Sci. Soc. Am. J.* 65:1618–1622.

Ogata, G. and L. A. Richards. 1957. Water content changes following irrigation of bare field soil that is protected from evaporation. *Soil Sci. Soc. Am. Proc.* 21:355–356.

Ong, S. K. and L. W. Lion. 1991. Effects of soil properties and moisture on the adsorption of trichloroethylene vapor. *Water Res.* 25:29–36.

Parker, J. J., Jr., and H. M. Taylor. 1965. Soil strength and seedling emergence relations: I. Soil type, moisture tension, temperature and planting depth effects. *Agron. J.* 57:289–291.

Parlange, J.-Y. and W. L. Hogarth. 1997. Comments on soil water diffusivity determination by general similarity theory. *Soil Sci.* 162:767–768.

Parlange, J.-Y. and R. E. Smith. 1976. Ponding time for variable rainfall rates. *Can. J. Soil Sci.* 56:121–123.

Parlange, M. B., A. T. Cahill, D. R. Nielsen, J. W. Hopmans, and O. Wendroth. 1998. Review of heat and water movement in field soils. *Soil Tillage Res.* 47:5–10.

Patten, H. E. 1909. *Heat Transference in Soils*. Bulletin 59. U.S. Department of Agriculture Bureau of Soils, Washington, DC.

Pauling, L. 1930. The structure of micas and related minerals. *Natl. Acad. Sci. Proc.* 16:123.

Pauling, L. 1948. *General Chemistry*. Freeman, San Francisco.

Peck, A. J. 1965. Moisture profile development and air compression during water uptake by bounded porous bodies: 3. *Vertical columns, Soil Sci.* 100:44–51.

Peck, A. J. 1983. Field variability of soil physical properties. In: D. I. Hillel (Ed.), *Advances in Irrigation*, Vol. 2. Academic Press, New York.

Penman, H. L. 1940a. Gas and vapor movements in the soil: I. The diffusion of vapors through porous solids. *J. Agric. Sci.* 30:437–462.

Penman, H. L. 1940b. Gas and vapor movements in the soil: II. The diffusion of carbon dioxide through porous solids. *J. Agric. Sci.* 30:570–581.

Penman, H. L. 1948. Natural evaporation from open water, bare soil, and grass. *Proc. R. Soc. Ser. A* 190:120–145.

Pennell, K. D. 2002. Aggregate specific surface area. pp. 295–315. In: J. Dane and G. Topp (Eds.), *Methods of Soil Analysis*, Part 4, *Physical Methods*. Soil Science Society of America, Madison, WI.

Pennell, K. D., S. A. Boyd, and L. Abriola. 1995. Surface area of soil organic matter reexamined. *Soil Sci. Soc. Am. J.* 59:1012–1018.

Perfect, E. and B. Kay. 1994. Statistical characterization of dry aggregate strength using rupture energy. *Soil Sci. Soc. Am. J.* 58:1804–1809.

Perroux, K. M. and I. White. 1988. Designs for disc permeameters. *Soil Sci. Soc. Am. J.* 52:1205–1215.

Philip, J. R. 1957a. The theory of infiltration: 1. The infiltration equation and its solution. *Soil Sci.* 83:345–357.

Philip, J. R. 1957b. The theory of infiltration: 2. The profile at infinity. *Soil Sci.* 83:435–448.

Philip, J. R. 1957c. The theory of infiltration: 3. Moisture profiles and relation to experiment. *Soil Sci.* 84:163–178.

Philip, J. R. 1957d. The theory of infiltration: 4. Sorptivity and algebraic infiltration equations. *Soil Sci.* 84:257–264.

Philip, J. R. 1957e. The theory of infiltration: 5. Influence of initial moisture content. *Soil Sci.* 84:329–339.

Philip, J. R. 1957f. Evaporation, and moisture and heat fields in the soil. *J. Met.* 14:354–366.

Philip, J. R. 1958a. The theory of infiltration: 6. Effect of water depth over soil. *Soil Sci.* 85:278–286.

Philip, J. R. 1958b. The theory of infiltration: 7. *Soil Sci.* 85:333–337.

Philip, J. R. 1968. Absorption and infiltration in two and three dimensional systems. pp. 503–525. In: *Proc. UNESCO Symp., Water in the Unsaturated Zone*, Vol. I. Wageningen, The Netherlands.

Philip, J. R. 1969. The theory of infiltration. *Adv. Hydrosci.* 5:215–296.

Philip, J. R. 1972. Reply to comments on "Hydrostatics and hydrodynamics in swelling soils." *Water Resour. Res.* 6:1248–1251.

Philip, J. R. 1975. Stability analysis of infiltration. *Soil Sci. Soc. Am. Proc.* 39:1042–1049.

Philip, J. R. 1983. Infiltration in one, two, and three dimensions. pp. 1–13. In: *Proc Natl. Conf. Advances in Infiltration.* American Society of Agricultural Engineers, St. Josephs, MI.

Philip, J. R. and D. A. de Vries. 1957. Moisture movement in soils under temperature gradients. *Trans. Am. Geophys. Union* 38:222–228.

Pratt, P. F., W. W. Jones, and V. E. Hunsaker. 1970. Nitrate in deep soil profiles in relation to fertilizer rates and leaching volumes. *J. Environ. Qual.* 1:97–102.

Pye, V., R. Patrick, and J. Quarles. 1983. *Ground Water Quality in the United States.* University of Pennsylvania Press, Philadelphia, PA.

Quisenberry, V. and R. E. Phillips. 1978. Displacement of soil water by simulated rainfall. *Soil Sci. Soc. Am. J.* 42:675–679.

Raats, P. A. C. 1973. Unstable wetting fronts in uniform and nonuniform soils. *Soil Sci. Soc. Am. Proc.* 37:681–685.

Raats, P. A. C. 1975. Distribution of salts in the crop root zone. *J. Hydrol.* 27:237–248.

Raats, P. A. C. and W. R. Gardner. 1975. Movement of water in the unsaturated zone near a water table. pp. 311–358. In: J. Van Schilfgaarde (Ed.), *Drainage for Agriculture.* Monograph 17. American Society of Agronomy, Madison, WI.

Radoslovich, E. W. 1960. The structure of muscovite. *Acta Cryst.* 13:919–932.

Radoslovich, E. W. and K. Norrish. 1962. The cell dimensions and symmetry of layer-lattice silicates. *Am. Mineral.* 47:599–616.

Raich, J. W. and A. Tufekcioglu. 2000. Vegetation and soil respiration: correlations and controls. *Biogeochemistry* 48:71–90.

Raney, W. A. 1949. Field measurement of oxygen diffusion through soil. *Soil Sci. Soc. Am. Proc.* 14:61–65.

Rao, P. S. C. and J. M. Davidson. 1979. Adsorption and movement of selected herbicides at high concentrations in soils. *Water Res.* 13:375–380.

Rao, P. S. C., A. G. Hornsby, and R. E. Jessup. 1985. Indices for ranking the potential for pesticide contamination of groundwater. *Proc. Soil Crop Soc. Fla.* 44:1–8.

Raven, P. and L. Berg. 2001. *Environment*, 3rd ed., Harcourt College Publishing, New York. 612 pp.

Rawlins, S. L. and G. S. Campbell. 1986. Water potential: thermocouple psychrometry. pp. 597–618. In: A. Klute (Ed.), *Methods of Soil Analysis*, Part 1. Monograph 9. American Society Agronomy, Madison, WI.

Reeves, M. J. 1980. Recharge of the English Chalk: a possible mechanism. *Eng. Geol.* 14:231–240.

Reeves, P. and M. Celia. 1996. A functional relationship between capillary pressure, saturation, and interfacial area as revealed by a pore-scale network model. *Water Resour. Res.* 32:2345–2358.

Reginato, R. J. 1974. Gamma radiation measurements of bulk density changes in soil pedon following irrigation. *Soil Sci. Soc. Am. Proc.* 38:24–29.

Ren, T., K. Noborio, and R. Horton. 1999. Measuring soil water content, electrical conductivity, and thermal properties with a thermo-TDR probe. *Soil Sci. Soc. Am. J.* 63:450–457.

Ren, T., G. J. Kluitenburg, and R. Horton. 2000. Determining soil water flux and pore water velocity by a heat pulse technique. *Soil Sci. Soc. Am. J.* 64:552–560.

Ren, T., T. E. Ochsner, R. Horton, and Z. Ju. 2003. Heat-pulse method for soil water content measurement: influence of the specific heat of the soil solids. *Soil Sci. Soc. Am. J.* 67:1631–1634.

Ressler, D. E., R. Horton, J. L. Baker, and T. C. Kaspar. 1997. Testing a nitrogen fertilizer applicator designed to reduce leaching losses. *Appl. Eng. Agric.* 13:345–350.

Ressler, D. E., R. Horton, J. L. Baker, and T. C. Kaspar. 1998. Evaluation of localized compaction and doming to reduce anion leaching losses using lysimeters. *J. Environ. Qual.* 27:910–916.

Ressler, D. E., R. Horton, T. C. Kaspar, and J. L. Baker. 1999. Crop response to localized compaction and doming. *Agron. J.* 90:747–752.

Reynolds, W. and D. Elrick. 1990. Ponded infiltration from a single ring: I. Analysis of steady flow. *Soil Sci. Soc. Am. J.* 54:1233–1241.

Reynolds, W., D. Elrick, E. Youngs, H. Booltink, and J. Bouma. 2002. Saturated and field-saturated water flow parameters: laboratory methods. pp. 802–804. In: J. Dane and G. Topp (Eds.), *Methods of Soil Analysis*, Part 4, *Physical Methods*. Soil Science Society of America, Madison, WI.

Rhue, R. D., P. S. C. Rao, and R. E. Smith. 1988. Vapor-phase adsorption of alkylbenzenes and water on soils and clays. *Chemosphere* 17:727–741.

Richards, L. A. 1953. Modulus of rupture as an index of crusting of soils. *Soil Sci. Soc. Am. Proc.* 17:321–323.

Ritchie, J. T. 1972. Model for predicting evaporation from a row crop at incomplete cover. *Water Resour. Res.* 8:1204–1213.

Robinson, R. A. and R. H. Stokes. 1959. *Electrolyte Solutions*. Butterworth, London.

Rogers, S. W. and S. K. Ong. 2000. Influence of porous media, airflow rate, and air channel spacing on benzene NAPL removal during air sparging. *Environ. Sci. Technol.* 34:764–770.

Rolston, D. E. and P. Moldrup. 2002. Gas diffusivity. pp. 1113–1139 In: G. C. Topp and J. H. Dane (Eds.), *Methods of Soil Analysis*, Part 4, *Physical Methods*. Soil Science Society of America, Madison, WI.

Romell, L. G. 1922. Luftväxlingen i marken som ekologisk faktor. *Medd. Statens Skogsfarsoksanstalt* 19:2.

Rose, C. W., W. R. Stem, and J. E. Drummond. 1965. Determination of hydraulic conductivity as a function of depth and water content for soil in situ. *Aust. J. Soil Res.* 3:1–9.

Ross, C. S. and E. V. Shannon, 1926. The minerals of bentonite and related clays and their physical properties. *J. Am. Ceram. Soc.* 9:77.

Roth, K., R. Shulin, H. Fluhler, and W. Attinger. 1990. Calibration of time domain reflectometry for water content measurement using a composite dielectric approach. *Water Resour. Res.* 26:2267–2274.

Rubin, J. 1968. Numerical analysis of ponded rainfall infiltration. pp. 440–450. In: *Proc. UNESCO Symp., Water in the Unsaturated Zone*, Vol. I. Wageningen, The Netherlands.

Russell, E. J. and A. Appleyard. 1915. The atmosphere of the soil, its composition, and causes of variation. *J. Agric. Sci.* 7:1–48.

Russell, E. W. 1973. *Soil Conditions and Plant Growth*, 10th ed. Longman, London.

Russo, D. and E. Bresler. 1980. Scaling soil hydraulic properties of a heterogeneous field. *Soil Sci. Soc. Am. J.* 44:681–684.

Russo, D. and E. Bresler. 1981. Soil hydraulic properties as stochastic processes: I. An analysis of spatial variability. *Soil Sci. Soc. Am. J.* 45:682–687.

Russo, D. and W. A. Jury. 1987a. A theoretical study of the estimation of the correlation length scale in spatially variable fields: 1. Stationary fields. *Water Resour. Res.* 23:1257–1268.

Russo, D. and W. A. Jury. 1987b. A theoretical study of the estimation of the correlation length scale in spatially variable fields: 2. Nonstationary fields. *Water Resour. Res.* 23:1269–1279.

Russo, D., W. A. Jury, and G. L. Butters. 1989. Numerical analysis of solute transport during transient irrigation: I. Effect of hysteresis and profile heterogeneity. *Water Resour. Res.* 25:2109–2118.

Ryden, J. C., L. J. Lund, and D. D. Focht. 1978. Direct in-field measurement of nitrous oxide flux from soils. *Soil Sci. Soc. Am. J.* 42:863–869.

Sander, G. C., I. Cunning, W. Hogarth, and J. Y. Parlange. 1991. Exact solution of nonlinear, nonhysteretic redistribution in vertical soil of finite depth. *Water Resour. Res.* 27:1529–1536.

Sauer, T. J. 2002. Heat flux density. pp. 1233–1248. In: G. C. Topp and J. H. Dane (Eds.), *Methods of Soil Analysis*, Part 4, *Physical Methods*. Soil Science Society of America, Madison, WI.

Sauer, T. J. and R. Horton. 2003. Soil heat flux. In: J. L. Hatfield (Ed.), *Micrometeorology in Agricultural Systems*. Monograph. American Society of Agronomy, Madison, WI (in press).

Sauer, T. J., J. L. Hatfield, J. H. Prueger, and J. M. Norman. 1998. Surface energy balance of a corn residue-covered field. *Agric. For. Meteorol.* 89:155–168.

Sawhney, B. L. 1988. Interstratification on layer silicates. In: J. B. Dixon and S. B. Weed (Eds.), *Minerals in Soil Environments*, 2nd ed. Soil Science Society of America, Madison, WI.

Scanlon, B. R., S. W. Tyler, and P. J. Wierenga. 1997. Hydrologic issues in arid, unsaturated systems and implications for contaminant transport. *Rev. Geophys.* 35(4):461–490.

Scarsbrook, G. E. 1965. Nitrogen availability. In: J. W. Bartholomew and F. E. Clark (Eds.), *Soil Nitrogen*. Monograph 10. American Society of Agronomy, Madison, WI.

Schofield, R. K. and H. R. Samson. 1954. Flocculation of kaolinite due to the attraction of oppositely-charged crystal faces. *Disc. Faraday Soc.* 18:135–145.

Schwertmann, U. and R. M. Taylor. 1988. Iron oxides. In: J. B. Dixon and S. B. Weed (Eds.), *Minerals in Soil Environments*, 2nd ed. Soil Science Society of America, Madison, WI.

Scotter, D. R., G. W. Thurtell, and P. A. C. Raats. 1967. Dispersion resulting from sinusoidal gas flow in porous materials. *Soil Sci.* 104:306–308.

Selker, J. S, T. S. Steenhuis, and J.Y. Parlange. 1992. Wetting front instability in homogeneous sandy soils under continuous infiltration. *Soil Sci. Soc. Am. J.* 56:1346–1350.

Shainberg, I. and W. D. Kemper. 1966. Hydration status of adsorbed cations. *Soil Sci. Soc. Am. Proc.* 30:707–713.

Shainberg, I. and W. D. Kemper. 1967. Ion exchange equilibria on montmorilinite. *Soil Sci.* 103:4–9.

Shao, M. and R. Horton. 1996. Soil water diffusivity determination by general similarity theory. *Soil Sci.* 161:727–734.

Shao, M. and R. Horton. 1997. Reply to comments on soil water diffusivity determination by general similarity theory. *Soil Sci.* 162:769–770.

Shao, M. and R. Horton. 1998. Integral method for estimating soil hydraulic properties. *Soil Sci. Soc. Am. J.* 62:585–592.

Shao, M. and R. Horton. 2000. Exact solution for horizontal water redistribution by general similarity. *Soil Sci. Soc. Am. J.* 64:561–564.

Shao, M., R. Horton, and D. B. Jaynes. 1998. Analytical solution for one-dimensional heat conduction–convection equation. *Soil Sci. Soc. Am. J.* 62:123–128.

Shouse, P., W. A. Jury, and L. H. Stolzy. 1982. Field measurement and modeling of cowpea water use and yield under stressed and well-watered growth conditions. *Hilgardia* 50:1–25.

Simmons, C. S. 1982. A stochastic-convective transport representation of dispersion in one dimensional porous media systems. *Water Resour. Res.* 18:1193–1214.

Simunek, J., K. Huang, M. Sejna, M. Th. van Genuchten, J. Majercak, V. Novak, and J. Sutor. 1997. *The HYDRUS-ET Software Package for Simulating the One-Dimensional Movement of Water, Heat, and Multiple Solutes in Variably-Saturated Media*, Version 1.1. USSL Research Report 133. Institute of Hydrology, Slovak Academy of Sciences, Bratislava, Slovakia. 184 pp.

Simunek, J., D. Wang, P. J. Shouse and M. Th. van Genuchten. 1998. Analysis of field tension disc infiltrometer data by parameter estimation. *Int. Agrophys.* 12:167–180.

Simunek, J., M. Th. van Genuchten, M. Sejna, N. Toride, and F. J. Leij. 1999. *The STANMOD Computer Software for Evaluating Solute Transport in Porous Media Using Analytical Solutions of Convection–Dispersion Equation*, Version 1.0. Report IGWMC-TPS 71. International Ground Water Modeling Center, Colorado School of Mines, Golden, CO. 32 pp.

Sleep, B. E. and P. D. McClure. 2001. Removal of volatile and semivolatile organic contamination from soil by air and steam flushing. *J. Contam. Hydrol.* 50:21–40.

Smith, A. 1932. Seasonal subsoil temperature variations. *J. Agric. Res.* 44:421–428.

Smith, W. O. and H. G. Byers. 1938. The thermal conductivity of dry soil of certain of the great soil groups. *Soil Sci. Soc. Am. Proc.* 3:13–19.

Smith, R., C. Corradini, and F. Melone. 1993. Modeling infiltration for multistorm runoff events. *Water Resour. Res.* 29:133–144.

Soil Survey Staff. 1999. *Soil Taxonomy: A Basic System of Soil Classification for Making and Interpreting Soil Surveys*, 2nd ed. Agriculture Handbook 436. USDA-NRCS, Washington, DC.

Spencer, W. F., M. M. Cliath, W. A. Jury, and Lian-Zhang Zhang. 1988. Volatilization of pesticides from soil as related to their Henry's law constants. *J. Environ. Qual.* 17:504–509.

Sposito, G. 1972. Volume changes in swelling soils. *Soil Sci.* 115:315–320.

Sposito, G. 1981. *The Thermodynamics of Soil Solutions*. Oxford University Press, Oxford.

Sposito, G. 1984. *The Surface Chemistry of Soils*. Oxford University Press, Oxford.

Stanhill, G. 1965. Observations on the reduction of soil temperature. *Agric. Meteorol.* 2:197–203.

Stolzy, L. H. and J. Letey. 1964. Characterizing soil oxygen conditions with a platinum electrode. *Adv. Agron.* 16:249–279.

Sullivan, P. J., G. Sposito, S. M. Strathouse, and C. L. Hanson. 1979. Geologic nitrogen and the occurrence of high nitrate soils in the western San Joaquin Valley. *Hilgardia* 47:15–31.

Swartzendruber, D. 1987. A quazi-solution of Richards equation for the downward infiltration of water in soil. *Water Resour. Res.* 23:809–817.

Swift, L. W. 1976. Algorithm for solar radiation on mountain slopes. *Water Resour. Res.* 12:108–112.

Takle, E. S. 2003. Soil management and conservation: windbreaks and shelterbelts. In D. Hillel et al. (Eds.), *Encyclopedia of Soil Science*. Elsevier, New York.

Takle, E. S., J. R. Brandle, R. A. Schmidt, R. Garcia, I. V. Litvina, W. J. Massman, X. Zhou, G. Doyle, and C. W. Rice. 2003. High-frequency pressure variations in the vicinity of a surface CO_2 flux chamber. *Agric. For. Meteorol.* 114:245–250.

Talsma, T. 1969. Infiltration from semi-circular furrows in the field. *Aust. J. Soil Res.* 7:277–284.

Tanner, C. B. 1968. Evaporation of water from plants and soil. In: T. Koslowski (Ed.), *Water Deficits and Plant Growth*. Academic Press, New York.

Taylor, G. I. 1953. The dispersion of soluble matter flowing through a capillary tube. *Proc. Math. Soc. London* 2:196–212.

Thomas, G. and R. E. Phillips. 1979. Consequences of water flow in macropores. *J. Environ. Qual.* 8:149–152.

Thomas, M. D. 1928. Aqueous vapor pressure in soils. *Soil Sci.* 25:485–493.

Thomson, N. R. and R. L. Johnson. 2000. Air distribution during in situ air sparging: an overview of mathematical modeling. *J. Hazard. Mater.* 72:265–282.

Tisdall, J. H. and R. D. Oades. 1982. Organic matter and water-stable aggregates in soils. *J. Soil Sci.* 33:141–163.

Topp, G. C. and P. A. Ferré. 2002a. Methods for measurement of soil water content. pp. 422–427. In: J. Dane and G. Topp (Eds.), *Methods of Soil Analysis*, Part 4, *Physical Methods*. Soil Science Society of America, Madison, WI.

Topp, G. C. and P. A. Ferré. 2002b. Water content: scope of methods and brief description. pp. 417–421. In: J. Dane and G. Topp (Eds.), *Methods of Soil Analysis*, Part 4, *Physical Methods*. Soil Science Society of America, Madison, WI.

Topp, G. C., J. L. Davis, and A. P. Annan. 1980. Electromagnetic determination of soil water content: measurement in coaxial transmission lines. *Water Resour. Res.* 16:574–582.

Troeh, F. R. and L. M. Thompson. 1993. *Soils and Soil Fertility*, 5th ed. Oxford University Press, New York. 462 pp.

Tuli, A. and J. Hopmans. 2003. Effect of degree of fluid saturation on transport coefficients in disturbed soils. *Eur. J. Soil Sci.* 54 (in press).

Tuller, M. and D. Or, 2001. Hydraulic conductivity of variably saturated porous media: film and corner flow in angular pore space. *Water Resour. Res.* 37:1257–1276.

Tuorilla, P. 1928. Uber Beziehungen zwischen Koagulation, elektrokinetischzen Wanderungs-geschwindigkeiten, Ionhydration und chemischer Beeinflussung. Kolloidchem. Beihefte 27:44–181.

Ulrich, R. 1894. Untersuchungen Fiber die Wärmekapacität der Bodenconstituenten. *Forsch. Geb. Agric. Phys.* 17:1–31.

U.S. Salinity Laboratory Staff. 1954. *Diagnosis and Improvement of Saline and Alkali Soils*. U.S. Department of Agriculture Handbook 60. USDA, Washington, DC.

Valocchi, A. J. 1985. Validity of the local equilibrium assumption for modeling sorbing solute transport through homogeneous soils. *Water Resour. Res.* 21:808–820.

Van Bavel, C. H. M. 1952. Gaseous diffusion and porosity in porous media. *Soil Sci.* 73:91–104.

Van Bavel, C. H. M. 1966. Potential evaporation: the combination concept and its experimental verification. *Water Resour. Res.* 2:455–467.

Van Bavel, C. H. M., N. Underwood, and R. W. Swanson. 1956. Soil moisture measurement by neutron moderation. *Soil Sci.* 82:29–41.

Van Duin, R. H. A. 1963. *The Influence of Soil Management on the Temperature Wave near the Surface*. Technical Bulletin 29. Institute for Land and Water Management Research, Wageningen, The Netherlands.

Van Genuchten, M. Th. 1980. A closed form equation for predicting the hydraulic conductivity of unsaturated soils. *Soil Sci. Soc. Am. J.* 44:892–898.

Van Genuchten, M. Th. and W. A. Jury. 1987. Progress in unsaturated flow and transport modeling. *IUGG Rev.* 25:135–140.

Van Genuchten, M. Th. and D. R. Nielsen. 1985. On describing and predicting the hydraulic properties of unsaturated soils. *Ann. Geophys.* 3:615–628.

Van Genuchten, M. Th. and P. J. Wierenga. 1976. Mass transfer studies in sorbing porous media: I. Analytical solutions. *Soil Sci. Soc. Am. J.* 40:473–480.

Van Genuchten, M. Th. and P. J. Wierenga. 1977. Mass transfer studies in sorbing porous media: II. Experimental evaluation with tritium. *Soil Sci. Soc. Am. J.* 41:272–278.

Van Olphen, H. 1963. *Clay Colloid Chemistry*. Interscience, New York.

Van Rooyen, M. and H. F. Winterkorn. 1959. Structural and textural influences on the thermal conductivity of soils. *Highway Res. Bd. Proc.* 38:576–621.

Van Wijk, W. R. 1963. *Physics of Plant Environment*. North-Holland, Amsterdam.

Vaneslow, A. P. 1932. Equilibria of the base exchange reaction of bentonites, permutites, soil colloids, and zeolites. *Soil Sci.* 33:95–113.

VDI [Verein Deutscher Ingenieure (Association of German Engineers)]. 1994. *Environmental Meteorology*: Interactions between atmosphere and surfaces; Calculation of shortwave and longwave radiation (in German and English). VDI Guideline. VDI-Richtlinie 3789, Part 2. Beuth Publishing Co., Berlin.

Wada, K. 1988. Allophane and imogolite. In: J. B. Dixon and S. B. Weed (Eds.), *Minerals in Soil Environments*, 2nd ed. Soil Science Society of America, Madison, WI.

Wang, F. and I. C. Ward. 2002. Radon entry, migration and reduction in houses with cellars. *Build. Environ.* 37:1153–1165.

Wang, H., E. S. Takle, and J. Shen. 2001. Shelterbelts and windbreaks: mathematical modeling and computer simulation of turbulent flows. *Annu. Rev. Fluid Mech.* 33:549–586.

Wang, Q., J. M. G. Shao, and R. Horton. 1999. Modified Green and Ampt models for layered soil infiltration and muddy water infiltration. *Soil Sci.* 164(7):445–453.

Wang, Q., R. Horton, and M. Shao. 2002a. Horizontal infiltration method for determining Brooks–Corey model parameters. *Soil Sci. Soc. Am. J.* 66:1733–1739.

Wang, Q., T. E. Ochsner, and R. Horton. 2002b. Mathematical analysis of heat pulse signals for soil water flux determination. *Water Resour. Res.* 38:10.1029/2001WR1089.

Wang, Z., J. Feyen, and D. E. Elrick. 1998. Prediction of fingering in porous media. *Water Resour. Res.* 34:2183–2190.

Wang, Z., L. Wu, and Q. J. Wu. 2000. Water-entry value as an alternative indicator of soil water-repellency and wettability. *J. Hydrol.* 231–232:76–83.

Wang, Z., A. Tuli, and W. A. Jury. 2003a. Unstable flow during redistribution in homogenous soil. *Vadose Zone J.* 2:52–60.

Wang, Z., L. Wu, T. Harter, J. Lu, and W. A. Jury. 2003b. A field study of unstable preferential flow during soil water redistribution. *Water Resour. Res.* 39, no. 4, 1075, doi:10.1029/ 2001wr000903.

Warrick, A. W. and D. R. Nielsen. 1980. Spatial variability of soil physical properties in the field. pp. 319–344. In: D. I. Hillel (Ed.), *Applications of Soil Physics*. Academic Press, New York.

Warrick, A. W., D. E. Myers, and D. R. Nielsen. 1986. Geostatistical methods applied to soil science. pp. 53–82. In: A. Klute (Ed.), *Methods of Soil Analysis*, Part 1. Monograph 9. American Society of Agronomy, Madison, WI.

Welch, S. M., G. J. Kluitenberg, and K. L. Bristow. 1996. Rapid numerical estimation of soil thermal properties for a broad class of heat-pulse emitter geometries. *Meas. Sci. Technol.* 7:932–938.

Westcott, D. W. and P. J. Wierenga. 1974. Transfer of heat by conduction and vapor movement in a closed soil system. *Soil Sci. Soc. Am. Proc.* 38:9–14.

Whisler, F. D., A. Klute, and D. B. Peters. 1968. Soil water diffusivity from horizontal infiltration. *Soil Sci. Soc. Am. Proc.* 32:6–11.

White, E. C. and D. G. Moore. 1972. Nitrates in South Dakota range soils. *J. Range Manag.* 25:27–32.

White, R. E. 1985. The influence of macropores on the transport of dissolved and suspended matter through soil. *Adv. Soil Sci.* 3:95–120.

White, R. E. 1987. A transfer function model for the prediction of nitrate leaching under field conditions. *J. Hydrol.* 92:207–222.

White, R. E., G. W. Thomas, and M. S. Smith. 1984. Modeling water flow through undisturbed soil cores using a transfer function model derived from 3HOH and Cl transport. *J. Soil Sci.* 35:159–168.

White, R. E., J. S. Dyson, R. A. Haigh, W. A. Jury, and G. Sposito. 1986. A transfer function model of solute movement through soil: II. Experimental applications. *Water Resour. Res.* 22:247–254.

Whitt, D. M. and L. D. Baver. 1937. Particle size in relation to base exchange and hydration properties of Putnam clay. *J. Am. Soc. Agron.* 29:905–916.

Wiegand, C. L. and E. R. Lemon. 1958. A field study of some plant–soil relations in aeration. Soil *Sci. Soc. Am. Proc.* 22:216–221.

Wiegner, G. 1925. *Dispersität und Basenaustauch.* Report of the Second Commission, 4th International Congress of Soil Science, Rome.

Wierenga, P. J. 1977. Solute distribution profiles computed with steady-state and transient flow models. *Soil Sci. Soc. Am. J.* 41:1050–1055.

Wiersum. L. K. 1960. Some experiences in soil aeration measurements and relationships to the depth of rooting. *Neth. J. Agric. Sci.* 8:245–252.

Willey, C. R. and C. B. Tanner. 1963. Membrane-covered electrode for measurement of oxygen concentration in soil. *Soil Sci. Soc. Am. Proc.* 27:511–515.

Williamson, R. E. and J. van Schilfgaarde. 1965. Studies of crop response to drainage: II. Lysimeters. *Trans. ASAE* 8:98–100, 102.

Wilson, M. J. (Ed.). 1994. *Clay Mineralogy: Spectroscopic and Chemical Determinative Methods.* Chapman & Hall, New York. 384 pp.

Wollney, E. 1878. Untersuchungen über den Einfluss der Exposition auf der Erwärmung des Bodens. *Forsch. Geb. Agric. Phys.* 1:263–294.

Wollney, E. 1883. Untersuchungen über den Einfluss der Pflanzendecke und der Beschttung auf die physikalischen Eigenschaften des Bodens. *Forsch. Geb. Agric. Phys.* 6:197–256.

Wooding, R. A. 1968. Steady infiltration from a shallow circular pond. *Water Resour. Res.* 4:1259–1273.

Yakuwa, R. 1945. Über die Bodentemperaturen in dem verschiedenem Bodenarten in Hokkaido. *Geophys. Mag. Tokyo* 14:1–12.

Yoder, R. E. 1937. The significance of soil structure in relation to the tilth problem. *Soil Sci. Soc. Am. Proc.* 2:21–23.

Young, M. 2002. Piezometry. pp. 547–574. In: J. Dane and G. Topp (Eds.), *Methods of Soil Analysis,* Part 4, *Physical Methods.* Soil Science Society of America, Madison, WI.

Young, M. and J. B. Sisson. 2002. Tensiometry. pp. 575–608. In: J. Dane and G. Topp (Eds.), *Methods of Soil Analysis,* Part 4, *Physical Methods.* Soil Science Society of America, Madison, WI.

Symbols

Lowercase

a	volumetric air content; air pressure potential head
b	overburden potential head
c	specific heat
c_a	specific heat of air
d	damping depth; stagnant boundary layer thickness
d_b	drainage breakthrough volume
d_w	cumulative drainage volume in breakthrough experiments
f_a	mass fraction for adsorbed phase
f_g	mass fraction for vapor phase
f_l	mass fraction for dissolved phase
f_L	leaching fraction
$f_L(I)$	net applied water pdf in transfer function model at depth L
$f_z(I)$	net applied water pdf in transfer function model at depth z
g	acceleration of gravity
h	matric potential head; lag distance
h_f	matric potential of wetting front in Green–Ampt model
h_H	transfer coefficient for sensible heat
h_T	total soil water potential head
h_v	transfer coefficient for water vapor
i	irrigation or infiltration rate
i_0, i_f	constants in Horton infiltration equation
$\hat{i}$	unit vector in x direction
$\hat{j}$	unit vector in y direction
$\hat{k}$	unit vector in z direction
K_H, k_H	Henry's constant
m	mean of a random variable
m_s	mass of solute; mass of solid
m_w	mass of water
p	hydrostatic pressure potential head
q	O_2 consumption rate in Lemon–Wiegand model
r_g	sink term in gas conservation equation
r_H	sink term in heat conservation equation
r_s	sink term in solute conservation equation
r_w	sink term in water conservation equation

s	solute potential head; slope of $P(T)$ curve; standard deviation
t_b	breakthrough time
t_{bR}	breakthrough time of chemical retarded by adsorption
t_c	time to end of first stage of evaporation
t_P	t-statistic at probability P
u	zero mean, unit variance normal random variate
z	gravitational potential head
z_0	reference elevation
z_{soil}	elevation of the location of interest in the soil
z_{wt}	water table elevation

Uppercase

A	area
B_λ	Planck energy density
C	heat capacity per unit volume
C^*	constant in BET equation
$C_1, /C_2$	constants in Planck radiation equation
C_a	concentration in adsorbed phase (mass/mass soil)
C_g^a	solute vapor concentration in atmosphere above soil surface
C_g	concentration in gas phase (mass/volume air)
C_{im}	solute concentration in immobile water zone
C_l	concentration in dissolved phase (mass/volume solution)
C_m	solute concentration in mobile water zone
C_p	O_2 concentration in dissolved phase at root-air interface
C_{soil}	soil heat capacity
C_T	total solute concentration (mass/volume soil)
$C_w(h)$	water capacity function
$\text{Cov}[X, Y]$	covariance of two random variables
CV	coefficient of variation
D	drainage rate; dispersion coefficient in CDE model; KS statistic
D_0	water diffusivity in Green–Ampt model
D_e	effective liquid diffusion–dispersion coefficient
D_E	effective liquid–vapor diffusion coefficient
D_g^a	diffusion coefficient of gas in air
D_g^s	diffusion coefficient of gas in soil
D_{lh}	hydrodynamic dispersion coefficient
D_l^s	liquid diffusion coefficient of solute in soil
D_l^w	liquid diffusion coefficient of solute in water
D_m	dispersion coefficient in mobile water zone
D_R	dispersion coefficient of chemical retarded by adsorption
D_{Tv}	thermal vapor diffusivity
D_w	water diffusivity
E	evaporation rate
$\text{E}[\cdot]$	expectation or ensemble average of a random variable

E_p	potential evaporation rate
ET	evapotranspiration rate
H	hydraulic head
H_f	latent heat of fusion
H_v	latent heat of vaporization
I	electrical current; cumulative infiltration; net applied water
I^*	integral scale
$\overline{I}$	mean water displacement volume
$\hat{I}$	median water displacement volume
J_g	soil gas flux
J_H	soil heat flux
J_{Hc}	conductive soil heat flux
J_{Hl}	latent soil heat flux
J_{lc}	convective chemical flux in dissolved phase
J_{ld}	diffusive chemical flux in dissolved phase
J_{lh}	dispersive chemical flux in dissolved phase
J_s	soil chemical flux; chemical flux in dissolved phase
J_{sb}	gaseous flux through surface boundary layer
J_w	soil water flux
$K(h)$	unsaturated hydraulic conductivity
K_0	hydraulic conductivity in Green–Ampt model
K_d	distribution coefficient in linear isotherm
K_f	Freundlich coefficient in Freundlich isotherm
K_H	dimensionless Henry's constant
K_{oc}	organic C partition coefficient
K_s	saturated hydraulic conductivity
K_T	thermal diffusivity
K_w	dissociation constant for water
L_c	capillary bundle length
M	solute mass in soil volume
M_a	molecular weight of air
M_s	molecular weight of solute
M_w	molecular weight of water
$N[u]$	normal curve of error
P	precipitation
$P(Z)$	cdf of a random variable Z
P_0	reference air pressure
P_a	air pressure
P_e	envelope pressure
P_l	liquid pressure
P_v	water vapor pressure
P_v^*	saturated water vapor pressure
Q	constant in Langmuir equation; volume flow rate
R	runoff rate; universal gas constant; retardation factor
R_a	phase partition coefficient for adsorbed phase

R_e	radius of root plus water film (Lemon–Wiegand model)
R_{earth}	long-wave thermal radiation from Earth
R_{ext}	extraterrestrial radiation
R_g	phase partition coefficient for gaseous phase
R_h	hydraulic resistance
R_l	phase partition coefficient for dissolved phase
R_N	net radiation
R_{nt}	net thermal radiation
R_s	solar radiation
R_{sky}	long-wave thermal radiation from sky
RH	relative humidity
S	sensible heat flux in atmosphere
T	temperature
T_A	annual average soil temperature
T_N	Nth travel-time moment
T_{ref}	reference temperature
V	volume; solute velocity in CDE equation
V_e	effective solute velocity
V_m	constant in BET equation; solute velocity in mobile zone
V_R	solute velocity of chemical retarded by adsorption
$\mathrm{Var}[Z]$	variance of a random variable Z
W	water storage
X	moles of chemical in solution
Y	regional variable
Z	random variable
$Z^*(x)$	estimation of $Z(x)$ at x
$\overline{Z}$	mean of Z
$\tilde{Z}$	median of Z
$\hat{Z}$	mode of Z

Lowercase Greek

α	constant in Kostiakov cquation; rate coefficient in MIM model
β	constant in Horton equation
γ	contact angle; constant in Kostiakov equation; psychrometer constant
$\gamma(h)$	variogram function
ϵ	emissivity
ϕ	porosity
ϕ_m	mobile water fraction
η	coefficient of viscosity
θ_g	gravimetric water content
θ_i	initial water content
θ_{im}	volumetric water content of immobile water zone
θ_m	volumetric water content of mobile water zone

θ_r	residual water content
θ_s	saturated water content
$\theta_v, /\theta$	volumetric water content
θ_0	surface water content in infiltration models
λ	thermal conductivity; dispersivity
λ^*	thermal conductivity of moist soil neglecting latent heat transfer
λ_{eff}	thermal conductivity including latent heat transfer
μ	solute degradation rate coefficient; parameter in lognormal pdf
μ_E	effective solute degradation rate coefficient
μ_T	chemical potential of soil water
ν_m	gamma-ray mass absorption coefficient of soil minerals
ν_w	gamma-ray mass absorption coefficient of water
ξ_g	soil gas diffusion tortuosity factor
ξ_l	soil liquid diffusion tortuosity factor
ξ'	modified gas diffusion tortuosity factor for water vapor
ρ	correlation coefficient
$\rho(h)$	autocorrelation function
ρ_a	air density
ρ_b	soil bulk density
ρ_p	soil particle density
ρ_v	water vapor density
ρ_v^*	saturated water vapor density
ρ_w	density of liquid water
σ	surface tension of water; constant in Stefan–Boltzmann equation; log variance
τ	period of a thermal wave
$\tau_{1/2}$	degradation half-life
ψ_T	total soil water potential
ψ_w	wetness potential
ψ_z	gravitational potential
ψ_s	solute potential
ψ_b	overburden potential
ψ_m	matric potential
ψ_a	air pressure potential
ψ_p	hydrostatic pressure potential
ψ_{tp}	tensiometer pressure potential
ω	ratio of gas-phase thermal gradient to average thermal gradient

Uppercase Greek

Θ	reduced water content
Σ	radiant energy flux
Σ_E	radiant energy flux at Earth's atmosphere

INDEX